Interdisciplinary Contributions to Archaeology

Archaeology stands alone among the sciences in its attempt to enlighten us about the entire record of humankind. To cover such a broad range of time and space, archaeologists must ensure that their findings are integrated into broader spheres of scientific knowledge. The IDCA series aims to highlight the collaborative and interdisciplinary nature of contemporary archaeological research.
Topics the series has covered include:

- Paleoecology
- Archaeological Landscapes
- Statistical Approaches
- Laboratory Methods
- Human Biological and Cultural Evolution
- Human Nutrition
- Emergence of Agriculture and Pastoralism

For a copy of the proposal form, please contact Christi Lue (christi.lue@springer.com). Initial proposals can be sent to the Series Editor, Jelmer Eerkens (jweerkens@ucdavis.edu). Proposals should include:

- A short synopsis of the work or the introduction chapter
- The proposed Table of Contents
- The CV of the lead author(s)
- If available: one sample chapter

We aim to make a first decision within 1 month of submission. In case of a positive first decision the work will be provisionally contracted: the final decision about publication will depend upon the result of the anonymous peer review of the complete manuscript. We aim to have the complete work peer-reviewed within 3 months of submission.

This book series is indexed in SCOPUS.

For more information, please contact the Series Editor at (jweerkens@ucdavis.edu).

More information about this series at https://link.springer.com/bookseries/6090

Agata Ulanowska • Karina Grömer
Ina Vanden Berghe • Magdalena Öhrman
Editors

Ancient Textile Production from an Interdisciplinary Perspective

Humanities and Natural Sciences Interwoven for our Understanding of Textiles

Springer

Editors
Agata Ulanowska
Faculty of Archaeology
University of Warsaw
Warsaw, Poland

Karina Grömer
Department of Prehistory
Naturhistorisches Museum
Vienna, Austria

Ina Vanden Berghe
Textile Research, Laboratory Department
Royal Institute for Cultural Heritage
Brussels, Belgium

Magdalena Öhrman
Faculty of Humanities
University of Wales, Trinity Saint David
Lampeter, UK

ISSN 1568-2722 ISSN 2730-6984 (electronic)
Interdisciplinary Contributions to Archaeology
ISBN 978-3-030-92169-9 ISBN 978-3-030-92170-5 (eBook)
https://doi.org/10.1007/978-3-030-92170-5

Cover illustration: Sassanian textiles of different colours, patterns and weave structures from the Chehrābād saltmine, Iran, 4th–6th c CE. Each box is a 2x2 cm detail of a textile (Image: K. Grömer, photos: F. Schapals, RUB/BDM)

This Springer imprint is published by the registered company Springer Nature Switzerland AG
The registered company address is: Gewerbestrasse 11, 6330 Cham, Switzerland

Contents

Contributors

Abolfazl Aali Archaeological Museum of Zanjan, Zanjan, Iran

Jennifer Beamer University of Leicester, Leicester, UK

Zlata Blažeska Independent Researcher, Skopje, North Macedonia

Tina Boloti Koukonisi Archaeological Project, Lemnos, Greece

Deborah Cassuto Martin (Szusz) Department of Land of Israel Studies and Archaeology, and the RIAB Minerva Center, Bar Ilan University, Ramat Gan, Israel

Albright Institute of Archaeological Research, Jerusalem, Israel

Lewis Ferrero University of Cambridge, Cambridge, UK

Julia Galliker University of Michigan, Ann Arbor, MI, USA

Karina Grömer Department of Prehistory, Naturhistorisches Museum, Vienna, Austria

Sabine Karg Free University of Berlin, Berlin, Germany

Zvi C. Koren The Edelstein Center for the Analysis of Ancient Artifacts, Department of Chemical Engineering, Shenkar College of Engineering, Design and Art, Ramat-Gan, Israel

Victor Kuzmichev Department of Clothes Design, Ivanovo State Polytechnic University, Ivanovo, The Russian Federation

Ronja Lau Ruhr-University Bochum, Bochum, Germany

Achim Lichtenberger Institut für Klassische Archäologie und Christliche Archäologie/Archäologisches Museum of Münster University, Münster, Germany

Gabriella Longhitano Department of Human Sciences, University of Catania, Catania, Italy

Ulla Mannering Ancient Cultures of Denmark and the Mediterranean, National Museum of Denmark, Copenhagen, Denmark

Ulrikka Mokdad Centre for Textile Research, University of Copenhagen and Expert Weaver, Copenhagen, Denmark

Aleksei Moskvin Department of Clothes Design and Technology, Saint Petersburg State University of Industrial Technologies and Design, Saint-Petersburg, The Russian Federation

Mariia Moskvina Department of Clothes Design and Technology, Saint Petersburg State University of Industrial Technologies and Design, Saint-Petersburg, The Russian Federation

Jasemin Nazim Curator Emerita, Museum of North Macedonia in Skopje, Skopje, North Macedonia

Marie-Louise Nosch Centre for Textile Research, Saxo Institute, University of Copenhagen, Copenhagen, Denmark

Magdalena Öhrman Faculty of Humanities, University of Wales, Trinity Saint David, Lampeter, UK

Andrea Orendi ArchaeoConnect GmbH, Department of Archaeological Science, Archaeobotany Laboratory, Tübingen, Germany

Helga Rösel-Mautendorfer Independent Scholar, Vienna, Austria

Itzhaq Shai Institute of Archaeology, Ariel University, Ariel, Israel

Orit Shamir Department of Museums and Exhibits of the Israel Antiquities Authority, Jerusalem, Israel

Thomas Stöllner Institute for Archaeological Science Ruhr-University Bochum, German Mining Museum Bochum, Bochum, Germany

Eva Andersson Strand Centre for Textile Research, Saxo Institute, University of Copenhagen, Copenhagen, Denmark

Oren Tal Department of Archaeology and Ancient Near Eastern Cultures of Tel Aviv University, Tel Aviv, Israel

Agata Ulanowska Faculty of Archaeology, University of Warsaw, Warsaw, Poland

Ina Vanden Berghe Textile Research, Laboratory Department, Royal Institute for Cultural Heritage, Brussels, Belgium

Elsa Yvanez Polish Center of Mediterranean Archaeology, University of Warsaw, Warsaw, Poland

Part I
About this Volume and Textile Research in 2020

the paper highlighted the importance of textiles and textile production in archaeology as matters closely associated with individuals but also the socio-cultural and economic life of any past society (cf. Andersson Strand et al. in this volume).

In the last decade, knowledge of textiles and textile production has become more widespread, appreciated and better understood by the archaeological scientific community, although textile studies have not completely entered mainstream archaeology. Similar developments have taken place in the fields of ancient and medieval history. Moreover, studies of textiles attract strong interest from the general public, especially in relation to reconstructions of fabrics, costumes and old textile techniques. In this period, the interdisciplinarity of textile research became a well-established standard in studies on fibres, dyes, fabrics, textile tools, techniques and technology, many of which developed from methods and approaches recommended and showcased in the 2010 paper.

The importance of textile studies is reflected by an impressive growth in number of research projects, overview publications and series of edited volumes and monographs related to textiles and textile production. To name but a few examples, cf. the North European Symposium for Archaeological Textiles (NESAT) and the PURPUREAE VESTES series of the conference proceedings, the Ancient Textiles Series by Oxbow Books and the CTR, the Bloomsbury series *A Cultural History of Dress and Fashion* and, currently in preparation, the *Bloomsbury Encyclopedia of World Textiles*. The number of specialist journals available is also growing, including *Textile History*, *Clothing and Textiles Research Journal, Archaeological Textiles Review*, and now also the recently created *Revue de l'Archéologie du Vêtement et du Costume*. The present volume originates from research presented at a session of the Annual Meeting of the European Archaeologists Association 2019, and retains the largely European focus of that conference session. However, textile studies are also well established and growing in global contexts with key publications emerging e.g. in North America, for example through Textile Society of America Symposium and its proceedings, and *The Textile Museum Journal*.

Scholarly work has been undertaken by individuals from universities and museums, as well as by major institutions with a specific focus on textile research. Examples from Europe include the CTR in Denmark, the Centre International d'Etude des Textiles Anciens [CIETA] in Lyon, France, the Textile Research Centre [TRC] in Leiden, Netherlands, the Landesamt für Denkmalpflege in Baden Württemberg and Bavaria, Germany: cf. Banck-Burgess 2019; Natural History Museum Vienna, Austria: cf. Grömer et al. 2013, Grömer 2015; or the Centre for Research on Ancient Technologies of the Polish Academy of Sciences, the University of Warsaw and other institutions in Poland: cf. Maik 2019. They have led research with a focus on regions such as Scandinavia, Central Europe, the Mediterranean and the Ancient Near East. Through their activities different methods and strategies have been gathered, and played a vital part in establishing standards for textile research, including both data mining and an analytical tool kit appropriate for the field.

Following the growth in textile studies in depth and breadth over the past decade, we aim to present a comprehensive update of the state-of-the-art. The diverse developments of the last decade, along with the increased recognition of the importance of textile studies in adjacent fields, now merit a dedicated, full-length publication entitled "Ancient Textile Production from an Interdisciplinary Perspective: Humanities and Natural Sciences Interwoven for our Understanding of Textiles". With this volume, we also wish to illustrate the current impact of textile archaeology on the scholarly perception of the past (not limited to archaeology alone). The volume presents new insights into the consumption, meaning, use and re-use of textiles and dyes, all of which are topics of growing importance in textile research. As indicated by the title, we demonstrate the continued importance of interdisciplinarity by showcasing several 'interwoven' approaches to environmental and archaeological remains, textual and iconographic sources, archaeological experiments and ethnographic data, from a large area covering Europe and the Mediterranean, Near East, Africa and Asia. The chronological span is deliberately wide, including materials dating from c. sixth millennium BCE to c. mid-fourteenth century CE.

1.1 Contents of the Volume

The idea of presenting the interdisciplinarity of textile studies as it stands in 2020 was initially discussed during session no. 133 at the 25th Annual Meeting of the European Archaeologists Association in Bern in 2019, convened under the same title as this edited volume, "Ancient Textile Production from an Interdisciplinary Approach: Humanities and Natural Sciences Interwoven for our Understanding of Textiles". The session was organised by Agata Ulanowska, Karina Grömer, Joanne Dyer and Ina Vanden Berghe. Contributions presented in that session make up the core of the present book,[1] although not all of the session's papers could be published here. We have also included contributions by colleagues who did not participate in the session but whose work captures or represents key developments in the field.

The interdisciplinarity of current textile research is reflected in the academic interests of the editors of this volume: Karina Grömer and Agata Ulanowska are archaeologists, Ina Vanden Berghe is a textile scientist and Magdalena Öhrman a philologist. Two of us, Karina Grömer and Ina Vanden Berghe, were active in the field as early as 2010, and collaborated, together with the CTR and others, in one of the first large-scale interdisciplinary projects on textiles.[2] Agata Ulanowska and Magdalena Öhrman are post-2010 newcomers to the field, attracted by the opportunities and multidisciplinary of textile research. Likewise, the contributors to this volume represent various areas of scholarly expertise and backgrounds. Among

[1] The full list of the contributors of the Session#133 is available in the EAA 2019 Programme Book: 332–335, https://www.e-a-a.org/EAA2019/Programme.aspx?WebsiteKey=92bb0c7f-79c0-4998-9804-b5db83f9a8b9andhkey=48f0a584-1ae8-4680-bcef-572c6fe82598andProgram=3#Program, accessed 6 Dec 2021.

[2] Clothing and Identities: New perspectives on textiles in the Roman Empire (DressID) – EU-culture programme (contract n.2007–1765/001–001 CTU COOPMU).

them, place has been given to the early career researchers who present their first research results in the field of textile studies, as well as to experienced and renowned colleagues. Through the composition of the editorial team and the featured authors, the volume showcases both the maturity and continued growth of textile research as a discipline.

The present volume is organised in five parts that, to our mind, reflect the main areas of the textile research in 2020. After the two introductory chapters (**Part I: About this Volume and Textile Research in 2020**), follow two chapters referring to dyes and dyeing technology in which analytical and material-based studies are linked to contextual sources (**Part II: Interdisciplinarity of Colour: Dye Analyses and Dyeing Technologies**). The six chapters of **Part III: Interdisciplinary Approaches to Textile Tools** discuss textiles and textile production starting from the analyses of tools, whether functional or as representative of technological developments or user identity. Archaeological and cultural contexts as well as textile traditions are the main topics of the six chapters in **Part IV: Traditions and Contexts: Fibres, Fabrics, Techniques, Uses and Meanings**. The two final chapters in **Part V: Digital Tools** refer to the use of digital tools in textile research, presenting two different case studies. The scope of this volume cannot fully convey the diversity of research on textiles at the beginning of the 2020s. However, we believe that it offers an overview of emerging innovative approaches and demonstrates the continued importance of established research standards and routines.

In Part I, the already mentioned Chap. 2: ***Old Textiles – New Possibilities. Ten Years on*** by Eva Andersson Strand, Ulla Mannering and Marie-Louise Nosch offers an inspiring, personal reflection by three great promoters of the field of textiles studies and its interdisciplinarity. The authors comment on the current state-of-the-art and future prospects, exemplifying their presentation with reference to studies and research activities based at the Centre for Textile Research in Copenhagen. They refer critically to the strong emphasis on production and functionality predominating in older studies, including their own, and advocate a change: "[T]he 2010 paper's definition of a textile as '*resources*, *technology*, and *society*' could now, ten years later, be challenged and alternatively defined as *individual choices*, *skills*, and *taste*." They also remind us that the "natural sciences are seen as objective quantifying, falsifiable, empirical, descriptive, factual and aiming at discovery, explanation, and solutions; the humanities, in contrast, promote the subjective, interpretative, theorizing, speculative, politicizing, and aesthetic, and aim at an understanding, intervention and at producing problems [...]. It is important that in textile research we include but also merge both these theoretical perspectives".

Part II starts with Chap. 3: ***Monobromoindigo: The Singular Chromatic Biomarker for the Identification of the Malacological Provenance of Archaeological Purple Pigments from* Hexaplex Trunculus *Species*** by Zvi C. Koren. It discusses the malacological provenance of archaeological purple pigments used in textile dyeing and in wall and stone paintings, using High Performance Liquid Chromatography (HPLC) and searching for two common components of all molluscan purple pigments: dibromoindigo and monobromoindigo, indicative of the molluscs' species. He argues that only *Hexaplex trunculus* species, whose shells are common finds in the Mediterranean since the Bronze Age, produces both reddish-purple as well as bluish-purple pigments having a considerable quantity of monobromoindigo. Moreover, *H. trunculus* contains the largest dye-content as compared to other species attested archaeologically as dye sources, being therefore "(...) the 'Snail of Snails' in the regal dye houses of antiquity".

HPLC is an analytical tool used also in the research presented in Chap. 4: ***Sassanid Dyes from Ancient Persia – Case Study Chehrābād in Northern Iran.*** Here, Ina Vanden Berghe and Karina Grömer discuss dyes and dyeing based on archaeological textiles. Due to the unique preservation conditions in the Chehrābād salt mine, excavated textiles provide new knowledge about dyeing traditions in northern Iran in the Sassanian period and possible trade in dyes between Iran and the Mediterranean. Most of the examined samples point to an established dyeing tradition, whereby textiles were commonly dyed with a combination of two or more dyes and by using technically different dye techniques. The closest parallels for these Sassanid dyes, as for well as the textile techniques (spinning and weaving) and the visual appearance of the Chehrābād fabrics, the authors find in Roman Egypt, Syrian Palmyra and north-east Iran.

In the first chapter of Part III, titled ***Textile Production in Some Early Neolithic Settlements in North Macedonia***, Zlata Blažeska and Jasemin Nazim examine the evidence for textile manufacture in sixth millennium BCE North Macedonia on the basis of textile imprints on clay: such impressions are known from vessel bases and from an anthropomorphic model house and a daub from the Velušina-Porodin and Anzabegovo-Vršnik Early Neolithic cultural groups. Most of the textile imprints identified by the authors show pieces of double weft-twining in S-direction, made of unspun fibres of plant origin. The uniformity of the impressed textiles (S-direction, similar thread count and weft-facing) implies an established knowledge of twining technology. Use-wear on several impressed textiles indicates re-cycling practices, with worn textiles used for modelling and drying pots. Potential textile tool finds from this period are scarce and problematic, however: a few weights and a single spool found in the Early Neolithic could have been used to produce similar twining textiles. Loom weights and spools only appear more frequently towards the end of the Middle Neolithic in contexts related to the expansion of Vinča culture. While rounded and perforated sherds may be considered as possible spindle whorls, the authors suggest that their function is likely to have been related to splicing rather than draft spinning.

In general, textile tools are among the artefacts of interest for textile studies which are available in quite high quantities from archaeological sites in Europe and the Mediterranean. Therefore, Chap. 6: ***Tradition versus Innovation? Horizontal (Ground-) and Vertical (Warp-weighted) Looms at Koukonisi, Lemnos*** by Tina Boloti, discusses an introduction of the warp-weighted loom at Koukonisi as a possible innovation in textile technology dated towards the end of the Middle Bronze Age. It is manifested by the appearance of loom weights, particularly Cretan-type discoidal weights, which seems to imply an adoption of the Cretan-origin technology and, possibly, a more specialised mode of production involving specialised weavers. The author argues that the warp-weighted loom might have gradually producing fabrics of qualities similar to the extant fragment. The recycling of textiles is once again considered, this time with regard to bleaching, a technique recognised in the Tell Iẓṭabba fragment, but otherwise attested in this period only in relation to linen fabrics used for garments and scroll wrappers.

Chapter 14, ***Unravelling the Threads of the Nubian Openworks. New Inquiries on a Unique Textile Tradition from Meroitic Sudan (c. 350 BCE–350 CE)***, by Elsa Yvanez and Ulrikka Mokdad brings the reader to the Meroitic kingdom and the Kushite people from present-day Sudan. The elaborate openwork borders of large cotton fabrics woven on a warp-weighted loom is a characteristic feature of the Meroitic decorative repertoire, present in excavated textiles, reflected in iconography, and attested by non-textile artefacts finished with tassels, pompoms, fringes and other hanging or dangling elements. The authors re-examine the pre-existing hypotheses on how openwork textiles were made, using an interdisciplinary approach based on the combination of textile studies with experimental archaeology. By an archaeologist and a weaver "bringing heads and hands together", this chapter offers a new opportunity for reconstructing and understanding the nature and history of this remarkable technique. Reuse and recycling, attested by excavated textiles, are also discussed as practices proving the notable price of the Meroitic openworks in the period.

With Chap. 15, ***Interdisciplinary Methods and New Perspectives on Inscribed Textiles: A Case Study of Christian 'Tiraz'***, we stay in Africa. The authors, Julia L. Galliker and Helga Rösel-Mautendorfer, discuss textiles with inscriptions in Greek and Coptic (a Christian version of '*tiraz*') dating to the Late Antique and early Islamic periods in Egypt. '*Tiraz*' is a term evolved to a generic designation for messages on cloth. The Christian form of '*tiraz*' comprises funerary textile fragments with Christian inscriptions, representing "a synthesis of Antique and Arabic religious cultures in Egypt under Muslim rule as a means to convey prayers and blessings". Through a case study based on an unpublished funerary shroud in the Kelsey Museum of Archaeology (KEL 88017a,b), the authors present their research methods including technical textile analysis, textual analysis, and experimental archaeology. This specific shroud, as suggested in the discussion "(..) was the product of a professional industry that produced luxury funerary shrouds for the Christian market".

The 16th and final chapter of Part IV, ***Weaving Experiments with the Rigid Heddle: Woven-in Metal Spirals from Siksälä and 'Auleja' Technique***, is authored by Ronja Lau. She discusses a rigid heddle, a simple and efficient tool for band weaving, in connection with bands and ornaments from the site of Siksälä in Estonia and the 'Auleja' weaving technique, dating to c. ninth–fifteenth century CE. According to her experiments with band weaving, a rigid heddle is an efficient tool for production of warp-picked up/floats patterns in Auleja technique, attested by numerous patterned bands found in Lithuania, Latvia and Russia. Its efficiency, however, is problematic when it comes to re-creating a band from Siksälä, which probably served as a headpiece, made of threads and bronze spirals. While the technique of production is the main concern of the author, written sources and folk beliefs as well as further ethnographic comparisons are referred to as data shedding more light on the symbolism of the bands and their patterns.

Part V is opened by Chap. 17, ***Investigating Relations between Textile Production and Seals and Sealing Practices in Bronze Age Greece: A Presentation of the New "Textile and Seals" Project Database*** by Agata Ulanowska. The online database presented has been designed and implemented as a tool facilitating tasks in the research project "Textiles and Seals" at the Faculty of Archaeology, University of Warsaw. By collecting various types of evidence, such as seal impressed textile tools, imprints of textile products on lumps of clay impressed by seals, and textile production-related imagery, the database and its search engines facilitate comparative studies of published evidence (seal impressed tools and iconography available also online in the *Corpus der minoischen und mykenischen Siegel* Arachne database) and new data coming from the first-hand analysis of the casts from the *Corpus der minoischen und mykenischen Siegel* Archive in Heidelberg (textile impressions). Although several correlations and cross-references among the registered data are theoretically possible to track without a new database, the amount of work required by such searches would potentially lead important observations unnoticed and the database will therefore allow swifter advances in this field of study.

In the last chapter of the volume, ***3D Visualization of the 2400-Year-Old Garments of Salt Man 4 from Chehrābād, Iran***, by Aleksei Moskvin, Karina Grömer, Mariia Moskvina, Victor Kuzmichev, Abolfazl Aali and Thomas Stöllner, readers are brought back to the Chehrābād salt mine, featured also in Chap. 2 in relation to dyestuffs. The authors present a digital reconstruction of the costume of the so-called Salt man 4 (died c. 405–380 BCE), a salt miner's mummy found in the mine. The limitations of the 3D-scanning technology for archaeological reconstructions led the authors to offer a new approach. The paper details how the entire process of clothes making and try-on was recreated in VR, on the basis of measurements, photographs, MRI (magnetic resonance imaging) scans and technical analysis of the find. Their reconstruction offers new ways of studying the costume in virtual reality, as well as an exceptional opportunity to observe the ergonomics of garments worn during heavy physical work.

1.2 Reflections and Outlook – A Multi-Dimensional Approach for Textile Research

The case studies presented here, arising from a decade of continual growth in archaeological textile and tool studies by the international scientific community, show the trends in recent scholarship and suggest directions for future research. The importance of considering textiles and textile tools from archaeological sites to advance our understanding of technical, economic and social aspects of past societies can no longer be overlooked. Archaeological textile research remains a heterogeneous field. The original textile finds are a significant source, assessed using different applied analytical methods. Nevertheless, basic research is still needed to gain statistically relevant data about archaeological textiles. In addition, textile tools, archaeological contexts and pictorial evidence as well as written sources are of relevance in understanding textile production and its economic and social implications in a specific area and time period.

As we have seen in the last decade, an establishment of specific research strategies was vital to gain basic data on textiles (e.g. Walton and Eastwood 1988; Barber 1991; Gillis and Nosch 2007; Peek 2013; Grömer 2014: 6–16; Andersson Strand and Nosch 2015; visualisation see Cybulska et al. 2010; Nowak-Böck and Voß 2015), dyes (e.g. Vanden Berghe et al. 2009; Vanden Berghe 2013) or textile tools (e.g. Mårtensson et al. 2009; Olofsson et al. 2015), as for the development of methodologies to interpret and contextualise such information. In modern research, this broad spectrum of methods and techniques (Fig. 1.1) enables a deeper insight into aspects of design, production and use of textiles than was possible in the past. In this volume, efforts in the area of dyestuff analysis in particular illustrate how technical development allows for more in-depth knowledge of materials and more accurate interpretation of data (cf. Koren in this volume; Vanden Berghe and Grömer in this volume).

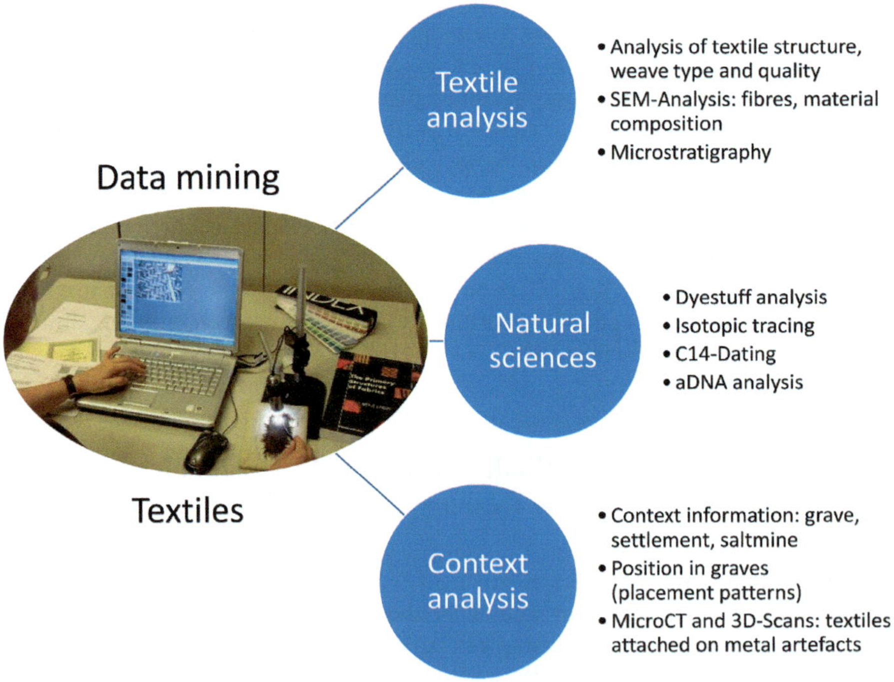

Fig. 1.1 Analytical tool kit used for the research on textiles from archaeological contexts. (Image: K. Grömer, NHM Vienna)

There are also attempts at integrating other analytical methods, ranging from aDNA, mass spectrometer-based protein sequencing and proteomics to strontium isotope analysis (Andersson Strand et al. 2010: 154–158 and Andersson Strand et al. in this volume). The implementation of promising new methods in archaeological research at the beginning of the twenty-first century has been defined by Kristian Kristiansen as the "Third Science Revolution" (2014: 19), leading to new theories and research fields. The application of those new methods in textile research opens new insights on the origin of raw materials and sheds light on textile trade routes, as well as mobility of textiles and of their individual users or producers. However, the use of novel methods raises new problems in data interpretation. Especially difficult to fully conceptualise at present are the relationships of such data and socio-economical processes (as the necessary background of these newly recognised phenomena) and the potential (un)awareness of people in the past about the revealed provenance/origin identifications.

For Kristiansen, the Third Science Revolution is also defined by the power of Big Data for systematic surveys via databases, as well as new methods of quantification and modelling. Digitalisation and making data accessible to the public via modern media are global challenges and important tasks in our postmodern world generally and particularly in the sciences and in heritage management. As Kristiansen points out, for archaeological research in the future, connecting Big Data from various studies and collections is the only way to extract new knowledge from them, through such recent innovative study fields as landscape modelling, agent-based modelling or the world system theory (2014: 20–22), which generates new understanding of our past. In textile research, a great deal remains to be done. However, putting together the results of data mining and the use of diverse analytical tools, Big Data about archaeological textiles and tools is already being generated (cf. a Digital Atlas in the EuroWeb project, see Nosch et al. 2020; Ulanowska in this volume).

Interdisciplinary research (Fig. 1.2) across different disciplines of the Humanities is nothing new in the twenty-first century. For archaeological textile research, a long tradition of integration of different studies of iconography and written sources has been vital for understanding textile production, dress and appearance. Different concepts and perspectives from a wide variety of historical disciplines (e.g. Prehistory, Classical Archaeology, Art History, Ancient History and History as well as Ethnography) have been combined with archaeological textile material since the very beginnings of textile studies. Multidisciplinary research also embraces cooperation with applied sciences such as chemistry or research in bio- and geosciences. For textile archaeology, both approaches have deep roots, as natural sciences such as Physical Anthropology, Geology, Biology and Zoology can be seen as traditional partners for archaeological research in general (e.g. Renfrew and Bahn 1991) and textile research in particular (e.g. Crowfood 1931; Pfister 1934, 1937; von Stokar 1938; Broholm and Hald 1940; Schlabow 1952; Hoffmann 1964) at least since the turn of the nineteenth to twentieth century. For textile research, archaeozoological and archaeobotanical data, together with data from landscape archaeology, geology, environmental studies and historical disciplines help to explain textile production

heritage: Documentation, preservation, and protection, 6th international conference, EuroMed 2016, Nicosia, Cyprus, October 31–November 5 2016, proceedings. Springer, Cham, p 11–15.
Barber, EJW (1991) *Prehistoric textiles. The development of cloth in the Neolithic and Bronze Age with special reference to the Aegean*. Princeton University Press, Princeton.
Banck-Burgess J (2019) THEFBO. The cultural-historical importance of textile production in the prehistoric wetland settlements. *Archaeological Textiles Review* 61:122–127.
Broholm HC, Hald M (1940) *Costumes of the Bronze Age in Denmark*. F. Brommer, Copenhagen.
Cybulska M, Florczak T, Maik J (2010) Visual reconstruction of archaeological textiles. Andersson Strand E, Gleba M, Mannering U, Munkholt C, Ringgaart M (eds) *North European symposium for archaeological textiles X*. NESAT X. Ancient Textiles Series 5. Oxbow Books, Oxford and Oakville, p 37–40.
Crowfoot GM (1931) *Methods of hand spinning in Egypt and the Sudan*. Bankfield Museum Note n°12. F. King & Sons Ltd., Halifax.
Gillis C, Nosch M-L (eds) (2007) *First aid for the excavation of archaeological textiles*. Oxbow Books, Oxford.
Grömer K (2014) *Römische Textilien in Noricum und Westpannonien – im Kontext der archäologischen Gewebefunde 2000 v. Chr.–500 n. Chr. in Österreich*. Austria Antiqua 5. Uni Press Graz Verlag, Graz.
Grömer K (2015) Textile research in Austria – an overview. In: Grömer K, Pritchard F (eds) *Aspects of the design, production and use of textiles and clothing from the Bronze Age to the Early Modern era. The North European symposium for archaeological textiles, 21st–24th May 2014 in Hallstatt, Austria*. NESAT XII. Archaeolingua 33, Budapest, p 13–24.
Grömer K (2016) *The art of prehistoric textile making. The development of craft traditions and clothing in Central Europe*. Veröffentlichungen der Prähistorischen Abteilung 5. Verlag des Naturhistorischen Museums Wien, Vienna.
Grömer K, Kern A, Reschreiter H, Rösel-Mautendorfer H (eds) (2013) *Textiles from Hallstatt. Weaving culture in Bronze and Iron Age salt mines. Textilien aus Hallstatt. Gewebte Kultur aus dem bronze- und eisenzeitlichen Salzbergwerk*. Archaeolingua 29, Budapest.
Hoffmann M (1964) *The warp-weighted loom*. Studia norvegica 1. Universitetsförlaget, Oslo.
Kristiansen K (2014) Towards a new paradigm? The Third Science Revolution and its possible consequences in archaeology. *Current Swedish Archaeology* 2:11–34.
Maik J (2019) 70 years of Polish research on archaeological textiles. *Przegląd Archeologiczny* 67:73–83.
Mårtensson L, Nosch M-L, Andersson Strand E (2009) Shape of things: Understanding a loom weight. *Oxford Journal of Archaeology* 28(4):373–398.
Nosch M-L, Ulanowska A, Yvanez E (2020) EuroWeb – a new European network and COST action 2020–2024. *Archaeological Textiles Review* 62:183–186.
Nowak-Böck B, Voß H (2015) Standardised mapping system for the digital documentation of organic materials on metal finds and *in-situ*-blocs. *Archaeological Textiles Review* 57:60–69.
Olofsson L, Andersson Strand E, Nosch M-L (2015) Experimental testing of Bronze Age textile tools. In: Andersson Strand E, Nosch M-L (eds) *Tools, textiles and contexts. Investigating textile production in the Aegean and Eastern Mediterranean Bronze Age*. Ancient Textiles Series 21. Oxbow Books, Oxford and Philadelphia, p 75–100.
Peek C (2013) Dokumentation organischer Bodenfunde. In: Banck-Burgess J, Nübold C (eds) *The North European symposium for archaeological textiles XI, 10-12 May 2011 in Esslingen am Neckar*. NESAT XI. Verlag Marie Leidorf, Rahden/Westfahlen, p 37–44.
Pfister R (1934) *Textiles de Palmyre*. Les Éditions d'Art et d'Histoire, Paris.
Pfister R (1937) *Nouveaux textiles de Palmyre*. Les Éditions d'Art et d'Histoire, Paris.
Renfrew C, Bahn P (1991) *Archaeology: Theories, methods and practice*. Thames and Hudson, London.
Schlabow K (1952) *Der Prachtmantel von Thorsberg, der Schlüssel zum altgermanischen Webstuhl*. Karl Wachholz Verlag, Neumünster.

Ulanowska, A (2016) Towards methodological principles for experience textile archaeology. *Prilozi Instituta za arheologiju u Zagrebu* 33:317–339.
Vanden Berghe I, Gleba M, Mannering U (2009) Towards the identification of dyestuffs in Early Iron Age Scandinavian peat bog textiles. *Journal of Archaeological Science* 36(9):1910–1921.
Vanden Berghe I (2013) Dye analysis of archaeological textile objects. In: Banck-Burgess J, Nübold C (eds) *The North European symposium for archaeological textiles XI, 10-12 May 2011 in Esslingen am Neckar*. NESAT XI. Verlag Marie Leidorf, Rahden/Westfahlen, p 57–62.
von Stokar W (1938) *Spinnen und Weben bei den Germanen. Eine vorgeschichtlich-Naturwissenschaftliche Untersuchungen*. Mannus-Bücherei 59. Curt Kabitzsch Verlag, Leipzig.
Walton P, Eastwood G (1988) *A brief guide to the cataloguing of archaeological textiles*. Textiles Research Associates, York.

Agata Ulanowska is Assistant Professor at the Faculty of Archaeology, University of Warsaw, Poland. Her research interests focus on textile production and technology, experimental archaeology, and seals and sealing practices in Bronze Age Greece. She was awarded two grants of the National Science Centre of Poland, for the projects: "Textile production in Bronze Age Greece – comparative studies of the Aegean weaving techniques" (2015–2017), and "Textiles and Seals. Relations between textile production and seals and sealing practices in Bronze Age Greece" (2018–2021). She is a chair of the COST Action CA 19131 "EuroWeb. Europe Through Textiles: Network for an integrated and interdisciplinary Humanities".

Karina Grömer is the Head of the Department of Prehistory, Natural History Museum Vienna. She studied prehistoric archaeology, ethnology and anthropology at the University of Vienna in Austria. Her habilitation thesis was entitled "Archaeological Textile Research – Technical, economic and social aspects of textile production and clothing from Neolithic to the Early Modern Era". She is specialized in interdisciplinary and integrated analysis of textiles, research on textile tools and reconstruction of prehistoric costume. Her research covers a timespan from c. 2500 BCE to 1000 CE and a geographical area from Central Europe to Iran, with research projects like "DressID – Clothing and Identity in Roman Empire", "CinBA – Creativity in the Bronze Age" and "Chehrābād Saltmummy and Saltmine Exploration Project". She also teaches at the Universities of Vienna (Austria) and Southampton (UK).

Ina Vanden Berghe is responsible for the Textile Research Laboratory of the Royal Institute for Cultural Heritage (KIK-IRPA) in Brussel, Belgium. She specializes in the study of materials, techniques and degradation aspects of organic materials, such as natural and synthetic organic dyes and protein fibres, as well as in studies on metal threads and plant-based fibres. Her special interest lies in archaeological textiles and, more specifically, in the degradation problems of fibres and dyes associated with different burial and conservation conditions. She has participated in several research projects such as "MODHT – Monitoring of the Damage in Historic Tapestries", "DressID – Clothing and Identity in Roman Empire', "Eu–Artech – Access, Research and Technology for the Conservation of the European Cultural Heritage"; "Charisma – Cultural heritage Advanced Research Infrastructures: Synergy for a multidisciplinary approach to conservation/restoration" and "Iperion – Integrated platform for European research infrastructures in the field of cultural heritage".

Magdalena Öhrman is Senior Lecturer in Classics at the University of Wales Trinity Saint David, UK. Trained as a classical philologist in Sweden and Germany, her current research interests are in the area of Graeco-Roman textiles and textile production, textile terminology and technology, and in sensory experience in past societies. In 2016–2018, she held a Marie Skłodowska-Curie fellowship at the Centre for Textile Research (University of Copenhagen) for the project "Textile Reflections: Multi-Sensory Representation of Textile Work in Latin Poetry and Prose".

Chapter 2
Old Textiles – New Possibilities. Ten Years on

Eva Andersson Strand, Ulla Mannering, and Marie-Louise Nosch

Abstract In 2010 we published a survey and state-of-the art paper, "Old Textiles – New Possibilities", on the position and advances of archaeological textile research. The paper reflected the research of the centre of excellence, the Centre for Textile Research (CTR), established in 2005 at the University of Copenhagen in Denmark, and our strategies to expand the field and develop new methodologies. In this review, we follow up the advancement of methods in archaeological textile research and discuss innovations since 2010. We also assess the ways in which textile research has expanded since 2010 and where there is still new ground to break.

Keywords Textile archaeology · Textile research · Textile tools · Textile technology

2.1 Introduction

In the years 2008–2009, we drafted the paper "Old Textiles – New Possibilities" (Andersson Strand et al. 2010) together with our colleagues Irene Skals, Margarita Gleba and Karin Margarita Frei. As authors, we were two prehistoric archaeologists (Mannering and Andersson Strand), one ancient historian (Nosch), one conservator (Skals), one classical archaeologist (Gleba) and one geologist/geochemist (Frei). At the time, we all worked closely together in the Centre for Textile Research (CTR) at

E. A. Strand (✉) · M.-L. Nosch
Centre for Textile Research, Saxo Institute, University of Copenhagen, Copenhagen, Denmark
e-mail: evaandersson@hum.ku.dk; nosch@hum.ku.dk

U. Mannering
Ancient Cultures of Denmark and the Mediterranean, National Museum of Denmark, Copenhagen, Denmark
e-mail: ulla.mannering@natmus.dk

A. Ulanowska et al. (eds.), *Ancient Textile Production from an Interdisciplinary Perspective*, Interdisciplinary Contributions to Archaeology,
https://doi.org/10.1007/978-3-030-92170-5_2

the University of Copenhagen and the National Museum of Denmark. CTR was founded in 2005 as a 10-year centre of excellence by the Danish National Research Foundation. The paper was driven by our desire to summarise the status of the field and to present the richness of methods in textile research to the broader archaeological community. Indeed, the introductory paragraph concluded:

> Textiles have an enormous potential in archaeological research, being able to tell about social, chronological, and cultural aspects of past societies, and at the same time giving us a unique opportunity to come very close to the prehistoric individual. (Andersson Strand et al. 2010: 150)

It was our aim to move research in textiles away from the specialised, auxiliary branches of archaeology and into the core and forefront of research. Moreover, CTR was expanding and we wished to gather our different disciplines into one landmark paper and present the essence of the interdisciplinary work taking place at the centre, and how the field of textile research could advance with such a combined effort.[1]

Looking back, we recall an exciting time of discussion and writing; the overall themes of the paper are clearly marked by the CTR scholars' main preoccupations of the time: to develop the CTR methodology of textile tools based on Andersson Strand's earlier works on Viking Age tools; Nosch's previous work on textile terminology; Frei's advancement of strontium analysis, Brandt's DNA and protein research as promising new tools in textile research; and the close examination of Danish archaeological Iron Age textiles by Mannering, Gleba and Skals.

In 2021, our paper has over hundred scholarly citations in Google Scholar. It has been adopted as part of many university module syllabi. In spite of its very generic statements, there seems to be a stable and even increasing interest in the paper, which possibly reflects the continued and growing interest in textile research in many related fields. In the abstract from 2010 we wrote:

> Textile research has become an important field of archaeology. Although the established analytical methods are often viewed as specialized, their integration with other interdisciplinary approaches allows us to deal with broader archaeological issues and provides the interpretational base for a much more comprehensive investigation of textiles in ancient times. Analyses of fibres, dyes, archaeobotanical and archaeozoological remains, as well as palaeoenvironmental and geochemical investigations, provide information about available resources, while tool studies, experimental testing, and visual grouping are approaches that explore the technology and techniques. Together, these approaches can provide new knowledge about textile production and consumption and, thereby, about people and society in ancient times. (Andersson Strand et al. 2010: 149)

The paper addresses textiles in the past but without defining this past geographically or chronologically. The time frame was set to 10,000 years: a deliberate choice to target communities of both prehistorians and classical archaeologists. This strategy can also be seen in the publication venue, *European Journal of Archaeology*, and in the paper's mixed use of references to the past as both *antiquity* and *prehistory*. This may explain the broad appeal of the paper. Overall, the paper argues strongly for

[1] We thank the editors for the opportunity to reflect upon our 2010 paper. In 2016, we followed up on the research in CTR (Andersson Strand et al. 2016).

inclusion of experts – both from laboratories and craft worlds. Hence, it argues implicitly for a greater interdisciplinarity and professionalization of the research field, something which has now happened.

In this new paper, we aim to summarise and discuss the development of textile research since the publication of our paper in 2010. It is clear that the field has grown dramatically in the last 10 years. The number of textile publications, conferences and sessions including textiles and textile research projects shows that it has certainly become 'the new black'. Furthermore, the field of archaeological textile research is now more integrated into archaeological and historical research than ever before, not only thanks to the work taking place at CTR, but through the genuine and persistent work undertaken by many individual researchers across the European university and museum landscape. It is therefore not possible to include all new initiatives and research here. This article should rather be seen as a contribution from our own individual perspectives and experiences, including the still ongoing work at CTR and National Museum of Denmark.

2.2 Textiles

The first part of our paper concerned archaeological textile analysis (2010: 151–159) with subsections on 'Basic analyses' and the ten basic descriptive features: *Dimensions, Condition, Colour, Textile structure, Yarn structure, Fibre, Decoration, Irregularities and faults, Wear, Construction.* This was followed by a section on more 'Specialized methods and approaches', leading to a detailed discussion on the two conventional methods, fibre and dye analyses, and two new methods, isotopic tracing and molecular analysis, assessing their status and potential. Fibre analysis in particular was considered a very promising field of research that could contribute to discussions on selective breeding, processing of raw materials, wear and degradation (Rast-Eicher 2008). At the same time, the 2010 paper criticised the evolutionary approach emerging from Michael Ryder's fibre analyses of archaeological textiles (Ryder 1983; for further references see also Rast-Eicher 2008), and argued that textiles are not direct evidence of fibre but the results of human choices of sorting and selection, as demonstrated by experienced craftspeople (2010: 154–155). Dye analyses were also discussed in detail, once again initially with a positive view of the new techniques, then with a critical assessment of their limitations and the need for expertise and contextualised knowledge. The paragraph on strontium analyses cautiously warned against over-interpretation of this new tool as more than a means of indicating local or non-local origin. Molecular analysis, also an emerging field in 2010, was evaluated more positively and with the optimistic conclusion:

> Hence, in the near future we should be able to use wool textiles to extract and analyse ancient DNA allowing us to approach the questions dealing with ancient sheep breeds and sheep domestication. (Andersson Strand et al. 2010: 158)

Radiocarbon, although not a new method, was briefly discussed because it was increasingly applied to textiles. We argued that textiles were particularly suited for 14C-analysis because of their shorter lifespan compared to other artefacts (a statement we would perhaps modify today), and for the potential for fine-grained chronological differences between a body and its associated garments. We also observed that careful pre-cleaning methods were imperative in order to avoid contamination. Today this is a standard procedure in 14C-analysis (Mannering et al. 2010). Visual grouping was a new concept introduced only 5 years earlier by Lena Hammarlund and Kathrine Vestergård Pedersen (2007), and here our paper takes a stance against its otherwise overall positivistic view by stating that "all the factors that create the final textile appearance cannot be quantified or qualified" (2010: 159). Further, it introduced the concept of 'look' but still hesitated to abandon the larger techno-cultural clusters for an individual view and landed in a middle ground assumption that visual appearance "permits investigation of regional and chronological trends in textile appearance" (2010: 159).

2.3 The Field of *Textiles* since 2010

Today it is recognised that textile studies can provide new and unforeseen perspectives, and that this material in its fragile organic structure is precious, worth preserving and analysing. Here, we would like to highlight some of the significant advancements in the field since 2010.

In terms of geographical focus, the anthology published in 2012 summarising the collections of archaeological textiles from 16 European countries highlighted textile research in a national and international perspective (Gleba and Mannering 2012). Despite its national structure, the book has very successfully shown the potential of conducting cross-regional research and of taking a longer perspective on textile craft, eschewing some conventional chronological divisions. It is hoped that within the next 10 years, other textile researchers will complete this work by including countries that still need to be documented in this way.

Regarding new methods, the strontium isotope method and its development on fibre materials (Frei et al. 2009) has yielded spectacular results since 2010. In 2015, the article about the iconic fourteenth-century-BCE Egtved girl (Fig. 2.1) and her clothing made headlines and received international press coverage (Frei et al. 2015, 2017). With this followed critical questions about the methodologies (Thomsen and Andreasen 2019; Frei et al. 2019; Kootker et al. 2020; Toxværd 2020) and about the archaeological interpretations (Bergerbrandt 2019).

Regrettably, the isotope discussion is difficult to follow for non-experts, but it will and must continue to be the subject of scholarly debate until more consensus about its applications and limitations is reached in the larger community of those working with isotopes and archaeology, including on other materials than textiles. In this way, textile research is at the forefront of a cutting-edge research field, and progress is made every year to understand its ramifications.

Fig. 2.1 The most iconic Danish Early Bronze Age grave, containing the remains the Egtved Girl, dated to 1370 BCE, in an outfit consisting of a short blouse, a belt and a corded skirt. When it was discovered in 1921, it created a furore among scholars and the general public, because it was unheard of that our female ancestors could have worn such daring outfits. Today the Egtved grave has provoked more controversy with its data on personal long-distance travel and exchange of raw materials. (Photo by R. Fortuna)

Likewise, large studies of aDNA and proteins have explored the possibility to secure species identifications and information about age, gender and, for instance, hair colour related to skin and textile production (Brandt et al. 2011; Brandt et al. 2014). The ongoing research by Luise Ørsted Brandt has facilitated increased integration and better understanding of the relationship between skin craft and textile craft in Scandinavia (Ebsen et al. 2019; Brandt et al. 2020; Brandt and Mannering 2020).

The field of fibre quality analysis has developed dramatically in the past 10 years, due to shared interests but independent efforts in conducting and publishing large scale analyses of South, Central European and North European textiles (see for instance Gleba 2012; Grömer et al. 2013; Rast-Eicher and Bender Jørgensen 2013; Skals and Mannering 2014; Bender Jørgensen and Rast-Eicher 2016; Rast-Eicher 2016; Skals et al. 2018). These studies have shown the potential of large-scale testing and how information about fibre quality can add important new perspectives to the understanding of wool production, provenance and trade, and also offer vital information about more tactile aspects such as textile feel and drape. It is hoped that this field will be further developed and highly integrated into textile research, preferably using a commonly accepted, standardised and a much more cost-efficient procedure.

Likewise, the field of dye analysis has shown great potential for visualising past textile production, and by using large-scale testing we can obtain a much clearer image of the integration of colours in past cultures (see for instance Vanden Berghe et al. 2009; Hofmann-de Keijzer et al. 2013; Vajanto 2015). Dye analysis and exploration of various ancient dyeing methods will doubtless play an increased role in archaeological textile research in the coming years.

Historical archaeology and in particular the archaeology of the Early Modern period (the sixteenth to eighteenth centuries) has expanded in the past decade, and

Fig. 2.2 Finely knitted wool stocking shaft found in the rubbish dumps excavated from the Copenhagen Town Hall Square (Rådhuspladsen) in Denmark dated to the seventeenth century. It shows clear signs of mending using silk yarn as well as repair with a wool twill patch on the heel and ankle part. The foot was cut off before the discarding. (Photo by Ch. Rimstad)

textile items have started playing a new role in scholarship on the period, challenging the dominance of written sources and rich iconography in the Early Modern period. In this way, textiles and clothing in historical archaeology enrich discussions of consumption and urbanisation history, and complement the elite iconography of clothing (see for instance Andersen 2017; Ringgaard 2007, 2017; Vestergård Pedersen and Nosch 2009) (Fig. 2.2).

Finally, the term "cloth cultures" introduced to textile archaeology by Susanna Harris (2008, 2012) has prompted a more inclusive and diverse understanding of dress technology. It has generated the possibility to perceive textiles and, for instance, skin technologies, on a more holistic and conceptual level which until recently was difficult to explain and understand, especially for non-textile specialists.

2.4 Beyond Textiles

In the second part of the 2010 article, *Beyond Textiles*, other sources were briefly mentioned but the main focus was on two areas highlighted by CTR, tool studies and experimental textile archaeology. This contributed to methods within these research fields becoming more clearly defined and standardised (Andersson Strand and Nosch 2015). In the first paragraphs on archaeobotany and archaeozoology, the blended nature of resources for textiles and for food were discussed (for example, linseed and flax stems, and domestic animals kept for their diverse products). We argued that this helps "locating and mapping the areas of exploitation in time and space, thereby advancing our knowledge of not only textile production but also of agriculture, husbandry, and ecology" (2010: 161). The focus was again on production and environmental impact, and textiles were described as part of agriculture, economy, and exchange.

Before 2010, tools studies were, with a very few exceptions, primarily based on typology such as shape and material, rather than the function of textile tools. In the 2010 paper, we therefore stressed the importance of also recording the tools' functional parameters, which led to a better understanding of their suitability for textile production. Analyses of the functional parameter recordings allowed an interpretation not just of the existence of textiles but also of the range of textile types produced.

> A study of number, distribution, and morphology of tools, based on statistical analyses of their physical parameters of preservation, size, weight, and use wear, can provide valuable information about technology, scale of production, and the importance of textile manufacture for a particular site, as well as about the raw materials and the final products (e.g., Andersson 2007; Walton Rogers 1997). (Andersson Strand et al. 2010: 161)

Furthermore, we concluded that to understand textile production fully, it is important to include invisible processes, techniques and tools of perishable material through experimental archaeology and ethnographic studies. It was our opinion that this approach would provide a more nuanced and comprehensive view of textile production and its significance. Despite a certain tradition of linking craft knowledge, locality and tradition in experimental and experiential archaeology, more essentialist interpretations are not present in the 2010 paper, and ethnographic research is used purely as a technical shortcut to assess textile tools, not as an essentialist cultural trait or style:

> Through ethnographic studies we often have knowledge about tool function and different processes such as fibre preparation, weaving techniques, and/or tools made of perishable materials. (Andersson Strand et al. 2010: 161)

2.5 The Field *Beyond Textiles* Since 2010

Today, the CTR approach to tool studies is indeed present is many publications, and generally adapted to textile tool studies in many other contexts. This can be seen in the many tool-centred articles published in *Archaeological Textile Review* and in conference proceedings and journals (cf. Meo 2014; Landenius Enegren 2015). Archaeologists outside the field of textile research have also understood the advantages of applying a more functional approach when studying textile tools, and how to include them in more general discussions on the impact of textile production in ancient societies (Andersson Strand and Heller 2017; Sarauw 2019).

Since 2010 the use of experimental archaeology has contributed to an important debate on the value of this branch of archaeology as a discipline, and particularly whether tests with tools really contribute to information on past production. Results from some tests indicate that potentially all textile craftspeople in ancient times could produce all types of products with all types of tools (Kania 2013). However, these tests do not consider the archaeological context, which is essential for any interpretation. In 2010 we stated that "tool studies can become even more informative when used in combination with experimental archaeology" (2010: 163), but only if including the information on the context as stated in the CTR guidelines for experimental testing of tools from the Bronze Age Aegean (Olofsson et al. 2015: 77). In order to contextualise the results, we suggested that:

- *The primary parameter to be investigated is function and the purpose is not to make reconstructions of ancient textiles;*
- *Tools must be reconstructed as precise copies of archaeological artefacts;*
- *Raw materials, such as wool and flax, must be selected according to our knowledge of the time and area under investigation;*
- *All processes must be performed by at least two skilled craftspeople.*

These guidelines are still valid today, but the use or mis-use of experimental archaeology and its combination with experiential archaeology as a method has demonstrated that specific guidelines must be developed for every single test. Every experiment is unique and guidelines are necessary to document how the tests are performed, and inform the interpretation of the results. To gain meaningful results, it is essential that the type of fibres, tools and techniques are the same as the fibres, tools and techniques from the period studied (Andersson Strand et al. forthcoming). If not, the results from an experiment can be directly misleading and cannot be used in the interpretation of the textile production from a specific context in the past (Fig. 2.3).

In sum, experimental archaeology as a method, both at CTR and as used by other scholars and craftspeople, has demonstrated that the suitability of textile tools has

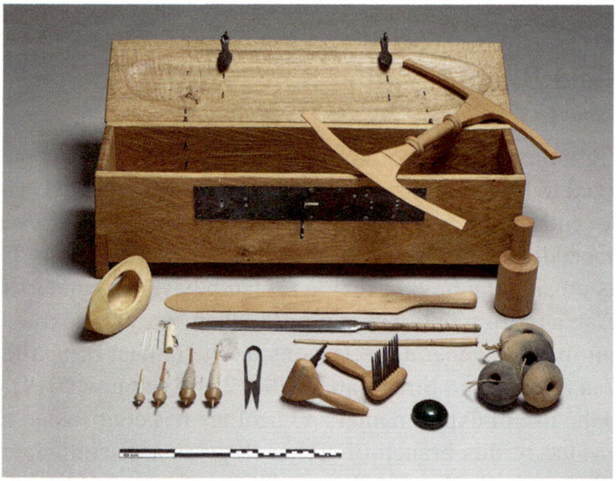

Fig. 2.3 Reconstructed Viking Age tool box. All textile tools are exact copies of textile tools from the Viking town Haithabu in North Germany and the Oseberg burial in Norway. The reconstructed tools have been used to reconstruct three Viking Age textiles based on fibre analyses of the originals, a wool chosen according to these results, and with the use of ancient techniques (Mannering 2018). The results from this project have clearly demonstrated the importance of working with the same parameters from a given context, region or time period in order to gain a product similar to the original. (Photo by Roberto Fortuna)

had an important impact on textile research but also on archaeological research in general, and given an understanding for the need of a detailed registration of textile tools (Belanová Štolcova and Grömer 2010).

While the focus in our paper was primarily placed on production, which is visible through the presence of tools and objects in the archaeological record, it has subsequently become clear that it is also necessary to include invisible production, such as non-preserved tools made of perishable materials, and other indirect data related to the production processes, in order to create a more complete picture (Andersson Strand and Nosch 2015). Ethnographic sources and knowledge of traditional textile crafts provide important insights and perspectives in relation to this aspect (see above).

However, in the last 10 years there has been an increasing awareness that it is not possible to understand and interpret the production of textiles in ancient societies based solely on ethnographic studies. It is clear that the production and use of textiles tools have changed considerably over time (Ebert et al. 2018). Therefore, the combination of ethnographic studies and contextualised experimental archaeology is essential.

Another innovative development in experimental archaeology is Motion Capture and 3D recording of movement, which allows us to come very close to the individual crafts people, spinners and weavers. This provides important new insights into how the movement of the individual craftsperson and the tools might influence the final product (Andersson Strand et al. 2018). Further, in combination with cognitive neuroscience, these methods can demonstrate both differences and similarities in people's actions and outcomes when comparing beginners and professionals.

In conclusion, in order to gain better insights and understanding through tool studies and experimental archaeology, it is necessary to include contextual perspectives in the discussion of the impact of textile production. It is also obvious that the geographical and chronological frameworks must be included in the interpretation of textile production and its impact on a specific past society. Each society and each individual are unique. How much does a society influence production and what choices did the individual craft person have? How was textile production organised and how much specialisation was needed? These questions need to be answered and discussions have already started (e.g., Grömer 2013; Andersson Strand and Mannering 2014; Bender Jørgensen et al. 2018). It is necessary to include textile research results in general archaeology but it is equally important that more textile researchers gain interest and knowledge that goes beyond textiles, tools and techniques.

2.6 Theoretical Approaches and Textile Research

At a distance of 10 years, it is now possible to position this paper theoretically within the cultural historical paradigm of archaeology. The opening phrase, "[t]extiles express who we are – our gender, age, family affiliation, social status, occupation, religion, and ethnicity" (2010: 150), set the interpretative framework: research on material culture is used to distinguish groups of peoples, their characteristics, and implicitly how they differ from each other. There is in this opening statement a clear reference to the twentieth century western historical and archaeological focus on cultures and peoples as reflected in their material culture. Even if the paper promises a rewarding "unique opportunity to come very close to the prehistoric individual" (2010: 150), our main concern was not the traces left by the individual but larger assemblages of traces leading to more generalised knowledge.

This same view is seen in the definition of a textile as "first and foremost a result of complex interactions between resources, technology, and society" as illustrated in Fig. 2.1 (2010: 151). One may even suggest that this mirrors the paper's theoretical foundation in traditional cultural history informed by (Marxist-inspired) materialism. In our paper, the three entities *resources*, *technology*, and *society* were seen as equally important yet overlapping, and were defined as three mutually dependent entities, which govern, control and enable each other and do not allow one entity to expand without stretching the others.

In our paper there is a strong emphasis on *production* (2010: 153: "gain a more integrated view of textile technology and production") that is also found in the scholarly literature inspired by historical materialism, craft studies, ethnology and folklore. By contrast, our paper focused less on consumption, use, or re-use, which are themes that have received more attention in consumer studies and identity studies such as sociology, anthropology, and studies of religion.

Regarding technology, our paper strongly emphasised functional typologies and functional analysis, ("function and efficiency", 2010: 163) and the use of functionality seemed to serve as leverage to assume a certain universality in the results, detached from cultures, peoples, and traditions, and adhering only to gravity and practicality.

Another significant theoretical feature of the paper is its assumption and ambition to speak of textiles *disregarding context, time and place*. "textiles are..." is a frequent catchphrase, and hence introduced a certain universality to the theme. There was clearly the ambition to establish general rules and universal laws, and textiles were naturally seen in the long perspective, with the only boundaries being the huge time span: from 10,000 years ago to industrialisation (2010: 150).

> Whatever the scale of society under investigation, the production of raw material for textiles requires long-term planning and the knowledge of animal husbandry, cultivation technology, and extensive exploitation and management of available natural resources. (2010: 165)

This global and *longue durée* perspective, is probably one of the reasons for our paper's impact. Hence, the paper became a useful guideline for colleagues studying all areas and times of the past.

Our paper's definition of a textile was illustrated in a figure as "*resources*, *technology*, and *society*" (p. 150). These three defining entities could now, 10 years later, be challenged. A textile could equally be defined as *individual choices*, *skills*, and *taste*. There is indeed a remarkable absence of people in the original paper: crafts, techniques and tools seem to move and act on their own, and the paper never ventured into exploring the bodies, genders, ages and minds of users.

Further, our paper carves out a new position for textile fibres as markers for cultures, tecnology and even chronology. The introduction of wool exploitation was coined 'fibre revolution' by J. McCorriston back in 1997. In the past decade, the fibre shift has generated a series of new concepts and interpreations, especially for the ancient Near East and the Meditteranean region, based on the economic, aestetic, technological and social meaning of wool; we now speak of wool economy, wool revolution, wool zone and wool age (Breniquet and Michel 2014; Nosch 2015; Becker et al. 2016; Andersson Strand and Nosch 2019; Bergerbrandt and Sabatini 2019).

Finally, our paper clearly saw great potential in the new analyses from the natural sciences, while also showing awareness of the risk of overstating their effects:

> Many of these more specialized methods and approaches are catalysts that can widen a microscopic view of a textile as an object or type of material leading towards a macroscopic vision of its function, meaning, and role in society. (2010: 154)

Our 2010 statement anticipated the seminal paper by Kristian Kristiansen (2014: 20) and its enthusiasm for new analytical methods. Kristiansen argued that we are entering the third revolution in archaeology, in which science offers an alternative to agency-based, contextual interpretation. This was not fully our position in 2010, as can be seen in our conclusion to the paper:

> Over the last 20 years, the field of textile archaeology has been enriched by a series of new scientific methods (isotopic tracing, radiocarbon dating, HPLC, SEM, X-ray spectroscopy, CAT-scanning, DNA analysis). The available methods, however, are not an end in themselves; rather, they are tools to be used in order to answer broader questions and are useful only when adapted to the aims of archaeology: to trace provenance, to understand chronology, to decode the craft, and ultimately to understand past societies. (2010: 164–165)

Ten years later, we have become more cautious. Indeed, as Tim Flohr Sørensen has criticised, currently in archaeology, these new and quite expensive tests are based on very little data but their results are treated as evidence for entire cultures and populations (Sørensen 2017). Kristiansen argues for a new theoretical paradigm in which the latest methods in science must play very important roles. However, natural sciences are seen as objective, quantifying, falsifiable, empirical, descriptive, factual and aiming at discovery, explanation, and solutions; the humanities, by contrast, promote the subjective, interpretative, theorising, speculative, politicising, and aesthetic, and aims at an understanding, intervention and at producing problems (Sørensen 2017). It is important that in textile research we include but also merge both these theoretical perspectives.

2.7 New Horizons

As mentioned in the introduction, the field of textile research has certainly expanded since 2010. The consolidation of several textile-specific journals, such as *Textile History, Clothing and Textiles Research Journal,* and *Ancient Textiles Review* has secured the rapid, free and peer-reviewed publication of many textile contributions. The *Ancient Textile Series* published by Oxbow Books in collaboration with CTR has issued more than 37 volumes since 2007 with more still to follow. The Berg/Bloomsbury series of *A Cultural History of Dress and Fashion* is an important medium for disseminating both research and standard knowledge of textile history. Many of these contributions could otherwise have been overlooked and forgotten in appendices and notes; instead, the increasing number of publications has made textile research more accessible and visible.

Since 2010 the number of textile research projects, individual as well as larger research programs has increased. One example are the many Marie Skłodowska-Curie fellowships that have contributed with money and research to the field of textile research at CTR and other institutions in Europe since 2011. These have also fostered internationalisation and mobility and contributed an invaluable pool of knowledge. In its essence the Marie Skłodowska-Curie Action programme is based on scholarly curiosity and exchange of knowhow. It has been an action been ideal for bringing together textile researchers otherwise scattered across Europe. ERC starting and consolidator grants have enabled important textile research projects, including the HERA project, *Creativity in the Bronze Age,* which has shown the potential of looking at textile craft in combination with other crafts (Bender Jørgensen et al. 2018). These grants enable aspects of innovation and training. We also see a great potential in putting more emphasis on aspects of cross-craft technologies in the future. The newly funded and ongoing COST Action – *EuroWeb* with its aim *to formulate a new vision of European history based on textiles* where textile researchers from most European countries come together may prove to be exactly such a hub, where exciting new research and unforeseen research field combinations can develop and flourish (Nosch et al. 2020).

In spite of the introduction of the many important analyses mentioned in 2010, it is mainly dye analysis that has succeeded in providing a commercially viable business, although the cost per sample is still high for those conducting textile studies at their own expense or without access to large-scale external funding. No doubt dye analysis and dyeing technology is a field that will receive increased attention in the near future due to its potential for visualising the past and its exciting cross-craft links to food production and conservation. Fibre analysis is still only performed in a few places in Europe, which is regrettable considering the research potential of fibre studies for visual appearance and provenance studies. However, scientific methods are constantly being developed and, hopefully, the future will bring new digital or less costly solutions that can change fibre analysis into an imperative standard tool in textile research. To broadening this field a new initiative ‘CTR TexLab:

Interdisciplinary hub for the study of past cloth cultures' has been initiated. TexLab is also a partner in the Danish Roadmap for Research Infrastructure 2020 E-RIHS.dk (Danish Research Infrastructure for Heritage Science), and together with our colleagues we look forward to developing this field from a scientific as well as theoretical perspective.

2.8 Conclusion

Textile research has shown that the field fully lives up to modern international scientific standards in terms of scientific conduct and publication of research. As a field, it is today at the frontline of interdisciplinary collaboration and project creation. These days, it is not an obstacle to an academic career to specialise in textile research, but neither must a scholar chose to focus their entire scholarly capital in this one field in order to be considered a specialist; instead, we observe the next generation of young scholars with more diverse professional profiles and with textile research as one of their strong fields among other areas of expertise. Textile research has thus benefitted from opening up the field; new scientific standards communicated transparently and disseminated internationally, as well as the development and use of new methods, as presented in the article from 2010, have made it easier for newcomers to enter textile research.

The future is unpredictable and textile research has already developed in so many unexpected ways. It is not our task to say which directions textile research should take but we hope this field will continue to grow. We are confident that the next generation of textile scholars will set new standards and investigate new fields. Their research will undoubtedly continue to demonstrate the important impact of textiles.

Acknowledgement We thank our 2010 co-authors Irene Skals, Margarita Gleba and Karin Margarita Frei for the collaboration, and we thank the editors for the opportunity to reflect upon our work 10 years later.

References

Andersen VL (2017) Old shoes in a new perspective – fashioning archaeology. *Fashion Practice* 9(2):168–182.

Andersson Strand E, Frei KM, Gleba M, Mannering U, Nosch M-L, Skals I (2010) Old textiles – new possibilities. *European Journal of Archaeology* 13(2):1–25.

Andersson Strand E, Heller S-G (2017) Production and distribution. In: Heller S-G (ed.) *Fashion in the Medieval Age (500–1450). A cultural history of dress and fashion.* Volume 2. Berg/Bloomsbury, London, p 29–52.

Andersson Strand E, Mannering U (2014) An exceptional woman from Birka. In: Bergerbrant S, Fossøy S H (eds) *A stitch in time. Essays in honour of Lise Bender Jørgensen.* GOTARC Series A. Department of Historical Studies, Gothenburg University, Gothenburg, p 301–316.

Andersson Strand E, Nosch M-L (eds) (2015) *Tools, textiles and contexts. Investigating textile production in the Aegean and Eastern Mediterranean Bronze Age.* Ancient Textile Series 21. Oxbow Books, Oxford and Philadelphia.
Andersson Strand E, Mannering U, Nosch M-L (2016) Mise en œuvre d'une approche globale des textiles anciens au Centre de recherche sur les textiles de Copenhague. *Perspective. La revue de l'INHA* 1:157–170.
Andersson Strand E, Lindgren S, Larsson C (2018) Motion capture and textile experimental archaeology a possible combination. In: Gleba M, Laurito R (eds) *Contextualising textile production in Italy in the 1st millennium BC.* ORIGINI, Prehistory and Protohistory of Ancient Civilisations XL 2017. Gangemi Editore International, Roma, p 129–140.
Andersson Strand E, Nosch M-L (2019) The wool zone in Prehistory and Proto-History. In Bergerbrandt S, Sabatini S (eds) *The textile revolution in Bronze Age Europe. Production, specialisation, consumption.* Cambridge University Press, Cambridge, p 15–38.
Andersson Strand E, Demant I, Wallenberg M et al. (forthcoming) Reconstructions of Viking Age textiles. From fibre analyses to finished product.
Becker C, Benecke N, Grabundžija A, Küchelmann H-C, Pollock S, Schier W, Schoch C, Schrakamp I, Schütt B, Schumacher M (2016) The Textile Revolution. Research into the origin and spread of wool production between the Near East and Central Europe. In: Graßhoff G and Meyer M (eds) *eTOPOI. Journal for Ancient Studies.* Special Volume 6:102–151.
Belanová Štolcova T, Grömer K (2010) Loom weights, spindles and textiles – textile production in central Europe from the Bronze and Iron Age. In: Andersson Strand E, Gleba M, Mannering U, Munkholt C, Ringgaart M (eds) *North European symposium for archaeological textiles X.* NESAT X. Ancient Textiles Series 5. Oxbow Books, Oxford and Oakville, p 9–20.
Bender Jørgensen L, Rast-Eicher A (2016) Innovations in European Bronze Age textiles. *Praehistorische Zeitschrift* 91(1):68–102.
Bender Jørgensen L, Sofaer J, Stig Sørensen ML (2018) *Creativity in the Bronze Age. Understanding innovation in pottery, textile, and metalwork production.* Cambridge University Press, Cambridge.
Bergerbrant S (2019) Revisiting "the Egtved girl". In: Berge R, Henriksen MM (eds) *Arkeologi og kulturhistorie fra norskekysten til Østersjøen. Festskrift til professor Birgitta Berglund.* NTNU, Trondheim, p 18–39.
Bergerbrandt S, Sabatini S (2019) (eds) *The textile revolution in Bronze Age Europe. Production, specialisation, consumption.* Cambridge University Press, Cambridge (2020).
Brandt LØ, Mannering U (2020) Taxonomic identification of Danish Viking Age shoes and skin objects by ZooMS (Zooarchaeology by mass spectrometry). *Journal of Proteomics* 231:104038.
Brandt LØ, Tranekjer LD, Mannering U, Ringgaard M, Frei KM, Willerslev E, Gleba M, Gilbert MTP (2011) Characterising the potential of sheep wool for Ancient DNA analyses. *Archaeological and Anthropological Sciences* 3:209–221.
Brandt LØ, Schmidt AL, Mannering U, Sarret M, Kelstrup CD, Olsen JV, Cappellini E (2014) Species identification of archaeological skin objects from Danish bogs: Comparison between Mass Spectrometry-based sequencing and microscopy-based methods. *PLOS ONE*, September 2014, Volume 9(9):1–10. http://journals.plos.org/plosone/article?id=10.1371/journal.pone.0106875, accessed 21 Jan 2021.
Brandt LØ, Ebsen JA, Haase K (2020) Leather shoes in early Danish cities: Choices of animal resources and specialization of crafts in Viking and medieval Denmark. *European Journal of Archaeology* 23(3):428–450.
Breniquet C, Michel C (2014) *Wool economy in the Ancient Near East and the Aegean: From the beginnings of sheep husbandry to institutional textile industry.* Ancient Textiles Series 17. Oxbow Books, Oxford and Oakville.
Ebert C, Frisch S, Harlow M, Andersson Strand, Bjerregaard L (eds) (2018) *Traditional textile craft – an intangible cultural heritage.* https://www.traditionaltextilecraft.dk/e-publication, accessed 21 Jan 2021.
Ebsen JA, Haase K, Larsen R, Sommer DVP, Brandt LØ (2019) Identifying archaeological leather – discussing the potential of grain pattern analysis and zooarchaeology by mass spec-

trometry (zooms) through a case study involving medieval shoe parts from Denmark. *Journal of Cultural Heritage* 39:21–31. https://doi.org/10.1016/j.culher.2019.04.008.

Frei KM, Frei R, Mannering U, Gleba M, Nosch M-L, Lyngstrøm H (2009) Provenance of ancient textiles – a pilot study evaluating the strontium isotope system in wool. *Archaeometry* 51(2):252–276.

Frei KM, Mannering U, Kristiansen K, Allentoft ME, Wilson AS, Skals I, Tridico S, Nosch M-L, Willerslev E, Clarke L, Frei R (2015) Tracing the dynamic life story of a Bronze Age female. *Scientific Reports* 5(10431). https://doi.org/10.1038/srep10431.

Frei KM, Mannering U, Vanden Berghe I, Kristiansen K (2017) Bronze Age wool; provenance and dye investigations of Danish textiles. *Antiquity* 91(357):640–654. https://doi.org/10.15184/aqy.2017.64.

Frei R, Frei KM, Jessen S (2019) Shallow retardation of the strontium isotope signal of agricultural liming – implications for isoscapes used in provenance studies. *Science of the Total Environment* 706(135710). https://doi.org/10.1016/j.scitotenv.2019.135710.

Gleba M (2012) From textiles to sheep: Investigating wool fibre development in pre-Roman Italy using scanning electron microscopy (SEM). *Journal of Archaeological Science* 39(12):3643–366. https://doi.org/10.1016/j.jas.2012.06.021.

Gleba M, Mannering U (eds) (2012) *Textiles and textile production in Europe from prehistory to AD 400*. Ancient Textiles Series 11. Oxbow Books, Oxford and Oakville.

Grömer K (2013) Discovering the people behind the textiles: Iron Age textile producers and their products in Austria. In: Gleba G, Pásztókai-Szeőke J (eds) *Making textiles in Pre-Roman and Roman times. People, places, identities*. Ancient Textiles Series 13. Oxbow Books, Oxford and Oakville, p 30–59.

Grömer K, Kern A, Reschreiter H, Rösel-Mautendorfer H (2013) *Textiles from Hallstatt. Weaving culture in Bronze Age and Iron Age salt mines. Textilen aus Hallstatt. Gewebe Kultur aus dem Bronze- und ersenzeitlichen Salzbergwerk.* Archeolingua, Budapest.

Hofmann-de-Keijzer R, van Bommel MR, Hartl A, Grömer K, Rösel-Mautendorfer H, Reschreiter H, Kania K, Joosten I, Proaño Gaibor AN, Erlach R, Lachner E, Wandl M, de Keijzer M (2013) Coloured Hallstatt textiles: 3500 year-old textile and dyeing techniques and their contemporary application. In: Banck-Burgess J, Nübold C (eds) *The North European symposium for archaeological textiles XI, 10. –13. Mai 2011 in Esslingen am Neckar*. NESAT XI. Verlag Marie Leidorf GmbH, Rahden/Westf, p 85–92.

Hammarlund L, Vestergård Pedersen K (2007) Textile appearance and visual expression – Craft knowledge applied to archaeological textiles. In: Rast-Eicher A, Windler R (eds) *Archäologische Textilfunde – Archaeological textiles. NESAT IX, Braunwald, 18.-21. Mai 2005*. NESAT IX. Archeo Text, Ennenda, p 213–219.

Harris S (2008) Textiles, cloth, and skins: The problem of terminology and relationship. *Textile* 6(3):222–237.

Harris S (2012) From the parochial to the universal: Comparing cloth cultures in the Bronze Age. *European Journal of Archaeology* 15(1):61–97.

Kania K (2013) Soft yarns, hard facts? Evaluating the results of a large-scale hand-spinning experiment. *Archaeological and Anthropological Sciences* 7(2015):113–130. https://doi.org/10.1007/s12520-013-0167-y.

Kootker LM, von Holstein ICC, Broeders J, Wescott DJ, Davies GR, Mickleburgh HL (2020) The effects of decomposition and environment on antemortem H-Pb-Sr isotope compositions and degradation of human scalp hair: actualistic Taphonomic Observations. *Forensic Science International* 312:110336. https://doi.org/10.1016/j.forsciint.2020.110336.

Kristiansen K (2014) Towards a new paradigm? The third science revolution and its possible consequences in archaeology. *Current Swedish Archaeology* 22:11–34.

Landenius Enegren H (2015) Loom weights in Archaic South Italy and Sicily. *Opuscula: Annual of the Swedish Institutes at Athens and Rome* 8:123–155. https://doi.org/10.30549/opathrom-08-06.

Mannering U, Possnert G, Heinemeier J, Gleba M (2010) Dating Danish textiles and skins from bog finds by means of 14C AMS. *Journal of Archaeological Science* 37:261–268.

Mannering U (2018) Fashioning the Viking Age. *Archaeological Textiles Review* 60:114–117.

McCorriston J (1997) The fiber revolution: textile extensification, alienation and social stratification in ancient Mesopotamia. *Current Anthropology* 38(4):517–549.
Meo F (2014) New archaeological Data for the Understanding of Weaving in Herakleia, Southern Basilicata, Italy. In: Harlow M, Nosch M L (eds) *Greek and Roman textiles and dress. An interdisciplinary anthology.* Ancient Textiles Series 19. Oxbow Books, Oxbow and Philadelphia, p 236–259.
Nosch M-L (2015) The Wool Age: Textile traditions and textile innovations. In: Ruppenstein F, Weilhartner F (eds) *Tradition and innovation in the Mycenaean palatial polities*. Mykenische Studien 34:167–201.
Nosch M-L, Ulanowska A, Yvanez E (2020) EuroWeb – a new European network and COST Action 2020–2024. *Archaeological Textiles Review* 62:183–186.
Olofsson L, Andersson Strand E, Nosch M-L (2015) Experimental testing of Bronze Age textile tools. In: Andersson Strand E, Nosch M-L (eds) *Tools, textiles and contexts. Investigating textile production in the Aegean and Eastern Mediterranean Bronze Age.* Ancient Textiles Series 21. Oxbow Books, Oxford and Philadelphia, p 75–100.
Ringgaard M (2007) A purple knitted silk among brown rags. Excavated textiles from an 18th century rubbish dump in Copenhagen. In: Johansen K (ed.) *Costume – Design and decoration. ICOM's Costume Committee. Proceedings from the 58th annual conference, 9–13th October, 2006, Copenhagen, Denmark and Lund, Sweden.* Costume Committee, Copenhagen, p 93–97.
Ringgaard M (2017) Framing early modern knitting. In: Welch E (ed.) *Fashioning the early modern: Dress, textiles, and innovation in Europe, 1500–1800.* Oxford University Press, Oxford, p 283–313.
Rast-Eicher A (2008) *Textilien, Wolle, Schafe der Eisenzeit in der Schweiz.* Antiqua 44. Archäologie Schweiz, Basel.
Rast-Eicher A (2016) *Fibres: Microscopy of archaeological textiles and furs.* Archaeolingua, Budapest.
Rast-Eicher A, Bender Jørgensen L (2013) Sheep wool in Bronze and Iron Age Europe. *Journal of Archaeological Science* 40:1224–1241.
Ryder M L (1983) *Sheep and man.* Duckworth, London.
Sarauw T (2019) *Bejsebakken: en nordjysk bebyggelse fra yngre jernalder og vikingetid.* Det Kongelige Nordiske Oldskriftselskab, Syddansk Universitetsforlag, København.
Skals I, Mannering U (2014) Investigating wool fibres from Danish prehistoric textiles. *Archaeological Textiles Review* 56:24–34.
Skals I, Gleba M, Taube M, Mannering U (2018) Wool textile and archaeometry: testing reliability of archaeological wool fibre diameter measurements. *Danish Journal of Archaeology* 7(2):1–19. https://doi.org/10.1080/21662282.2018.1495917.
Sørensen TF (2017) The two cultures and a world apart: Archaeology and science at a new crossroads. *Norwegian Archaeological Review* 50(2):101–115.
Thomsen E, Andreasen R (2019) Agricultural lime disturbs natural strontium isotope variations: implications for provenance and migration studies. *Science Advances* 5(eaav8083):1–11.
Toxværd S (2020) The strontium isotope ratio 87Sr/86Sr in archaeological organic matter conserved in acidic anaerobic environments is hard to interpret. *Journal of Archaeological Science: Reports* 31:102379. https://doi.org/10.1016/j.jasrep.2020.102379.
Vajanto K (2015) *Dyes and dyeing methods in Late Iron Age Finland.* Unpublished dissertation, University of Helsinki.
Vanden Berghe I, Gleba M, Mannering U (2009) Towards the identification of dyestuffs in Early Iron Age Scandinavian peat bog textiles. *Journal of Archaeological Science* 36:1910–1921.
Vestergård Pedersen K, Nosch M-L (2009) *The medieval broadcloth: Changing trends in fashions, manufacturing and consumption.* Ancient Textiles Series 6. Oxbow Books, Oxford and Oakville.

Eva Andersson Strand is Associate Professor in Textile Archaeology. She is one of the three founders of the Centre for Textile Research in Copenhagen, Denmark. Her research focus is the integration of the impact of textiles into a wider economic and culture discussion of ancient societies. She has worked on a number of time periods, such as Viking Age Scandinavia, the Bronze Age Aegean, and the Ancient Near East. Her research has combined comprehensive context analysis and experimental and digital archaeology. Since 2017, Eva Andersson Strand is the Director of the Centre for Textile Research.

Ulla Mannering is Research Professor (MSO) at the National Museum of Denmark, and has responsibility for the unique Danish collection of archaeological textiles. She is one of the three founders of the Centre for Textile Research in Copenhagen. Ulla Mannering has dedicated years of research to the study of the development of Bronze and Iron Age textile design, as well as costume design, iconography and textile production in Late Iron and Viking Age Scandinavia.

Marie-Louise Nosch is Professor in Ancient Greek History at the SAXO Institute at University of Copenhagen, where she was the Director of the Centre for Textile Research in 2005–2016. Her research focuses on Mycenaean and ancient Greek production and use of textiles, with emphasis on technology, terminology and gender. Marie–Louise Nosch is currently President of the Royal Danish Academy of Sciences and Letters.

Part II
Interdisciplinarity of Colour: Dye Analyses and Dyeing Technologies

Chapter 3
Monobromoindigo: The Singular Chromatic Biomarker for the Identification of the Malacological Provenance of Archaeological Purple Pigments from *Hexaplex Trunculus* Species

Zvi C. Koren

Abstract The determination of the malacological provenance of archaeological purple pigments used in textile dyeings and in wall and stone paintings is possible due to the chromatic fingerprinting of the pigment into its individual components via High Performance Liquid Chromatography. The common component of all molluskan purple pigments from all oceans and seas is the doubly-brominated indigo dye, dibromoindigo (DBI). However, the singularity of the *Hexaplex trunculus* sea snails that sets them apart from all other purple-producing whelks is that they contain colorants unique to this species. Besides DBI, *H. trunculus* pigments also contain monobromoindigo (MBI) as well as indigo (IND), where the latter two are negligible in other species. While indigo can also be extracted from certain plant sources, the chromatic biomarker that is unique to *H. trunculus* pigments is the existence of a significant quantity of MBI. All archaeological pigments analyzed to date contain a sizable amount of MBI dye, and hence all purple pigments were produced in antiquity from the *H. trunculus* species. Further, two varieties of this species exist: one produces DBI-rich reddish-purple pigments, and the other yields IND-rich bluish-purple pigments. Other reddish *Muricidae* pigments were generally only used as additions to reddish-purple *H. trunculus* pigments.

Keywords Pigments · Dibromoindigo · Monobromoindigo · Indigo · *Hexaplex trunculus* · Purple · HPLC

Z. C. Koren (✉)
The Edelstein Center for the Analysis of Ancient Artifacts, Department of Chemical Engineering, Shenkar College of Engineering, Design and Art, Ramat-Gan, Israel
e-mail: zvi@shenkar.ac.il

A. Ulanowska et al. (eds.), *Ancient Textile Production from an Interdisciplinary Perspective*, Interdisciplinary Contributions to Archaeology,
https://doi.org/10.1007/978-3-030-92170-5_3

3.1 Introduction

Once upon a time, a bit over a century ago, it was believed that the colorant present in the purple pigment produced from molluskan sources was essentially the doubly-brominated indigo dye, chemically known as 6,6′-dibromoindigo (abbreviated as DBI). This determination was based on the laborious work of Friedländer when he identified DBI as the colorant in the pigment produced from thousands of sea snails (Friedländer 1909). This dye was previously synthetically produced by the chemists Sachs and Kempf (1903). More than two decades after Friedländer's discovery, Pfister (1935) devised a series of colorimetric tests in order to detect the possible presence of this DBI dye – as well as dyes from other sources – in purple-colored archaeological textiles by adding various solvents to archaeological samples and comparing the resultant colors with those using a standard sample of DBI.

In order to break away from the subjectivity of visual color comparisons for the identifications of all types of historical dyes from flora and fauna sources, relatively objective instrumental methods were utilized. One of the first such techniques that was used for the analysis of ancient dyes was infrared (IR) spectroscopy performed by Edelstein and Abrahams who analyzed second century CE archaeological textile dyeings found in the Judean Desert of Ancient Israel (Abrahams and Edelstein 1963, 1964). Instrumental dye analyses were also performed via visible spectrophotometry by analyzing solutions of the investigated dyes, and a review of this technique was given by Saltzman (1992).

An example of the application of visible spectrophotometry specifically for the identification of a malacological purple pigment is from a dyeing vat assembly from the sixth century BCE Phoenician site in Tel Dor in modern north-central Israel (Koren 1993). The analysis of a dimethyl formamide (DMF) extract of a dark residue on a small pebble-sized limestone found in this complex centered on the ability of detecting the presence of DBI in the stain, and if so, then that pigment is from a molluskan source. Figure 3.1 shows the visible spectra of DMF solutions of standard indigo (IND) and dibromoindigo (DBI) dyes, as well as of the dark residue from the pigmented Tel Dor limestone. Each pigment was dissolved in DMF at 100 °C for 5 min. Interestingly, one cannot visually distinguish between DBI and IND in solution since DMF solutions of the reddish-purple DBI are blue and, to the naked eye, virtually identical to the blue color of indigo solutions. As seen spectrophotometrically, both of these two solutions absorb visible light at around 600 nm. However, the Tel Dor pigment solution shows a wavelength at maximum absorption, λ_{max}, that is nearly identical to that of DBI (600 vs. 598 nm, respectively). This clearly indicates that the dyestuff source of this pigment is malacological. Further, as also noted in that publication (Koren 1993), the fact that the characteristic wavelength of the archaeological pigment is slightly above that of DBI, implies that the colorants in the pigment also consists of MBI and/or IND, which show λ_{max} values that are somewhat higher than that of DBI.

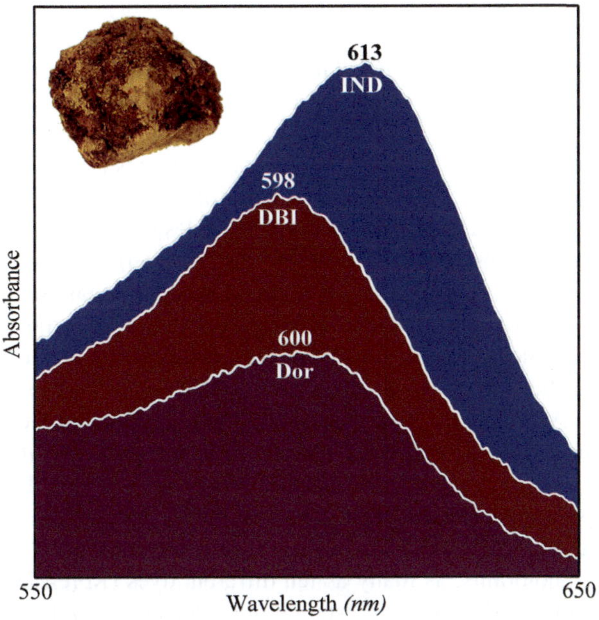

Fig. 3.1 Visible spectra of DMF solutions of indigo (IND), dibromoindigo (DBI), and the purple residue on a small Limestone (also shown) from sixth century BCE Tel Dor, Israel. (© Zvi Koren)

In order to understand fully the detailed composition of molluskan purple pigments, the micro-separation method of High Performance Liquid Chromatography (HPLC) was utilized.

3.2 HPLC Analyses of Purple Pigments

The optimal analytical technique for the study of multicomponent mixtures, such as possibly in the purple pigment, is HPLC. It was first used for the analysis of molluskan purple pigments in the early 1990s (Wouters and Verhecken 1991; Wouters 1992).

Prior to an HPLC analysis, the molluskan pigment must be extracted with an appropriate organic solvent at elevated temperatures. Hot pyridine was first used for this dissolution (Wouters and Verhecken 1991; Wouters 1992); however, the noxious fumes of that solvent, even when performed in a fume hood, resulted in its disuse. Shortly following these publications, the first use of hot DMF for the detection of a malacological purple was reported (Koren 1993). Subsequently, it was observed that hot DMSO (dimethyl sulfoxide) was very effective in extracting the red carthamin component from safflower and its dyeings (Koren 2001). Following the successful application of this solvent on safflower, it was then investigated for the extraction of the various brominated indigoids found in purple pigments. Indeed, hot DMSO was found to be the optimal solvent for such extractions and its first use on molluskan pigments was published in 2008 (Koren). The effectiveness of DMSO for such extractions has also been later re-confirmed by Karapanagiotis et al. (2013).

The results of HPLC analyses show that the proposition that DBI is overwhelmingly the dominant colorant in molluskan purple pigments is only partially correct. It is factual to say that the common denominator of all purple pigments from all sea snails is indeed the presence of DBI, and this includes the mollusks from all seas of the world. These mollusks can be found in the Mediterranean waters of various regions: the Levantine countries in the Eastern Mediterranean, or in the northern Mediterranean area of, for example, Greece and Italy, or to the south in North Africa. Purple-producing sea snails can also be found well beyond the Mediterranean, for example, off the coast of Japan, along the Pacific Ocean of Mexico and Central America, or the Atlantic Ocean of France and the UK. Because the DBI dye is not found in any plant or insect dyes, its detection automatically signifies that the pigment in which it is present is from a marine source. However, HPLC analyses performed in the past three decades have shown that while some of the purple pigments produced from these snails contain nearly only DBI, other pigments can contain many additional colorants, as many as ten different dyes (Koren 2008). In fact, in the purple pigment produced from certain molluskan species, DBI may not even be the major constituent. Table 3.1 lists the common and abbreviated names of the known dyes. The chemistry associated with them has been previously given (Cooksey 2001, 2013; Karapanagiotis 2019).

The primary question regarding these HPLC analyses was whether such investigations can point to specific molluskan species that were used to produce purple pigments in antiquity. A review of the major murex-purple producing workshops in the Mediterranean has recently been given (Kalaitzaki et al. 2017). Such pigments were found to be used as early as nearly four millennia ago in Minoan wall paintings and later also for painting of objects (Karapanagiotis 2019), as well as for transforming the insoluble pigment to a soluble dye for textile dyeing, more than three and a half millennia ago (Koren 2013, 2020; Breniquet et al. 2018). Towards this

Table 3.1 Common and abbreviated names of the known dyes in molluskan purple pigments

Abbreviated name	Common name
IS	Isatin
6BIS	6-Bromoisatin
IND	Indigo (also referred to as indigotin)
MBI	6-Monobromoindigo
DBI	6,6′-Dibromoindigo
INR	Indirubin
6MBIR	6-Monobromoindirubin
6’MBIR	6′-Monobromoindirubin
DBIR	6,6′-Dibromoindirubin

goal, HPLC analyses of pigments produced from various Mollusca from around the world were performed. Since the material culture of purple usage was dominant in the Mediterranean basin, especially in the Eastern Mediterranean region, special focus was given to the three *Muricidae* species that inhabit these waters: *Hexaplex (= Murex) trunculus, Bolinus (= Murex) brandaris,* and *Stramonita (= Purpura) haemastoma.*

The results of the HPLC analyses are given in Figs. 3.2, 3.3 and 3.4. The chromatograms are shown in Fig. 3.2, the respective spectra and molecular structures of the dye components in Fig. 3.3, and the semi-quantitative constitutions of the pigments in Fig. 3.4.

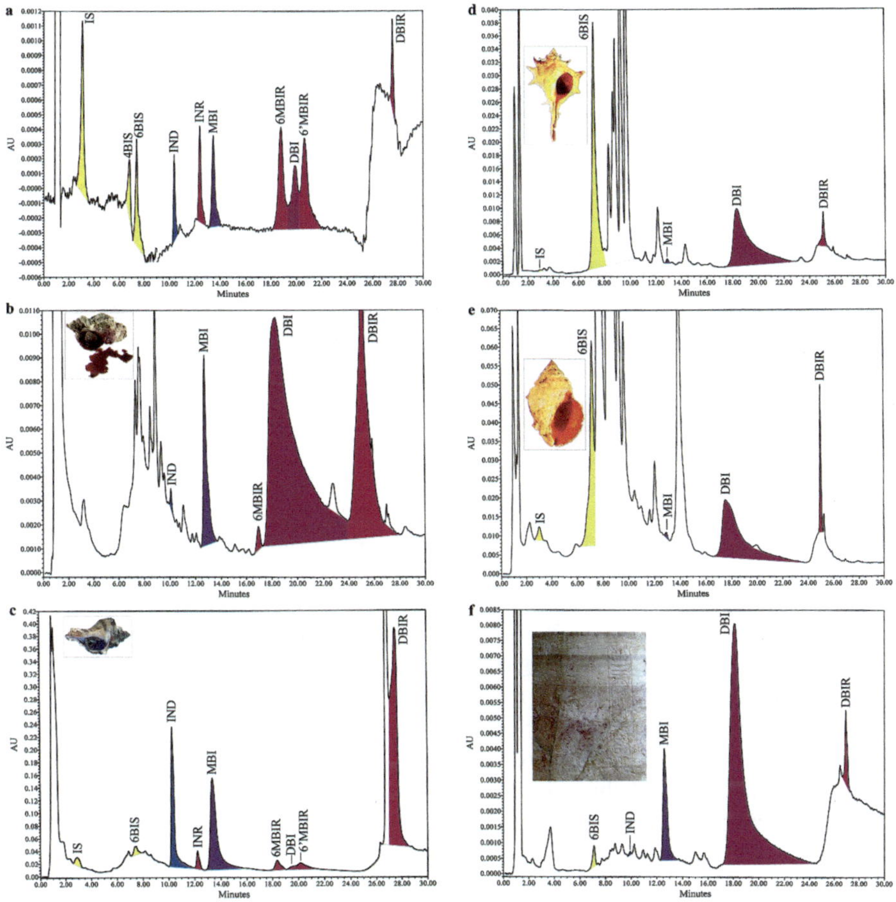

Fig. 3.2 Chromatograms of DMSO solutions of (**a**) 10 standard dyes, (**b**) red-purple pigment from *Hexaplex trunculus* (at 290 nm), (**c**) blue-purple pigment from *H. trunculus* (at 300 nm), (**d**) *Bolinus brandaris* (at 288 nm), (**e**) *Stramonita haemastoma* (at 288 nm), (**f**) purple pigment on a 2500-year old king Darius marble jar (at 300 nm); wavelengths chosen for visualization purposes only. (© Zvi Koren)

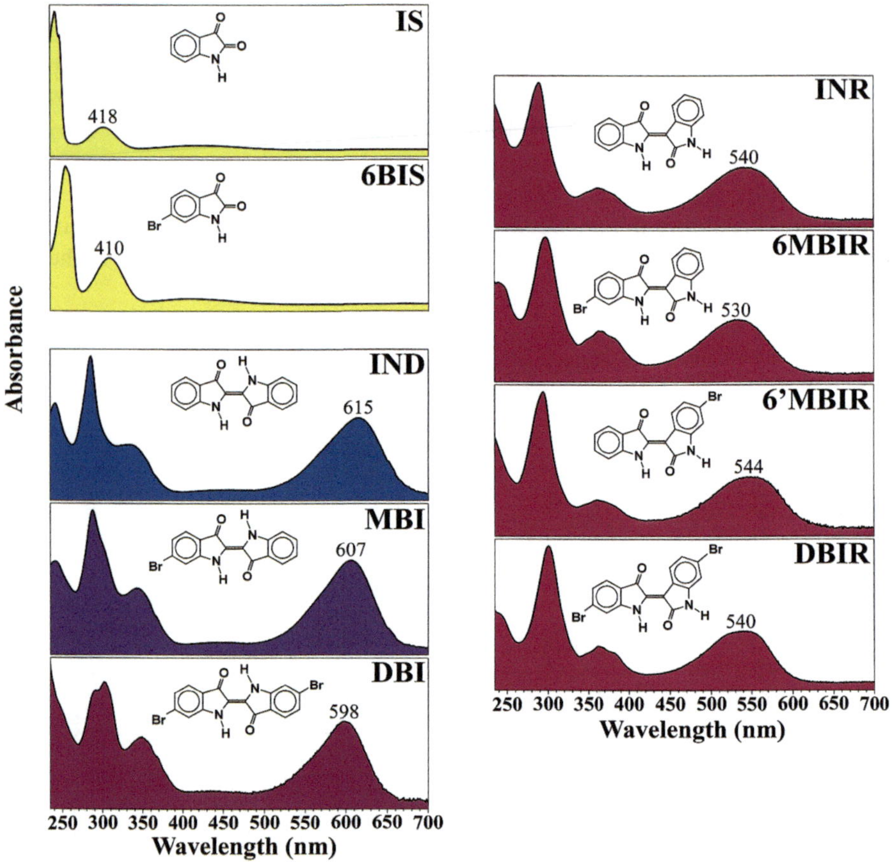

Fig. 3.3 UV/Vis spectra and visible wavelengths at maximum absorption (λ_{max}) of DMSO solutions of nine standard dyes (with their molecular structures) as produced by the photo-diode array (PDA) detector of the HPLC, and their respective colors in their solid state. (© Zvi Koren)

Figure 3.2 shows the chromatograms and identified peaks of DMSO solutions of the colorants under investigation. Figure 3.2a shows the separation of the standard reference dyes utilizing the efficient chromatographic elution system published elsewhere (Koren 2008). The chromatograms of DMSO solutions of purple pigments from each of the three sea snail species (Figs. 3.2b–e), as well as for an archaeological pigment painted on an early fifth century BCE marble jar ascribed to the Persian King Darius the Great (Fig. 3.2f), are also shown. In addition, it was visually observed that the *H. trunculus* species could produce two varieties of purple pigments – one that is reddish-purple while the other is bluish-purple (violet) – and these two types and their respective chromatograms are shown in Fig. 3.2b and c.

The respective UV wavelengths chosen for displaying the chromatograms of Fig. 3.2 are the ones that clearly show all the detectable colorants, and are not the ones that are used for quantifying the various constituents of each pigment, as

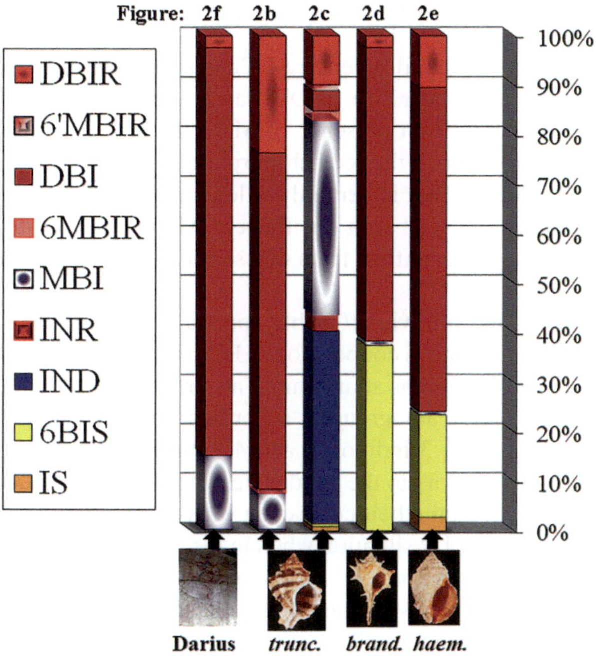

Fig. 3.4 Relative integrated peak areas measured at 288 nm of the nine dyes from the pigments of Fig. 3.2: archaeological Darius pigment, and modern red-purple *H. trunculus*, blue-purple *H. trunculus*, *B. brandaris*, and *S. haemastoma* pigments. (© Zvi Koren)

mentioned below. The other peaks appearing in the chromatograms only absorb in the UV region and thus are non-colorants.

In all of the chromatograms, the presence of DBI is detectable, as expected, though in some pigments it may not be the major component, and can even be an extremely minor constituent. This is seen, for example, in the violet pigment from *H. trunculus* in Fig. 3.2c, which has a major IND content. On the other hand, DBI is the dominant component, as expected, of reddish-purple pigments produced from the other variety of *H. trunculus* (Fig. 3.2b). This interplay between IND and DBI in the two *H. trunculus* varieties is very interesting, as these two dye components appear at opposite constitutional ends. With the presence of a relatively large quantity of DBI (as in Fig. 3.2b), there is only a trace of IND present; whereas with a significant quantity of IND present (Fig. 3.2c), there is only a trace of DBI. The amount of violet MBI dye is substantial in only *H. trunculus* snails – both varieties – and negligible in the other two species – *B. brandaris* and *S. haemastoma*. The nearly exclusive dye in the latter two species is DBI and, as expected, they thus only produce reddish-purple pigments. Of the four possible indirubinoids that may be present in a purple pigment, and listed in Table 3.1, it is mainly DBIR – DBI's isomer – that is dominant in the raw pigments of all types of snails investigated, and also usually present in relatively small quantities in archaeological pigments and dyeings, as observed, e.g., in various HPLC analyses (James et al. 2009; Karapanagiotis 2019; Koren 2014).

The chromatogram of the fifth century BCE pigment on the Persian Darius jar (Fig. 3.2f), shows a preponderance of DBI, as is indicative of its reddish-purple coloration. With the significant presence of MBI, this pigment's chromatographic profile most resembles that of the reddish-purple pigment from the Mediterranean *H. trunculus* (Fig. 3.2b). This is certainly a logical geographic possibility as the Persian Empire extended well into the Eastern Mediterranean and beyond, and undoubtedly used the purple extract from that sea snail species as a decorative paint material.

Figure 3.3 shows the ultraviolet/visible spectra produced by the Photo-Diode Array (PDA) detector, which is coupled to the HPLC, of the relevant dissolved dyes and their respective wavelengths at maximum absorbance, λ_{max}, in the visible range. In the figure, the spectra are grouped according to their molecular functional groups – 2 isatinoids dyes, 3 indigoids, and 4 indirubinoids – together with their respective approximate colors in the solid state. The yellow isatinoids and reddish-purple indirubinoids maintain their color even when in solution. However, the indigoids present a more complex situation when in the dissolved state. While blue IND retains its color when dissolved, as mentioned above, both the violet MBI and reddish-purple DBI form blue solutions, similar to the color of IND, and the visible spectra of their solutions are thus all similar, with all absorbing at a λ_{max} in the same region at a rounded value of about 600 nm.

Figure 3.4 depicts the relative constitution of each colorant in a pigment, based on the integrated peak areas obtained from each of the five chromatograms, as measured at the accepted standard wavelength of 288 nm. This uniform wavelength was chosen because all the dyes in this study have significant absorptions at this value. Though these peak areas are not strictly concentrations – because these values would need to be adjusted according to the absorptivity coefficient of each dye at that wavelength – nevertheless, these semi-quantitative values can be used for comparative interpretations.

Of all the various isatinoids, indigoids, and indirubinoids that could be present in the purple pigment, the main dyes found in archaeological pigments and dyeings are the three indigoids. Their common molecular skeleton and approximate colors on dyed wool are shown in Fig. 3.5. These dyes have been detected in HPLC analyses of archaeological samples from Israel, Greece, Italy, and beyond, as indicated, for example, in the analyses mentioned above. Often some reddish DBIR is also contained in an archaeological pigment, which adds a little more reddish color to that which would have been obtained with the presence of DBI. Though the isatinoids can exhibit a significant presence in the raw pigments from *B. brandaris* and *S. haemastoma*, these isatinoids nevertheless possess at most a trace quantity in archaeological pigments and dyes.

The violet color of solid MBI can be perceived as in-between the colors of blue IND and reddish-purple DBI, and hence the color of a 'purple' pigment from *H. trunculus* mollusks – whether reddish or bluish – would be generally influenced by the relative quantities of DBI vs. IND. The relative peak area values shown in

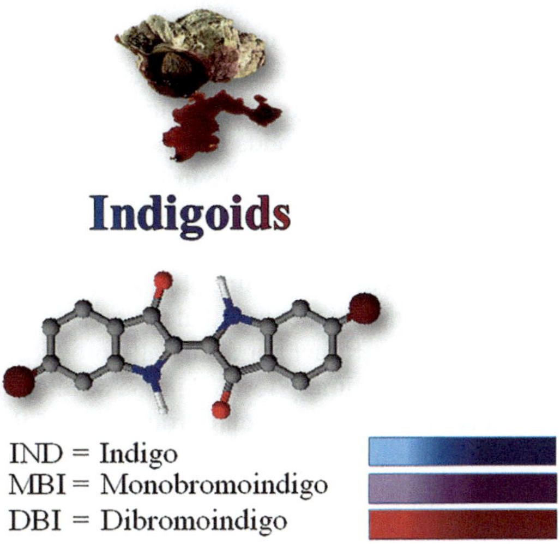

Fig. 3.5 The three main indigoid dyes from *H. trunculus* pigments. (© Zvi Koren)

Fig. 3.4 logically match the visual appearance of the two colors of the pigments from *H. trunculus* species as seen in the inserts of the chromatograms of Fig. 3.2b and c. Thus, it is clear that the reddish-purple colored pigment of Fig. 3.2b is due to the high content of DBI (relative to IND) and that the bluish-purple coloration in the other pigment (Fig. 3.2c) is due to a higher IND content relative to DBI. These two varieties of *H. trunculus* sea snails – and they may even be 'sub-species' of the *H. trunculus* species – are denoted as 'DBI-rich' and 'IND-rich' mollusks, respectively. These two types could have been identified, and separately collected, based on one or more of the three 'S' properties of the snails: sex, size, site. A preliminary investigation of these factors was previously performed (Boesken Kanold and Koren 2009).

3.3 Malacological Provenance of Purple Pigments

Based on the results presented in the previous sections, it is now possible to determine the zoological provenance of malacological purples in archaeological samples. From the chromatograms and composition diagrams of Figs. 3.3 and 3.4, it can be observed, as noted above, that only *H. trunculus* species – both those producing reddish-purple pigments as well as bluish-purple ones – have a considerable quantity of MBI. Analyses of all other sea snail pigments, from all waters, have shown that this dye is negligible or not present at all. This then is the chromatic biomarker that would set *H. trunculus* apart from all the other snails.

It is tempting to point to the presence of a significant quantity of IND in a pigment as originating from the *H. trunculus* species, since in other sea snails, the quantity of IND is nearly non-existent. However, the detection of the presence of IND in an archaeological dyeing cannot be the proverbial 'smoking gun' evidence that indicates that the malacological source of that pigment was the *H. trunculus* snail species. The IND dye originating from indigo-producing plants is of course identical to the IND from a molluskan source. Thus, it is not 'beyond a reasonable doubt' that, for whatever coloristic reason desired by the dyer, indigo from a flora source may have been added to the molluskan purple. There is a number of examples whereby other dyes were in fact added to a molluskan purple, so that we may never know if the detection of IND in an archaeological purple is integral to that molluskan source or was an externally added colorant. Hence, the detection of IND cannot be used as a determiner for the malacological source of the pigment.

It is important to mention that the presence of MBI and IND in *H. trunculus* pigments is not an artifact nor artificially produced in the reduced dye bath, but rather, these dyes are naturally formed in the development of this pigment as performed in antiquity. Today, the artificial production of the MBI dye, as well as IND, can be performed from DBI by simply irradiating a reduced DBI solution with sunlight (or artificial light), which would result in photo-debromination – the successive cleavages of Br from the parent DBI molecule. Nevertheless, this would not have been possible in the ancient processing of the molluskan pigment because of two factors: in antiquity, the dyeing was performed in non-transparent clay vats (not in glass vessels), which were also covered in order to eliminate air (oxygen) from entering the dye bath during the anaerobic reductive fermentation that was occurring (Koren 2013).

Hence, finding the appreciable presence of MBI in an archaeological pigment clearly indicates that a mollusk belonging to one of the two varieties of *H. trunculus* species was used for this dyeing. It may be possible that the pigments from other snails were additionally used, but the presence of MBI signifies that, at least, *H. trunculus* was utilized. To date, all archaeological molluskan purple pigments and dyes that have been investigated by fully extracting them with the appropriate solvent – e.g., DMSO, DMF, or pyridine – and then analyzed via HPLC have detected significant levels of MBI. This is an exceptional result as it signifies that all archaeological molluskan purple pigments and dyes were produced from *H. trunculus* species! Further, both reddish-purple and bluish-purple dyeings would have been produced by the same *H. trunculus* species, through the use of the two varieties of this species mentioned above. Of course, these two color types could have been mixed to produce dyeing colors that are in-between the reddish and bluish colorations obtained with the use of each of these respective types alone.

The other Mediterranean purple-producing molluskan species, *B. brandaris* and *S. haemastoma*, were also in use for dyeing – but not alone. As mentioned above, these two species mainly contain the reddish-purple DBI dye, and thus produce similarly colored reddish dyeings. As the dye content that each of these two species contains is considerably less than from *H. trunculus* species, the latter mollusks would have been the preferred ones to be used. For the production of bluish-colored dyeings, only the IND-rich *H. trunculus* species would have been utilized. Though the other DBI-rich *H. trunculus* species would produce the reddish dyeings, if a redder coloration were desired then additions from *B. brandaris* and/or *S. haemastoma* would have been made. Thus, the roles of the latter two species were seemingly as color additions to the red-purple dyeings that were already produced from the 'reddish' DBI-rich *H. trunculus*. This occurrence was observed in a Late Roman period Egyptian textile whereby two differently colored reddish yarns were observed under magnification (Koren and Verhecken-Lammens 2013). While HPLC analyses showed the appreciable presence of MBI in both reddish yarns – which indicates that the *H. trunculus* species was used for both dyeings – some of the yarns were noticeably redder than the others. The HPLC analyses indicated that these redder dyeings contained a relatively greater amount of DBI than that obtainable from *H. trunculus* alone. This indicates that a species such as *B. brandaris* or *S. haemastoma*, which contains mainly DBI, was added to the DBI content from *H. trunculus* species.

The possibility of finding the use of a non-*trunculus* species in a historical setting – rather than an ancient one – is from late twelfth century Manas (France) and from early twentieth century Leeds (UK). In the first case, a brown wall-paint pigment from a church in the Drôme region of France, probably built during the time of the third Crusade (1187–1192), was reported to have a DBI:MBI 'molar ratio' of 98:2, respectively (March et al. 2011). The authors indicated that the molluskan source of the pigment could have been *B. brandaris*, or an equivalent purple-producing snail, such as *Nucella lapillus*. However, since they did not report on integrated peak area values of the relative components in the HPC chromatogram, a DBI-rich *trunculus* cannot be ruled out.

The most modern – yet, historical – sample of a molluskan purple dyeing is from Leeds of the early 1900s (Whitworth and Koren 2016). This century-old item consisted of purple-dyed twisted cotton yarns from a textile archive belonging to the Bedford family of dyers in Leeds. The floral handwriting on the envelope containing these strings noted its contents as "Tyrian Purple. Dyed with the bodies of shell-fish found on the African Coast." HPLC analyses have shown that the source of the dyeing is indeed molluskan, but not from *H. trunculus*, and based on the geographic locations the malacological source could have been one of the other species mentioned above: *B. brandaris*, *S. haemastoma*, or *N. lapillus*.

Though recently the statistical principal component analysis (PCA) method was used in attempting to identify the molluskan source of a purple pigment (Karapanagiotis et al. 2013; Papliaka et al. 2017), a simpler quantitative property that can resolve the identity of the type of *H. trunculus* pigment used was developed earlier (Koren 2008). This parameter is the Di-Mono Index (DMI), which quantifies the chromatographic peak area ratios (calculated at 288 nm) of the doubly brominated indigo to the singly brominated one. These dyes are unique to the marine world and are thus used in the calculation, since IND is not unique to snails, as mentioned above, and therefore is not used in the evaluated index. Using published HPLC results at 288 nm of modern and archaeological molluskan pigments (Koren 2008; Papliaka et al. 2017), the calculated DMI values fall into an orderly pattern:

DMI < 0.6, IND-rich *H. trunculus*
DMI 0.8–9, DBI-rich *H. trunculus*
DMI > 14 (usually >30), *B. brandaris* and *S. haemastoma*

For archaeological pigments that show a relatively low DMI value of about 1, which is somewhat on the borderline between the IND-rich and DBI-rich *H. trunculus* mollusks, it is possible that the two types of these snails were mixed, as mentioned above.

3.4 Conclusions

The findings in this study regarding the use of the three *Muricidae* in textile dyeings can be schematically summarized in Fig. 3.6. Of all the species of purple-producing sea snails, only the pigments from *Hexaplex trunculus* species contain a sizable quantity of the MBI dye, which can, thus, be used as a chromatic-marker specific to that species. To date, all analyzed molluskan archaeological purples contain an appreciable quantity of MBI colorant, which therefore indicates that they were all produced from *H. trunculus* sea snails! The archaeological proof that *H. trunculus* species were probably always used for the production of the purple pigment and dye is evidenced from the overwhelming accumulations of whole or broken *H. trunculus* shells – typically much more than other species – found in practically all sites associated with molluskan purple (Reese forthcoming). This species consists of two

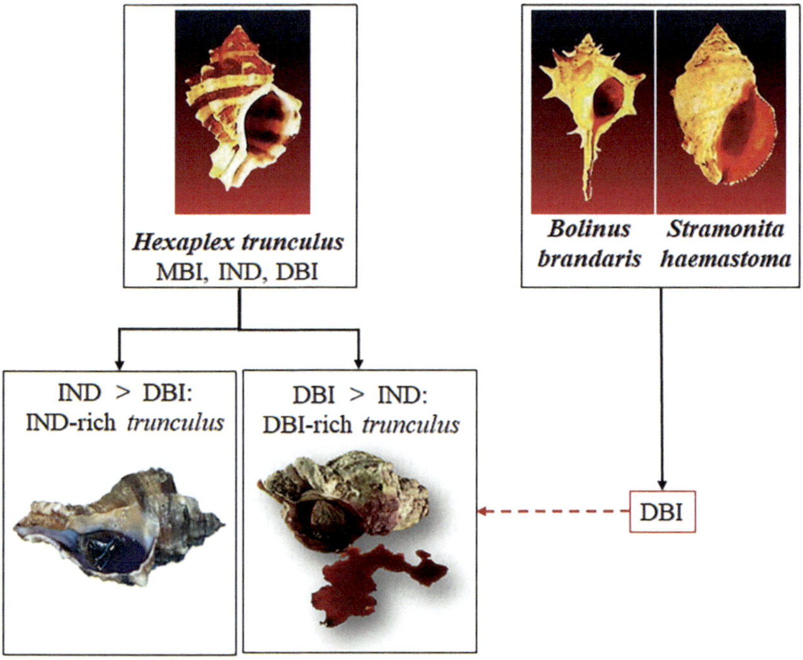

Fig. 3.6 The main compositions of blue-purple pigments and of red-purple pigments from *H. trunculus* species, and of *B. brandaris* and *S. haemastoma* species. (© Zvi Koren)

types – one that is IND-rich and produces the bluish-purple (violet) dyeings, and the other variety is DBI-rich and produces the reddish-purple dyeings, whereby mixing of these two varieties to yield in-between colors was possible.

The other two molluskan species, *B. brandaris* and *S. haemastoma*, were not used alone, but rather were added to the reddish producing *H. trunculus* dyeings in order to produce even redder dyeings.

In conclusion, while in ancient times Royal Purple was the 'Dye of Dyes', the chromatically versatile *Hexaplex trunculus* mollusk was the 'Snail of Snails' in the regal dye houses of antiquity.

Acknowledgments The author is sincerely grateful to the Sidney and Mildred Edelstein Foundation for the support of this research.

References

Abrahams D, Edelstein S (1963) A study of the textiles from the color standpoint. In: Yadin Y (ed.) *The finds from the Bar Kokhba period in the Cave of Letters*. Israel Exploration Society, Jerusalem, p 270–279.

Abrahams D, Edelstein S (1964) A new method for the analysis of ancient textiles. *American Dyestuff Reporter* 53(1):19–25.

Boesken Kanold I, Koren Z (2009) Hexaplex snail sex and the purple pigment. Paper presented at DHA28, the 28th Meeting of Dyes in History and Archaeology. Institute of Natural Fibres and Medicinal Plants, Poznań, Poland, 22–23 October 2009.

Breniquet C, Desrosiers S, Nowik W, Rast-Eicher A (2018) Les textiles découverts dans les tombes de l'âge du Bronze moyen à Chagar Bazar (Syrie). In: Tunca Ö, Baghdo A (eds) *Chagar Bazar (Syrie) VIII. Les tombes ordinaires de l'âge du Bronze ancien et moyen des chantiers D-F-H-I (1999-2011)*. Études diverses. Mission archéologique de l'Université de Liège en Syrie. Peeters, Leuven, p 11–47.

Cooksey C (2001) Tyrian purple: 6,6′-dibromoindigo and related compounds. *Molecules* 6:736–769.

Cooksey C (2013) Tyrian purple: The first four thousand years. *Science Progress* 96:171–186.

Friedländer P (1909) Über den Farbstoff des antiken Purpurs aus murex brandaris. *Berichte der Deutschen Chemischen Gesellschaft* 42:765–770.

James M, Reifarth N, Mukherjee A, Crump M, Gates P, Sandor P, Robertson F, Pfalzner P, Evershed R (2009) High prestige Royal Purple dyed textiles from the Bronze Age Royal Tomb at Qatna, Syria. *Antiquity* 83:1109–1118.

Kalaitzaki A, Vafiadou A, Frony A, Reese D, Drivaliari A, Liritzis I (2017) PO-PU-RE: Workshops, use and archaeometric analysis in pre-Roman central eastern Mediterranean. *Mediterranean Archaeology and Archaeometry* 17:103–130.

Karapanagiotis I, Mantzouris D, Cooksey C, Mubarak MS, Tsiamyrtzis P (2013) An improved HPLC method coupled to PCA for the identification of Tyrian purple in archaeological and historical samples. *Microchemical Journal* 110:70–80.

Karapanagiotis I (2019) A Review on the archaeological chemistry of shellfish purple. *Sustainability* 11:3595. https://doi.org/10.3390/su11133595.

Koren Z (1993) Methods of dye analysis used at the Shenkar College Edelstein Center in Israel. *Dyes in History and Archaeology* 11:25–33.

Koren Z (2001) A successful Talmudic-flavored High-Performance Liquid Chromatographic analysis of carthamin from red safflower dyeings. *Dyes in History and Archaeology* 16/17:158–166.

Koren Z (2008) Archaeo-chemical analysis of Royal Purple on a Darius I stone jar. *Microchimica Acta* 162:381–392.
Koren Z (2013) New chemical insights into the ancient molluskan purple dyeing process. In: Armitage RA, Burton JH (eds) *Archaeological chemistry* VIII. ACS Symposium Series 1147. American Chemical Society, Washington DC, p 43–67.
Koren Z (2014) Scientific study tour of Ancient Israel. In: Orna MV (ed.) *Science history: A traveler's guide*. ACS Symposium Series 1179. American Chemical Society, Washington DC, p 319–351.
Koren Z (2020) Archaeological shades of purple from flora and fauna from the Ancient Near East. In: Orna MV, Rasmussen SC (eds) *Archaeological chemistry: A multidisciplinary analysis of the past*. Cambridge Scholars Publishing, Newcastle upon Tyne, p 256–300.
Koren Z, Verhecken-Lammens C (2013) Microscopic and chromatographic analyses of molluskan purple yarns in a Late Roman period textile. *e-Preservation Science* 10:27–34.
March R, Papanastasiou M, McMahon A, Allen N (2011) An investigation of paint from a mural in the church of Sainte Madeleine, Manas, France. *Journal of Mass Spectrometry* 46:816–820.
Papliaka Z, Konstanta A, Karapanagiotis I, Karadag R, Akyol A, Mantzouris D, Tsiamyrtzis P (2017) FTIR imaging and HPLC reveal ancient painting and dyeing techniques of molluskan purple. *Archaeological and Anthropological Sciences* 9:197–208.
Pfister R (1935) Teinture et alchimie dans l'Orient hellénistique. In: *Seminarium Kondakovianum. Recueil d'études. Archéologie. Histoire de l'art.* Études byzantines VII. Institut Kondakov, Prague, p 1–59.
Reese DS (forthcoming) Murex use in the Eastern Mediterranean. In: Mylona D, Brogan T, Iakovou M, Eaby E (eds) *The materiality of purple dye production and use in Cyprus and the Aegean from Prehistory to the Late Roman period.* Studies in Mediterranean Archaeology. Åströms Editions, Uppsala.
Sachs F, Kempf R (1903) Über p-Halogen-o-nitrobenzaldehyde. *Berichte der Deutschen Chemische Gesellschaft* 36:3299–3303.
Saltzman M (1992) Identifying dyes in textiles. *American Scientist* 80:474–481.
Whitworth I, Koren Z (2016) Orchil and Tyrian Purple: Two centuries of Bedfords from Leeds. *Ambix* 63:244–267.
Wouters J (1992) A new method for the analysis of blue and purple dyes in textiles. *Dyes in History and Archaeology* 10:17–21.
Wouters J, Verhecken A (1991) High-performance liquid chromatography of blue and purple indigoid natural dyes. *Journal of the Society of Dyers and Colourists* 107:266–269.

Zvi C. Koren (formerly Kornblum) is Professor Emeritus at the Shenkar College of Engineering, Design and Art in Ramat-Gan, Israel. Koren received the BS degree (cum laude) in chemistry and mathematics from Brooklyn College and his PhD in physical chemistry from the City University of New York (CUNY). At Shenkar, he has been the Director of The Edelstein Center for the Analysis of Ancient Artifacts since the research center's inception in 1991, and also served as the Head of the Department of Chemical Engineering. In Israel, Koren pioneered the research area involving chromatographic analyses of archaeological and historic colorants from flora and fauna sources. He has also served as Chairman of the Department of Chemistry at The Cooper Union for the Advancement of Science and Art in Manhattan.

Chapter 4
Sassanid Dyes from Ancient Persia – Case Study Chehrābād in Northern Iran

Ina Vanden Berghe and **Karina Grömer**

Abstract The Douzlākh salt mine Chehrābād in northern Iran offers good preservation conditions for organic finds of all kinds (wood, botanical remains, human faeces, textiles, ropes, leather and fur objects, human mummies). More than 15% of all artefacts found in the waste layers of the salt mine are textiles, most of them covering roughly a timespan between the fifth century BCE to the sixth century CE, which refers to the Achaemenid and Sassanid periods of the Persian Empires.

Within the framework of a recent research project, our focus lies on interdisciplinary textile research, including diverse analytical tools to gain basic data, radiocarbon dating, isotopic tracing, fibre analysis and wool measurements, dyestuff analysis and advanced conservation methods. The textiles from Chehrābād are an important source for our understanding of textile technology of the Ancient Near East and the Persian Empires.

This chapter deals with the dyeing research of selected Sassanid textiles (approx. 400–600 CE) and discusses how the outcome of the research, in terms of the biological dye sources applied in the material, can contribute to the geographical and historical context of the finds. This will provide knowledge about the dyeing tradition in Iranian territory and possible trade in dyes and textiles between Iran and other contemporary Mediterranean cultures.

Keywords Chehrābād salt mine · Dye preservation · Sassanid textiles · HPLC-DAD · Organic dyes

---

I. Vanden Berghe
Textile Research, Laboratory Department, Royal Institute for Cultural Heritage, Brussels, Belgium
e-mail: ina.vandenberghe@kikirpa.be

K. Grömer (✉)
Department of Prehistory, Naturhistorisches Museum, Vienna, Austria
e-mail: karina-groemer@nhm-wien.ac.at

A. Ulanowska et al. (eds.), *Ancient Textile Production from an Interdisciplinary Perspective*, Interdisciplinary Contributions to Archaeology,
https://doi.org/10.1007/978-3-030-92170-5_4

4.1 Introduction

Textiles are organic materials and easily disintegrate, especially if buried in the ground. Salt mines provide excellent preservation conditions, with plant material surviving, as well as animal fibres, leather and skins. In a salty environment, single-celled bacteria dry out and die. This prevents the decomposition process of organic materials due to bacterial activity. Unlike other environments such as bogs or waterlogged contexts, the salt conserves everything without exception. So, we know of artefacts made from animal, as well as of plant origins. The textiles are so well preserved in their organic matter that they are still colourful, elastic and supple when recovered.

Important sites with salt mines dating to prehistory and antiquity are located in Austria (Hallstatt and Dürrnberg), and Iran. In 1994, a mummified body, as well as other archaeological artefacts were found by chance at a site called Douzlākh (salt ground) near Chehrābād in northern Iran, and regular excavations started in 2004 (Aali and Stöllner 2015). In the salt mine, at least six mummies and more than 1000 textiles have so far been found in the Achaemenid (c. 500–400 BCE), Sassanid (c. 400–600 CE), Safavid and Quajar layers (late seventeenth to early twentieth century CE, for details on dating see Pollard et al. 2008). The large number of textiles known from the site, covering a wide time-span, and intensive on-site studies make it possible to answer different research questions, such as the development of textile technology, imports and trade.

Within the framework of the research projects "Saltmen Iran 2" (Deutsche Forschungsgesellschaft) and "Patrimonies Project Gerda Henkel Stiftung", textile analysis was done with the expertise of salt mine textile research from the Natural History Museum Vienna (Grömer and Bagherpour 2018). Specialised research in terms of dyestuff and fibre analysis focus on the Achaemenid and Sassanid textiles, as these are vital for the understanding of the mechanisms of ancient textile production, the use of raw material and fibres.

4.2 Textiles from Chehrābād

Some of the textiles found in the salt mine Chehrābād have already been published (Hadian et al. 2012; Krug-Ochmann 2014; Mouri et al. 2014), others have been investigated within the recent projects (e.g., Grömer and Aali 2016, 2020). From the Achaemenid layers, the garments associated with salt men 3, 4 and 5, are among the most prominent textiles (cf. Moskvin et al. in this volume). They mainly consist of light, undyed weft-faced tabbies in sheep wool. Remarkable are the red dyed cords attached to the garments and a multi-coloured belt from salt man 3. A cotton textile was also found together with salt man 3. Most of the other textile fragments found in Achaemenid layers are medium fine textiles in the basic weave type tabby and variants, the textiles are usually of light colour and undyed. Some few have patterning techniques such as stripes and weft-floating patterns.

The majority of textiles dated by context are the Sassanid textiles. The most impressive among these are the textile fragments associated with salt man 1: complete short trousers and diverse patterned textiles (kept at the National Museum Tehran, Iran). Important is also the complete upper garment (kept at the Zolfaghari-Museum Zanjan, Iran) and diverse, more or less well-preserved parts of trousers. The textile fragments from Sassanid layers are mainly monochrome and of very fine to coarse tabby and warp-faced tabby, and, more rarely, basket weaves. As raw material, mostly sheep wool could be detected, but also goat hair and cotton. Additionally, there are various patterned and dyed textiles, such as textiles with compound-weave techniques; warp-based patterning (‘*jajim*’) kelim-techniques (tapestry); stripes, and textiles with weft-floating pattern. Textiles have also been excavated from the younger layers of the mine, dating to the Safavid and Quajar period (seventeenth–twentieth century CE). There are hand-made textiles as well as machine-made (industrial) ones. The patterned textiles include chequered textiles and colour prints. In respect of raw material, they are made of wool, goat hair, cotton and synthetic fibres.

Both fibre and dyestuff analysis started very early, when the first textiles were found by chance during modern construction works at the salt mine in 1994. Some dyestuff analysis from prominent items have been published (Vatandoust and Hadian Dehkordi 2005; Hadian et al. 2012: 156–157). One Achaemenid and two Sassanid samples have been analysed for dyestuffs by a team from the Boston University (USA), Department of Chemistry (Mouri et al. 2014).

The recent interdisciplinary research project gave us the opportunity to analyse the dyes in the textile fragments from the Sassanid period in a more systematic way and thus to obtain a more representative picture of the dyeing traditions and the economic context at that time. Detailed information about the organic dye molecules preserved was obtained with chromatography.

4.3 Dye Investigation

The organic dyes were analysed using High Performance Liquid Chromatography with photo diode array detection (HPLC-DAD) by which dye compound identification is done based on the characteristic UV-Vis absorbance spectra and retention time of each dye molecule. This is a micro-destructive technique, requiring samples of about 4–6 mm of coloured yarn. The focus of the dye identification in the current study of 2018–2019 is on the Sassanid fabrics. The reason for this is that Achaemenid textiles found in the salt mine are more or less monochrome and one of the few textiles with colour has been sampled previously. The fourteen investigated samples belong to five textiles stored at the Zolfaghari-Museum Zanjan.

4.3.1 Analytical Procedure

The chromatography equipment consists of an Alliance HPLC instrument with automatic injection and autosampler (Waters Chromatography BV) coupled to a photodiode array detector (model 996), scanning the absorbance within the wavelength range between 200 and 800 nm. A LiChrosorb RP-18 end capped column was used for the separation. The mobile phase is composed of eluent A: pure methanol (grade: for HPLC >99.8%), eluent B: a mixture of methanol and Milliq. water in the volumetric ratio of 1/9 and eluent C: a 5% phosphoric acid solution (85 wt% *pro analisi*). The mobile phase was delivered at a flow rate of 0.2 ml/min in gradient mode: 0–3 min: 23A/67B/10C, 3–29 min: linear gradient to 90A/0B/10C, 30–35 min: 23A/67B/10C. The applied software system for data treatment was Empower 3. All equipment was from Waters Chromatography BV. Previously to analysis, the colorants are recovered from the textile fibres by soft acidic extraction during 15 min at 80 °C in a solution of 250 μl methanol/acetone/water/2,1 M oxalic acid (30:30:40:1, v/v/v/v). After centrifugation and vacuum evaporation of the solution, the residue is dissolved in 60 μl methanol/water (1/1, v/v) from which 20 μl is injected for analysis. (Vanden Berghe et al. 2009).

Detected chromophores of dye extracts are identified by comparison with spectra from a user-generated reference spectra database developed of commercially and non-commercially available references of natural dye sources. The database covers a large group of mordant, direct and indigoid dye compounds belonging to a wide variety of plant and animal dye sources.

4.3.2 Corpus: Investigated Sassanid Textiles

The first Sassanid textile investigated is a large fragment of a woollen weft-faced tabby (No. 152–1) of medium quality (thread count: 9 warp threads, 28 weft threads) with colour stripes in six different colours, from which red, green, blue and yellow threads (samples #3030–3033) have been taken (Fig. 4.1). The colour stripe pattern on a beige base web consists of a regular sequence of wider dark or light brown stripes (2.5–3 cm), between them an 11.5 cm wide pattern zone with two groups of three colour stripes (3–3.5 cm wide). The groups always consist of two colours: one colour inside, the other colour surrounding it. The colour stripes are between 0.6 and 0.7 cm wide. The textile was made of 0.7–0.9 mm S-plied yarn in warp, while the yarns used for the wet differ a bit according to the different threads used, e.g., 0.4–0.5 mm Z-single yarn for the red, yellow and blue threads, 0.6–0.7 mm for green and brown and 0.8 mm Z-single yarn for beige. This might be a hint that ready-dyed yarns were combined from different sources. On this textile, a selvedge on one side also survived, a simple returning of the weft thread into the web with bundles of three threads at the edges forming a more stable and thicker selvedge.

The second Sassanid textile examined is a beige weft-faced tabby with a band of colour pattern with cross motif in tapestry technique (No. 2415-0, Fig. 4.2). It has

No. of sample	Location on fabric	Sample image (stereomicroscope Stemi, Zeiss)
#3030	 *Find no. 152-1*	
#3031		
#3032		
#3033		

Fig. 4.1 Striped Sassanid textile, find no. 152-1 and colour thread samples. (Photos: RUB/DBM/MFZ, by Vanden Berghe)

been sampled for red, yellow and blue (samples #3034–3036). The textile is a fragment of a larger one, and only parts of the former pattern are visible: there is a dark blue band surrounded by c. 4 mm wide yellow stripes, followed by red threads. The dark blue band is patterned with large cross-shaped elements (3.2 cm wide) in red, white and yellow; between them there are white dots with red filling. The textile is flat and even, with weft threads densely packed, varying depending of the pattern. The base web has a count of 9 threads in warp and 28–32 threads in weft. The threads used are 0.6–0.7 mm S-plied yarn for warp and 0.4–0.5 mm Z-single yarn in weft.

A larger textile fragment with an embroidered colour stripe and ram pattern (No. 2477-0) served for sampling of green, red and blue threads (samples #3037–3038, Fig. 4.3). The base web is a medium fine beige tabby weave, made of S-plied yarn in one thread system and Z-single yarn in the other with a thread count between 10–12 threads per cm and 0.6–0.7 mm thick yarns. The pattern in detail shows bands in red and blue surrounding very abstract and geometric standing rams, whose horns are shown in frontal view, while the body is displayed from the side. The pattern is stitched very carefully following the thread system of the base web. The stitch type used for the embroidery is a running stitch and the coloured threads employed for this pattern are blue, red and green 1 mm S-plied yarns.

No. of sample	Location on fabric	Sample image (stereomicroscope Stemi, Zeiss)
#3034	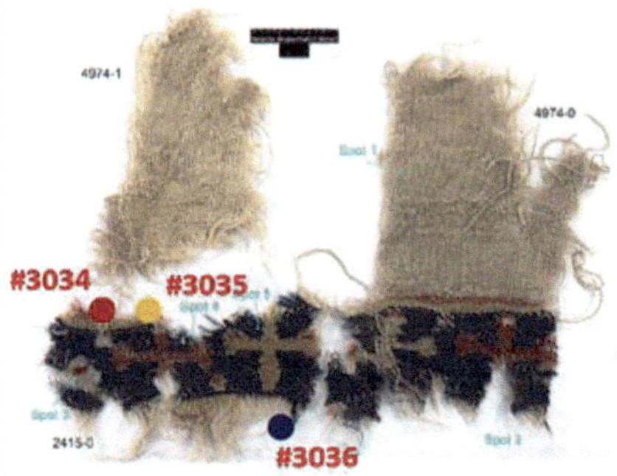	
#3035		
#3036	*Find no. 2415-0*	

Fig. 4.2 Sassanid textile in tapestry technique, find no. 2415-0 and colour thread samples. (Photos: RUB/DBM/MFZ, I. Vanden Berghe)

Further blue as well as dark brown fibres (samples #3039–3040) have been collected from a blue felt fragment which was wrapped and tied together to a package with dark brown and off-white threads (No. 3942-0, Fig. 4.4). The 2–3 mm thick felt is sewn together with a seam from two similar felted parts. Originally, the felt was made from off-white wool and then dyed blue; sometimes the dye did not reach the whole item or did not penetrate it totally.

Finally, a textile with geometric pattern in tapestry technique (No. 5529-0) served to select light blue and red shades (samples #3014–3042). It is a fragment of a medium fine beige tabby textile in wool (Fig. 4.5). It was made of 0.6 mm S-plied warp and 0.4–0.5 mm z-single yarn in weft, the weft is densely packed to 20–22 threads per cm. The 7.4 cm wide pattern consists of stripes that surround bands of geometric motifs in red, yellow light blue and dark blue. The motif bands are of diagonal chessboard design in red and beige, another band consists of a dark blue stripe with yellow star-like elements, between them red circles with yellow inside.

4.4 Dye Composition and Biological Sources

In total, an average representation of colours present in the Sassanid layers from the Douzlākh salt mine near Chehrābād have been sampled, from red, yellow/orange shades to light blue, darker blue to green. The dye composition of each of the

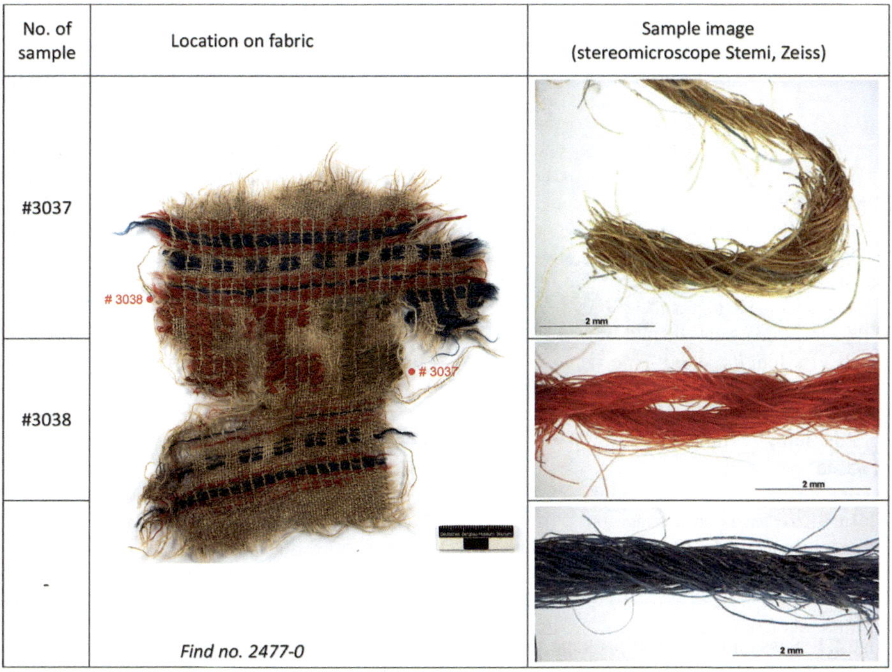

Fig. 4.3 Embroidered Sassanid textile, find no. 2477-0 and colour thread samples. (Photos: RUB/DBM/MFZ, I. Vanden Berghe)

extracts is shown in Table 4.1, indicating the most likely applied biological dye source in the third column.

The embedding in saline conditions has a positive impact on the preservation of fibres and organic dyes, as previously discovered in the finds from the prehistoric salt mine Hallstatt in Austria (Hofmann-de Keijzer et al. 2013) and equally applicable to the textiles selected from the Douzlākh mine. Not only are the colours of the textiles still very vivid, but the dye molecules are still clearly preserved in the protein fibres. Also striking is the resemblance in hue of the sampled yarns of the five fragments, especially in the red and blue samples. This similarity becomes even more evident in the discussion of the results of the dye identification per colour group. Finally, as noted above, dark brown thread was included in the analyses, although it has been presumed to consist of naturally pigmented wool. This assumption was confirmed by the HPLC analyses as no evidence was found for the use of any organic dye molecule.

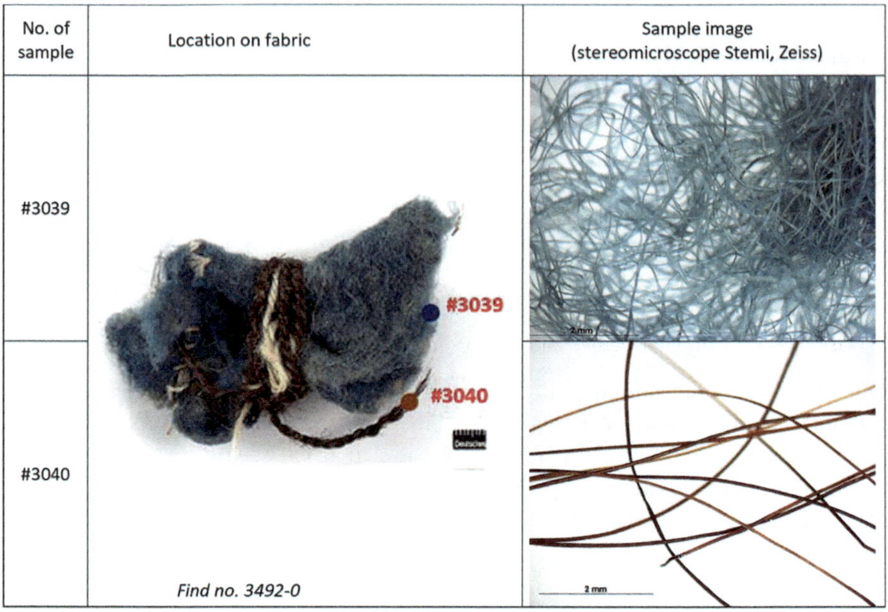

Fig. 4.4 Sassanid dyed felt package, find no. 3492-0 and colour thread samples. (Photos: RUB/DBM/MFZ, I. Vanden Berghe)

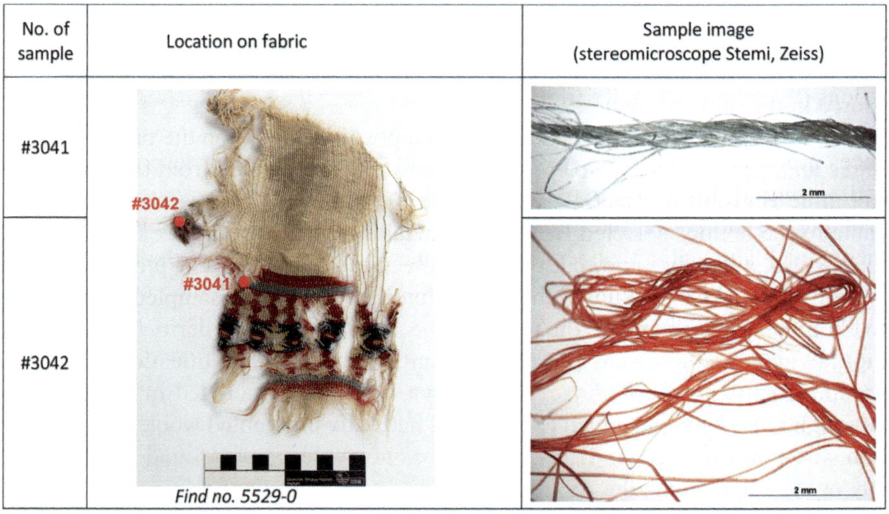

Fig. 4.5 Sassanid textile in tapestry technique, find no. 5529-0 and colour thread samples. (Photos: RUB/DBM/MFZ, I. Vanden Berghe)

Table 4.1 Result table HPLC-DAD analyses of the Sassanid textiles samples from the Chehrābād salt mine Biological dye source(s) and dye composition

No. of sample	Colour	Biological dye source(s)	Detected dye molecules
Fabric 152-1			
#3030	Red	Madder	90 alizarin, 8 purpurin, 1 anthrapurpurin, (nordamnacanthal, anthragallol, xanthopurpurin)
#3031	Green	Luteolin yellow Indigo/woad Madder	77 luteolin-7-0-glucoside and derivative, 12 alizarin, 4 apigenin-7-0-glucoside, 2 unknown compound (rt8 min, 414 nm), 2 indigotin, 2 isatin, 1 luteolin, (indirubin)
#3032	Blue	Indigo/woad Madder	65 alizarin, 15 indigotin, 8 indirubin, 6 isatin, 6 rubiadin
#3033	Yellow	Luteolin yellow Madder Tannin	45 alizarin, 16 ellagic acid, 12 flavonoid (rt3.3 min, 350 nm), 10 luteolin-7-0-glucoside + derivative, 8 flavonoid (rt12.7 min, 370 nm), 8 purpurin
Fabric 2415-0			
#3034	Red	Madder Tannin (trace)	81 alizarin, 13 purpurin, 6 nordamnacanthal, (anthragallol, ellagic acid)
#3035	Yellow	Luteolin yellow Madder	70 luteolin-7-0-glucoside and derivative, 16 alizarin, 5 purpurin, 4 apigenin-7-0-glucoside, 3 flavonoid (rt20.5 min, 345 nm), 2 luteolin, 1 apigenin
#3036	Blue	Indigo/woad Madder	68 alizarin, 18 indigotin, 7 rubiadin, 5 indirubin, 2 isatin
Fabric 2477-0			
#3037	Green	Luteolin yellow Indigo/woad Tannin Madder	58 luteolin-7-0-glucoside + derivative, 18 luteolin, 16 ellagic acid, 4 alizarin, 4 indigotin
#3038	Red	Madder Tannin (trace)	80 alizarin, 12.5 purpurin, 5 nordamnacanthal, 1 ellagic acid, (anthragallol, anthrapurpurin, xanthopurpurin)
–	Blue	Indigo/woad Tannin Madder	58 alizarin, 21 indigotin, 6 indirubin, 5 rubiadin, 4 purpurin, 3 isatin, 3 ellagic acid
Fragment 3942-0			
#3039	Blue	Indigo/woad Madder	52 alizarin, 28 indigotin, 8 indirubin, 8 isatin, 4 rubiadin
#3040	Dark brown	–	No dyes detected
Fabric 5529-0			
#3041	Light blue	Indigo/woad Tannin Madder	46 alizarin, 31 indigotin, 23 ellagic acid, (isatin)
#3042	Red	Madder Tannin (trace)	86 alizarin, 13 purpurin, (anthrapurpurin, anthragallol, ellagic acid)

4.4.1 Red

The principal red dye molecule detected in the acid extracts from the four red samples is alizarin. It was found in the presence of other minor anthraquinone compounds such as purpurin, nordamnacanthal, anthrapurpurin, anthragallol and xanthopurpurin. Such a combination of dye compounds is indicative of the use of the roots from the dyer's madder *Rubia tinctorum* L. (Cardon 2007; Mouri and Laursen 2012). Dyer's madder is native to the Middle East and the eastern parts of the Mediterranean, but the geographical distribution reaches to the south of Europe on the one side, up to the far East on the other (Cardon 2007: 107–121).

Small amounts of ellagic acid were found in samples #3034, #3038 and #3042, suggesting the use of tannins as mordants.

4.4.2 Blue

Blue is, along with red, one of the most common colours in the Sassanid textiles from the salt mine. Indigotin is the major marker molecule for the use of an indigoid dye plant. It has been found in the presence of its isomer indirubin and isatin. Possible biological dye sources are an oriental indigo plant (*Indigofera* or *Polygonum* species) or the woad plant (*Isatis tinctoria* L.). As the same marker dye molecules are present in the extracts of textiles dyed with both types of plants, distinction between these sources based on compound detection is not possible (Hofenk de Graaff 2004: 240–257). Nor can either plant species be excluded on the basis of the geographical context of the finds. The *Indigofera* genus comprises approximately seven hundred species, occurring both wild and cultivated in a very wide geographical area including Asia and the territory of present-day Iran, but woad is also found growing in areas ranging from Turkey and the countries of central Asia to the Himalayas and the Far East (Cardon 2007: 354–367).

The analyses of the blue yarns are further characterised by the presence of anthraquinone dye molecules. In particular, all these blue samples contain alizarin and a small amount of rubiadin. The second important marker compound for cultured madder, purpurin, was found only in the blue sample from textiles No. 2477–0. The different composition of the red dye molecules detected in the red and the blue yarns respectively could suggest that another plant source from the *Rubiaceae* family may have been used in both dyeings, even though it cannot be excluded that other parameters during the dyeing processes are responsible for the different composition. On the other hand, the red dye molecules found in the blue coloured yarns may also refer to the use of madder roots as an additive in the indigo fermentation bath and as such are not used to obtain a more purple or violet hue. However, this assumption seems less likely and is not fully supported by the expected relative ratios between alizarin and rubiadin in this case (Hartl et al. 2015).

In conclusion, all the blue yarns analysed from the Sassanid textiles were obtained by vat dyeing with an indigo or woad plant as dye source for blue dyes, in combination with the roots of a plant from the *Rubiaceae* family (dyer's madder or another), either used as additive or as source of red mordant dyes.

4.4.3 Yellow and Orange

Yellow and orange shades are less common among the textiles in the Sassanid layers from Chehrābād. The analyses of the two yellow samples allowed the identification of the flavone dye molecules luteolin-7-O-glucoside and luteolin, as well as apigenin-7-O-glucoside and apigenin, in smaller amount. Apart from that, some other flavonoid compounds with absorbance maxima between 345–370 nm were detected in very small amounts.

Luteolin refers to the use of a yellow luteolin containing dye source. A broad range of plants are among the possible sources, such as weld (*Reseda luteola* L.), sawwort (*Serratula tinctoria* L.), dyer's greenweed (*Genista tinctoria* L.) or types of chamomile (*Anthemis sp.*) or other equivalents. Among the central Asian sources are Eurasian sawwort (*Serratula coronata* L.) (Cardon 2007: 177) and the chamomile species (Schweppe 1993: 350). Apart from the flavonoid components, alizarin and purpurin were systematically detected, indicating the mixture of the yellow dyes with anthraquinone dyes from madder to obtain an orange shade. Ellagic acid was only found in sample #3033.

4.4.4 Green

Green yarns have only been found in very few cases, among the hundreds of textiles from Sassanid Chehrābād. The two green shades analysed were obtained with a combination of an indigoid dye source (indigo or woad) and the yellow luteolin dye source (weld, chamomile, sawwort, etc.). Also, a low amount of alizarin is present in both samples, indicative of the use of a *Rubiaceae* dye plant.

The first, sample #3031 from Textile No. 152-1 (see Fig. 4.1), is homogeneously green dyed, while the second, sample #3037 from Textile No. 2477-0 (see Fig. 4.3), is a yarn composed of yellow and blue fibres, hence giving clear evidence of a fleece dyeing process (Fig. 4.6). In addition, tannin is also present in that sample.

4.5 Dyes from Chehrābād in the Context of the Fourth to Sixth Century CE

From the archaeological sites of the modern territory of present-day Iran, almost no textiles from 600 BC to 600 CE have been analysed in terms of dyestuffs (for the few exceptions, see: Vogelsang-Eastwood 2006: 236, for Germi; 237, for Shahr-i

Fig. 4.6 Sassanid textile, find no. 2477-0: Detail of the pattern with green yarn and #3037, composed of yellow and blue dyed fibres. (Photo by I. Vanden Berghe)

Qumis). Therefore, the studies about Chehrābād are an important contribution to basic research in that area.

Dyer's madder, indigo or woad, as well as luteolin based yellow plants (likely sawwort, weld or chamomile species) are among the oldest sources of dye used for dyeing wool with red, blue and yellow dyestuff respectively. They were used for this purpose over a very extensive geographical area that easily covers the entire Sassanid Empire at its widest (Cardon 2007). Yellow flavonoid dyes are very rarely detected in archaeological samples, as their chemical structure is generally less resistant to burial conditions than that of many red and blue dye molecules. The very exceptional embedding conditions in the salt mines, which are most advantageous for organic compounds in general, have therefore also ensured that these more unstable compounds have not been completely lost over time. As such, this gives us important rare information about the use of yellow dyes in that era.

Most of the wool samples examined were dyed using a combination of two of these plant-based sources, either two sources of mordant dyes (such as madder and luteolin yellow) or two sources from technically very different dyes (namely a vat dye, indigo/woad, and a mordant dye). Such mixtures of dyes indicate a thorough knowledge of the dyeing process and thus clearly indicates evidence of an established dyeing tradition.

Furthermore, the plant dye sources identified in this study used to dye the different woollen weft threads, fibres of the selected tabby woven textiles and the fleece fragment (all Sassanid period from Chehrābād) show a very close resemblance to the results obtained previously from two such textiles, from the same period and the same mine (Hadian et al. 2012; Mouri et al. 2014). Dyer's madder and indigo or woad have been identified in both studies as the natural sources for respectively red and blue dyestuffs. In fact, dyer's madder was also detected in that study in a much older woollen tabby fragment (No 10031) from the same salt mine dated to the Achaemenid period.

The main difference between the Sassanid textiles is seen in the yellow dyes used. The current study identified a luteolin based yellow dye source, likely weld (*Reseda luteola* L.), Eurasian sawwort (*Serratula* sp.) or chamomile (*Anthemis* sp.)

(see 5.4.3) in the yellow and green yarns, based on the detection of luteolin-7-O-glucoside and apigenin-7-O-glucoside and their aglycones. This is in contrast to the previous analysed Sassanid textiles from Chehrābād, where also a species of tamarisk has been reported as the yellow source used in textile no. 10050. This was based on the detection of quercetin-3-O-sulfate and quercetin-3-O-glucuronide and other flavonols and derivatives. In one of the previous studies, a second, "unknown yellow" in the yellow and green yarns of fragment no. 10141 was also found, based on the detection of quercetin-3-O-glucuronide and other 3-O-flavonol glycosides (Mouri et al. 2014). Tamarisk (*Tamarix* sp.) is mainly known as oriental tannin-rich shrub or small tree whose gallnuts were used to dye black. In the western regions of India, it was also used to dye cotton brown, or as an additive in the dye bath to make the colour darker (Mohanti et al. 1987; Cardon 2007: 145).

In order to gain a better insight into the dye tradition on the Iranian territory and possible trade of dyes and textiles between Iran, Rome and Egypt, more far-reaching comparisons with dye sources used in the ancient world are required through large-scale dyestuff studies. Only from the Egyptian territory have a substantial number of textiles been preserved and dye analyses systematically undertaken over the last decades. Madder, indigo/woad and a luteolin based dye source (weld, saw-wort, etc.) have been found as the predominant dye sources in Roman times. To a lesser extent, animal dye sources were identified considering the few evidences found of the use of murex purple in combination with kermes (*Kermes vermillio* Planchon). From 600 CE onwards, the influence from the East in these textiles from Egyptian burial sites becomes clear from the identification of wood dyes such as sappanwood (*Caesalpinia sappan* L.) and young fustic (*Cotinus coggygria* Scop), as well as red insect dyes Armenian cochineal (*Porphyrophora hamelii* Brandt) and Indian lac (*Kerria lacca* Kerr) (Cardon et al. 2004; Wouters et al. 2008; Wouters 2009; Vanden Berghe 2011).

The results of the current study of the dyes in the Sassanid textiles coincides well with the dyes used in the textiles found in the Egyptian territory in Roman times. Also, in terms of some pattern and weave types, we see clear comparative material. For instance, this is the case for textiles with clavi found fragmented in the salt mine Chehrābād (Hadian et al. 2012: Fig. 7) that resemble artefacts from the eastern part of the Roman empire, e.g., Egyptian ones (Pennick Morgan 2018: e.g., Figs. 4.2 and 4.3). In addition, similarities can be seen in the dominance of weft-faced tabbies, the composition of threads (use of S-plied warps) and thread counts that are comparable. This tapestry technique is quite common in Late Antiquity (e.g. Falck and Lichtwarck 1996; Paetz gen. Schieck 2005) and might be a Roman influence.

Clear similarities are also found with textiles from Palmyra, in central Syria, dating to100–300 CE (Schmidt-Colinet et al. 2000), in terms of weaving technology, being woollen weft-faced tabby fragments, as well as in terms of dyes used. Madder and indigo have been reported for red and blue colours and a yellow had been determined as 'yellow weed'. In addition, insect reds have been identified, such as kermes (*Kermes vermillio* Planchon), used alone for red and combined with indigo in purple stripes in woollen fabrics, and cochineal, presumably the Ararat or Armenian species (*Porphyrophora hamelii* Brandt). Even lac dye (*Kerria lacca* KERR) was identified in many of the Palmyra textiles, hence clearly demonstrating the trade connections with India in the first and the second century CE (Böhmer and Karadag 2003). However, their article does not specify whether it was used in woollen textiles in particular.

Other contemporaneous comparable weft-faced tabby fragments have been found in the Iranian site Shahr-i Qumis in Damghan (north-east Iran) (Kajitani, mentioned in Hadian et al. 2012). However, no information has so far been found on the dyes used.

Despite some promising similarities, it is currently not possible to obtain accurate information on the origin of the textile finds, the dyeing tradition and the possible dye or textile trade between Iran and the surrounding regions on the basis of the dyestuffs identified in the Sassanid textiles from the Chehrābād salt mine. The red and blue dyes found in the textiles analysed so far, are too widespread (both in space and time) to make this possible, and any refinement severely hampered by the lack of comparanda. In addition, the comparisons described above with other relevant textile finds are largely limited to the use of the red dyes, as, so far, no yellow dye sources have been identified in other contemporaneous woollen textiles. A further study of the dyes in a larger number of textile fragments of the Chehrābād salt mine will make it possible to determine to what extent the dye palette now obtained, with one red (dyer's madder), one blue (indigo or woad) and at least two yellow dye sources (a luteolin based yellow (sawwort, weld, chamomile, etc.), tamarisk and the 'unknown yellow', is representative for this site in that period.

Acknowledgments This research was carried out within the framework of two main projects in cooperation with the Iranian authorities (Miras Farhangi Iran, ICAR): Project Gerda Henkel Stiftung 2018–2020: "Die Salzmänner Irans – Das Kulturerbe des Salzmumien-Museums in Zanjan", Project leader: Thomas Stöllner (Ruhr-Universität Bochum, Deutsches Bergbaumuseum Bochum, Deutschland), Abolfazl Aali (Archaeological Museum Zanjan, Iran), Markus Egg (Römisch-Germanisches Zentralmuseum Mainz), Wolfgang David (Museum Frankfurt), Grant Number: AZ 07/BE/17. Deutsche Forschungsgesellschaft DFG Project 2015-2017: "Saltmen Iran – Chehrabad Saltmummy & Saltmine Exploration Project", Project manager: Abolfazl Aali (Archaeological Museum Zanjan, Iran), Thomas Stöllner (Ruhr-Universität Bochum, Deutsches Bergbaumuseum Bochum, Germany), Grant Number STO 458/12-2. Finally, we would like to thank Alexia Coudray from the textile lab for her cooperation in this project.

References

Aali A, Stöllner T (eds) (2015) *The archaeology of the salt miners. Interdisciplinary research 2010–2014*. Metalla 21(1–2)/2014. Bochum.

Böhmer H, Karadag R (2003) New research on Palmyra textiles. *Dyes in History and Archaeology* 19:88–93.

Cardon D, Wouters J, Vanden Berghe I, Richard G, Breniaux R, (2004) Dye analyses of selected textiles from Maximianon, Krokodilo and Didymoi (Egypt). In: Alfaro C, Wild JP, Costa B (eds) *Purpureae Vestes, Actas del I Symposium internacional sobre Textiles y tintes del Mediterrano en e'poca romana*. PURPUREAE VESTES I. Consell Insular de Eivissa i Formentera, Universitat de València, València, p 145–154.

Cardon D (2007) *Natural dyes. Sources, tradition, technology and science*. Archetype Publications Ltd., London.

Grömer K, Aali A (2016) Durchdacht gemacht. Die Kleidung eines Salzmannes aus Chehrabad, Iran. *Universum Magazin* 3:108–109.

Grömer K, Aali A (2020) How to make a Sassanian tunic? Understanding handcraft skills based on a find from the salt mine in Chehrābād, Iran. In: Quillien L, Sarri K (eds) *Textile workers. Skills, labour and status of textile craftspeople between the prehistoric Aegean and the ancient Near East. Proceedings of the workshop held at 10th ICAANE in Vienna, April 2016*. Oriental and European Archaeology 13. Austrian Academy of Sciences Press, Vienna, p 59–74.

Grömer K, Bagherpour N (2018) "The saltmummies of Zanjan". Textile research in Iran in a conservation and exhibition project. *Archaeological Textiles Review* 60:110–113.

Hadian M, Good I, Pollard AM, Zhang X, Laursen R (2012) Textiles from Douzlakh salt Mine in Chehr Abad, Iran: A technical and contextual study of late Pre-Islamic Iranian textiles. *International Journal of Humanities* 19(3):152–167.

Hartl A, Néss Proano Gaibor A, van Bommel M, Hofmann-de-Keijzer R (2015) Searching for blue: Experiments with woad fermentation vats and an explanation of the colours through dye analysis. *Journal of Archaeological Science Reports* 2: 9–39.

Hofenk de Graaff J (2004) *The colourful past. Origins, chemistry and identification of natural dyestuffs*. Abegg-Stiftung & Archetype Publications Ltd., Riggisberg and London, p 240–257.

Hofmann-de Keijzer R, van Bommel M, Hartl A, Grömer K, Rösel-Mautendorfer H, Reschreiter H, Kania K, Joosten I, Proano Gaibor AN, Erlach R, Lachner E, Wandl M, de Keijzer M (2013) Coloured Hallstatt textiles: 3,500-year-old textile and dyeing techniques and their contemporary application. In: Banck-Burgess J, Nübold C (eds) *The North European symposium for archaeological textiles XI, 10–13 May 2011 in Esslingen am Neckar*. NESAT XI. Verlag Marie Leidorf GmbH, Rahden/Westfalen, p 125–130.

Krug-Ochmann J (2014) Achaemenid and Sassanian trousers. A short technical description from Douzlakh Salt Mine at Chehr Abad, Iran. *Archaeological Textiles Review* 56:60–64.

Mohanti BC, Chandramouli KV, Naik HD (1987) *Natural dyeing processes of India*. Calico Museum of Textiles, Sarabhai Foundation, Ahmedabad.

Mouri C, Aali A, Zhang X, Laursen R (2014) Analysis of dyes in textiles from the Chehrābād salt mine in Iran. *Heritage Science* 2:20.

Mouri C, Laursen R (2012) Identification of anthraquinone markers for distinguishing Rubia species in madder-dyed textiles by HPLC. *Microchim Acta* 179:105–113.

Paetz gen. Schieck A (2005) *Die koptischen Textilien, Gewebe und Gewänder des ersten Jahrtausends aus Ägypten*. Kolumba 19. Köln.

Pennick Morgan F (2018) *Dress and personal appearance in Late Antiquity. The clothing of the middle and lower classes*. Brill, Leiden and Boston.

Pollard AM, Brothwell DR, Aali A, Buckley S, Fazeli H, Hadian Dehkordi M, Holden T, Jones AKG, Shokouhi JJ, Vatandoust R, Wilson AS (2008) Below the salt. A preliminary study of the dating and biology of five salt-preserved bodies from Zanjan province, Iran. *Iran* 46:135–150.

Schmidt-Colinet A, Stauffer A, Al-As'ad K (2000) *Die Textilien aus Palmyra: Neue und alte Funde*. Damaszener Forschungen 8. Philipp von Zabern, Mainz.

Schweppe H (1993) *Handbuch der Naturfarbstoffe, Vorkommen, Verwendung, Nachweis*. Nikol Verlagsgesellschaft mbH & Co., KG Hamburg, p 346–357.

Vanden Berghe I (2011) Dye palette of the supplementary weft patterned linen fabrics from the Katoen Natie collection. In: De Moor A, Fluck C (eds) *Dress accessories of the 1st millennium AD from Egypt and neighbouring countries. Proceedings of the sixth Textiles from the Nile Valley conference (Antwerp, Belgium 2–3 October 2009)*. Lannoo Publishers, Tielt, p 286–293.

Vanden Berghe I, Gleba M, Mannering U (2009) Towards the identification of dyestuffs in Early Iron Age Scandinavian peat bog textiles. *Journal of Archaeological Science* 36:1910–1921.

Vatandoust A, Hadian Dehkordi M (2005) "Saltman" a new archaeological discovery: scientific investigation and conservation. *Journal of Biological Research* 80:236–242.

Vogelsang-Eastwood G (2006) A preliminary survey of Iranian archaeological sites with textiles. In: Schrenk S (ed.) *Textiles in situ. Their find spots in Egypt and neighbouring countries in the first millennium CE.* Riggisberger Berichte 13, Riggisberg, p 221–254.
von Falck M, Lichtwark F (eds) (1996) *Ägypten. Schätze aus dem Wüstensand. Kunst und Kultur der Christen am Nil.* Katalog zur Ausstellung. Ludwig Reichert Verlag, Wiesbaden.
Wouters J, Vanden Berghe I, Richard G, Breniaux R, Cardon D (2008) Dye analysis of selected textiles from three Roman sites in the Eastern Desert of Egypt: A hypothesis on the dyeing technology in Roman and Coptic Egypt. *Dyes in History and Archaeology* 21:1–16.
Wouters J (2009) Red and purple dyes in Roman and 'Coptic' Egypt. In: De Moor A, Fluck C (eds) *Clothing the house. Furnishing textiles of the 1st millennium AD from Egypt and neighbouring countries. Proceedings of the fifth Textiles from the Nile Valley conference (Antwerp, 6–7 October 2007).* Lannoo Publishers, Tielt, p 182–185.

Ina Vanden Berghe is responsible for the Textile Research Laboratory of the Royal Institute for Cultural Heritage (KIK-IRPA) in Brussel, Belgium. She specializes in the study of materials, techniques and degradation aspects of organic materials, such as natural and synthetic organic dyes and protein fibres, as well as in studies on metal threads and plant-based fibres. Her special interest lies in archaeological textiles and, more specifically, in the degradation problems of fibres and dyes associated with different burial and conservation conditions. She has participated in several research projects such as "MODHT – Monitoring of the Damage in Historic Tapestries", "DressID – Clothing and Identity in Roman Empire', "Eu–Artech – Access, Research and Technology for the Conservation of the European Cultural Heritage"; "Charisma – Cultural heritage Advanced Research Infrastructures: Synergy for a multidisciplinary approach to conservation/restoration" and "Iperion – Integrated platform for European research infrastructures in the field of cultural heritage".

Karina Grömer is the Head of the Department of Prehistory, Natural History Museum Vienna. She studied prehistoric archaeology, ethnology and anthropology at the University of Vienna in Austria. Her habilitation thesis was entitled "Archaeological Textile Research – Technical, economic and social aspects of textile production and clothing from Neolithic to the Early Modern Era". She is specialized in interdisciplinary and integrated analysis of textiles, research on textile tools and reconstruction of prehistoric costume. Her research covers a timespan from c. 2500 BCE to 1000 CE and a geographical area from Central Europe to Iran, with research projects like "DressID – Clothing and Identity in Roman Empire", "CinBA – Creativity in the Bronze Age" and "Chehrābād Saltmummy and Saltmine Exploration Project". She also teaches at the Universities of Vienna (Austria) and Southampton (UK).

Part III
Interdisciplinary Approaches to Textile Tools

Chapter 5
Textile Production in Some Early Neolithic Settlements in North Macedonia

Zlata Blažeska and Jasemin Nazim

Abstract The most conclusive evidence for the existence of textile production in North Macedonia during the first quarter of the sixth millennium BC are textile imprints on pottery bases. We analyzed 45 imprints from six settlements of two geographically close Early Neolithic cultural groups, Velušina-Porodin and Amzabegovo-Vršnik. We identified the textile structures, measured the yarn diameters and their twist directions and angles, as well as the thread count. The close double weft twining in S direction and the spliced fibres from decorticated plant stems are common features of all of these textiles. The weft rows are either very close or tightly packed, producing weft-faced and flexible fabrics. Fraying and holes attest secondary use of the worn-out textiles during the construction of the vessels. In these settlements, there was a persistent and uniform community of practice in the production of twined textiles, differing from the twining structures practiced by the other Neolithic groups in the Balkans. Textile tools are represented by 22 perforated rounded pottery sherds, probably used as spindle whorls, while only a few objects can be recognized as loom weights: one dumbbell weight and five discoid weights.

Keywords Early Neolithic · North Macedonia · Textile imprints · Twining · Textile tools

Z. Blažeska (✉)
Independent Researcher, Skopje, North Macedonia

J. Nazim
Curator Emerita, Museum of North Macedonia in Skopje, Skopje, North Macedonia

A. Ulanowska et al. (eds.), *Ancient Textile Production from an Interdisciplinary Perspective*, Interdisciplinary Contributions to Archaeology,
https://doi.org/10.1007/978-3-030-92170-5_5

5.1 Introduction

The Neolithic settlements in North Macedonia, existing approximately from the end of the seventh to the fifth millennium BC, have been continuously researched since the second half of the twentieth century. Numerous excavations with a large number of finds contributed to the understanding of various aspects of the local Neolithic cultures. However, the industries producing perishable goods, such as textiles, cannot be directly attested, as ancient fabrics have not been preserved due to the properties of the soil and the climate. The function of the tools identified by archaeologists as loom weights, spools and spindle whorls was not established with certainty. However, conclusive evidence is provided by textile imprints on clay. As archaeologists in North Macedonia have thus far paid limited attention to this production, we were challenged to start the project "The Textile Production in the Territory of North Macedonia in Prehistoric Times". Through the joint efforts of an archeologist and a museum curator for folk textiles, we have been researching the archaeological collections in museums housing finds from Macedonia. We are grateful to the archaeologists from several museums in North Macedonia who provided us access to the museum collections and information on the finds. Unfortunately, some of the collections were not completely available to us and the information held on precise archaeological contexts, layers and dating is sometimes limited: some finds were originally discovered on the surface or in trial trenches, and not all archaeological sites concerned were systematically researched and published. However, we could gather a considerable amount of data, including tools for producing textiles and imprints on clay of textiles and straw mats from the Neolithic and Chalcolithic periods and textile tools from Bronze and Iron Ages. The rising interest in prehistoric textile production among archaeologists in the Balkans gave us an opportunity to compare the data and follow, as far as possible, the geographical and chronological distribution of the textile technology during the Neolithic in the Balkans and Anatolia.

In this article we focus on the analysis of the textile manufacture in several closely connected Early Neolithic settlements, where a considerable number of textile imprints were discovered during previous excavations. Furthermore, there are still ongoing excavations with new valuable data provided generously to us by the archaeologists Aleksandar Mitkoski and Goce Naumov for Vrbjanska Čuka and Elena Kanzurova for Kanli Čair.

5.2 Analytical Procedures

Pottery sherds with negative imprints of fabrics were pressed upon dentist paste in order to obtain a positive cast. We photographed both the sherds and the positive imprints with a camera and observed them with hand lens. Not having any sophisticated equipment, we magnified the photographs on the computer screen, where we could observe the properties of the yarns, as well as the damages on the textiles. We visually identified the structure of the textiles, and measured the yarn diameters, their twist directions, and the thread count. The structure of the clay and the irregularities and damages on the surface of the pottery bases were observed and compared to the appearances of the textiles on the imprints. We also observed how the yarn diameters and thread counts were affected by the uneven surfaces of the vessel bases and by the damages on the actual fabrics. The tools were measured and classified according their forms, size and materials.

5.3 Technology and Tools

The analysis of the textiles is based on 45 imprints. The majority of these (42 items) are preserved on the outer sides of the fragments of pottery bases, one on a fragment of an anthropomorphic house model and two on burnt daub wall fragments. The imprints originate from six Early Neolithic settlements from North Macedonia (Fig. 5.1), dating approximately to the first centuries of the sixth millennium BC (Mitrevski 2013). Vrbjanska Čuka, Pešterica, Čuka Topolčani and Ġumušica are

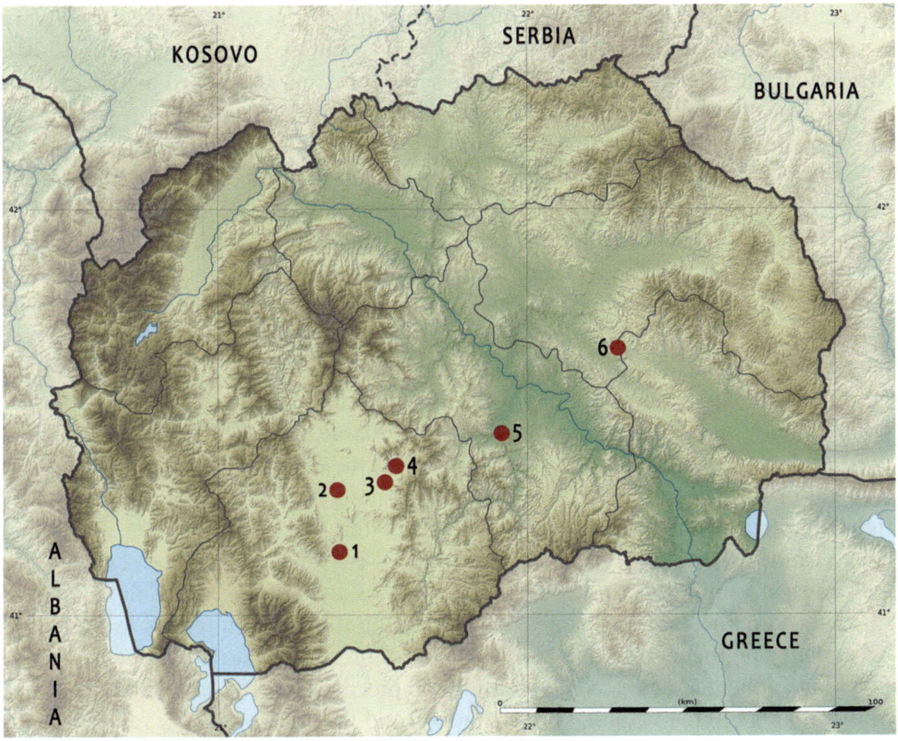

Fig. 5.1 Map of the Republic of North Macedonia with the Neolithic sites mentioned in the text: 1 Čuka Topolčani, 2 Vrbjanska Čuka, 3 Ġumušica, 4 Pešterica, 5 Čakovec, 6 Kanli Čair. (Courtesy of G. Naumov)

situated in the northern part of the Pelagonia Plain, near the town of Prilep, and belong to the Early Neolithic Velušina-Porodin cultural group. Calibrated C^{14} dates are available only for Čuka Topolčani (Layer I 7050–6200 BC; Layer II 6260–5500 BC) (Fidanoski 2009b) and Vrbjanska Čuka (6000–5700 BC)(Naumov et al. 2021). Čakovec is located to the northeast of Prilep in the Middle Vardar River Valley, with dominant influence from the Velušina-Porodin and some elements of the Anzabegovo-Vršnik cultural group (Sanev 1987; Mitrevski 2013). Anzabegovo-Vršnik is chronologically older, its earliest phases dating roughly from 6500–5700 cal. BC (Fidanoski 2009b). At this stage of research, the cultural affiliations and influences at the multilayered Neolithic settlement Kanli Čair near Radoviš, in Southeastern Macedonia, are not fully clear, as it was most probably culturally associated with several contemporary neighbouring Neolithic groups: to the northwest with the latest phases of the Early Neolithic in Anzabegovo-Vršnik and to the southwest with the Early Neolithic Velušina-Porodin (Sanev 1989; Kanzurova 2017).

The vessels with imprints are mostly made of coarse clay, but some of them are made of fine clay fabrics. Very often their bases are uneven, with protruding and sunken parts, full of bumps, holes and cracks. The textiles imprinted on the pottery bases were most probably pieces of worn-out ragged pieces of cloths, used either to protect the pottery from contamination by dirt during the modeling and drying, or as a support, a kind of primitive variant of a rotational device (Özdemir 2007; Mazăre 2012). Plaited straw mats had the same function, as attested by the imprints of mats on the pottery sherds from Vrbjanska Čuka and Ġumušica, as well from other Neolithic sites in North Macedonia.

Twenty seven of the imprints were discovered at Vrbjanska Čuka, eight at Kanli Čair, four at Čuka Topolčani, four at Ġumušica, one at Čakovec and one at Pešterica. We identified the process of craftsmanship and the features of the textiles, limiting the full investigation to finds where such features could be recognized and measured on the clay imprints and on their positives. We used the classification model and code for each structure by Adovasio and Maslowski (1988) and Mazăre (2014) as they were most compatible for the analyses of the Neolithic twining structures from North Macedonia. The weave analysis and technological classification (Table 5.1) point to only one technique: a two-thread or double weft twining with S twining twist of the weft, or class II-s, where the warp is the passive and the weft the active element of the structure. We identified two variants of this class: 36 imprints with tight simple S-twist twining (II_2-s-A1) where the rows of the wefts are tightly packed and seven imprints with closed simple S-twist twining (II_2-s-A2) where the weft rows are slightly separated. All of the fabrics are weft faced and flexible.

Identifying the exact structures and measuring the textile imprints from the two burnt daub fragments found in Vrbjanska Čuka presented particular challenges. At first glance, there appears to be several fabric pieces with different features on each of them, as the thread count, weft diameters and the weft row gaps vary considerably. However, by following the weft rows, we noticed that they run parallel all over the extremely coarse and uneven surfaces of the daub imprints. When the fabric was pressed upon the bumps on the daub, the weft diameters appeared larger and the

Table 5.1 Yarn measurements and structural features of the twined textiles impressed on ceramic vessels from six Early Neolithic settlements in North Macedonia

Sample number	Museum number	Archeological group	Settlement	Textile structure	Thread diameter (mm) active/passive	Final twist active/passive	Thread count
1	MP 398	Velušina-Porodin	Vrbjanska Čuka	II2-s-A1	1.8–2.8 /	S /	12 (14)
2	MP 382	Velušina-Porodin	Vrbjanska Čuka	II2-s-A2	1.0–2.0 /	/ /	/
3	MP 1073	Velušina-Porodin	Vrbjanska Čuka	II2-s-A2	1.5–2.0 /	S /	15 (17)
4	MP 413	Velušina-Porodin	Vrbjanska Čuka	II2-s-A1	3.5–4.0 /	S /	7 (9)
5	MP 228	Velušina-Porodin	Vrbjanska Čuka	II2-s-A1	1.5–2.5 /	S /	14
6	MP 22	Velušina-Porodin	Vrbjanska Čuka	II2-s-A1	1.9–2.5 /	/ /	12
7	MP 393	Velušina-Porodin	Vrbjanska Čuka	II2-s-A1	1.5–2.1 /	/ /	14 (16)
8	MP 1251	Velušina-Porodin	Vrbjanska Čuka	II2-s-A1	2.2–2.9 /	/ /	11
9	MP 372	Velušina-Porodin	Vrbjanska Čuka	II2-s-A1	2.7–3.5 /	S /	9 (11)
10	MP 164	Velušina-Porodin	Vrbjanska Čuka	II2-s-A1	1.8–2.5 /	S /	12 (14)
11	MP 147	Velušina-Porodin	Vrbjanska Čuka	II2-s-A1	2.2–3.2 /	S /	10 (12)
12	MP 156	Velušina-Porodin	Vrbjanska Čuka	II2-s-A1	2.5–3.2 /	S /	10
13	MP 57	Velušina-Porodin	Vrbjanska Čuka	II2-s-A1	1.8–2.2 /	S /	14 (16)
14	MP 185	Velušina-Porodin	Vrbjanska Čuka	II2-s-A1	3.0–3.7 /	S /	9
15	MP 18	Velušina-Porodin	Vrbjanska Čuka	II2-s-A2	1.8–2.2 /	S /	14
16	MP 400	Velušina-Porodin	Vrbjanska Čuka	II2-s-A1	2.8–3.5 /	S /	9
17	MP 412	Velušina-Porodin	Vrbjanska Čuka	II2-s-A1	1.8–2.1 /	/ /	15 (17)
18	MP 250	Velušina-Porodin	Vrbjanska Čuka	II2-s-A1	2.2–3.0 /	/ /	11 (13)
19	MP 87	Velušina-Porodin	Vrbjanska Čuka	II2-s-A2	1.5 /	/ /	/
20	MP 88	Velušina-Porodin	Vrbjanska Čuka	II2-s-A1	1.8–2.5 /	/ /	11 (13)
21	MP 89	Velušina-Porodin	Vrbjanska Čuka	II2-s-A1	2.0–2.8 /	/ /	12 (14)

(continued)

Table 5.1 (continued)

Sample number	Museum number	Archeological group	Settlement	Textile structure	Thread diameter (mm) active/ passive	Final twist active/ passive	Thread count
22	MP 145	Velušina-Porodin	Vrbjanska Čuka	II2-s-A1	2.8–3.2 /	S /	10 (12)
23	MP 207	Velušina-Porodin	Vrbjanska Čuka	II2-s-A1	/ /	/ /	/
24	MP 140	Velušina-Porodin	Vrbjanska Čuka	II2-s-A1	/ /	/ /	/
25	MP 85	Velušina-Porodin	Vrbjanska Čuka	II2-s-A1	/ /	/ /	/
26	MP 132a	Velušina-Porodin	Vrbjanska Čuka	/	/ /	/ /	/
27	MP 132b	Velušina-Porodin	Vrbjanska Čuka	/	/ /	/ /	/
28	MP 15	Velušina-Porodin	Ġumušica	II2-s-A2	2.2–3.4 /	S /	10 (12)
29	MP 16	Velušina-Porodin	Ġumušica	II2-s-A1	2.2–3.4 /	S /	10 (12)
30	MP 43	Velušina-Porodin	Ġumušica	II2-s-A1	1.5–1.8 /	/ /	14
31	MP 50	Velušina-Porodin	Ġumušica	II2-s-A1	2.0–2.8 /	S /	12
32	MP 995	Velušina-Porodin	Čuka Topolčani	II2-s-A1	1.6–2.3 /	/ /	12 (14)
33	MP 982	Velušina-Porodin	Čuka Topolčani	II2-s-A1	1.8–3.2 /	S /	12 (14)
34	MP 993	Velušina-Porodin	Čuka Topolčani	II2-s-A1	2.7–3.4 /	S /	9
35	MP 996	Velušina-Porodin	Čuka Topolčani	II2-s-A2	1.8–2.3 /	/ /	15
36	AMNM /	Velušina-Porodin	Čakovec	II2-s-A1	1.6–2.3 /	S /	15
37	MP 466	Velušina-Porodin	Pešterica	II2-s-A2	1.8–2.3 /	S /	11
38	AMNM /	Anzabegovo-Vršnik I	Kanli Čair	II2-s-A1	1.8–2.6 1.8–2.0	S S	11
39	AMNM /	Anzabegovo-Vršnik I	Kanli Čair	II2-s-A1	1.7–2.8 /	S /	12
40	AMNM /	Anzabegovo-Vršnik I	Kanli Čair	II2-s-A1	1.2–1.8 /	/ /	18
41	AMNM /	Anzabegovo-Vršnik I	Kanli Čair	II2-s-A1	2.2–3.2 /	S /	11
42	AMNM /	Anzabegovo-Vršnik I	Kanli Čair	II2-s-A1	1.2–2.2 /	S /	15 (17)

(continued)

Table 5.1 (continued)

Sample number	Museum number	Archeological group	Settlement	Textile structure	Thread diameter (mm) active/ passive	Final twist active/ passive	Thread count
43	AMNM /	Anzabegovo-Vršnik I	Kanli Čair	II2-s-A1	3.0–5.0 /	S /	9
44	AMNM /	Anzabegovo-Vršnik I	Kanli Čair	II2-s-A1	1.8–2.8 /	S /	11
45	AMNM 267	Anzabegovo-Vršnik I	Kanli Čair	II2-s-B2	2.0–5.0 1.0–2.0	S S	9

Samples are from two museum collections, distinguished in the second column by initials preceding each museum number, with AMNM used for the Archeological Museum of North Macedonia and MP for the Institute for Protection of Cultural Monuments and Natural Rarities and Museum – Prilep

Fig. 5.2 A burnt daub of a house wall with textile imprint, Vrbjanska Čuka, Slavej, Prilep, Early Neolithic. (Photo by A. Mitkoski)

weft rows separated, while on the sunken parts, the weft diameters were smaller and the weft rows tightly packed. We concluded that on both of the daub fragments single pieces of fabric were imprinted, with the same simple S-twist twining, common to the other textiles, only distorted by the configuration of the daub surface and stretching of the flexible fabrics (Fig. 5.2).

On most of the imprints the structure of the weft yarns is quite visible and even the strips of plant fibres could be recognized (Fig. 5.3). The individual fibres are most clearly seen in frayed and damaged textiles (Fig. 5.3 [7]). At some points the fibres run almost parallel, as if the weft is not spun at all, however a slight S-twist of 5°–15° could be measured on some of the weft yarns. The thickness of the weft on all of the imprints ranges from less than 1–5 mm in diameter (Table 5.1), although the original diameters of the threads could be altered by the shrinking of the original imprint after drying and baking of the pots. The diameters of the weft threads on each single imprint also vary significantly (Table 5.1). These features of the weft are

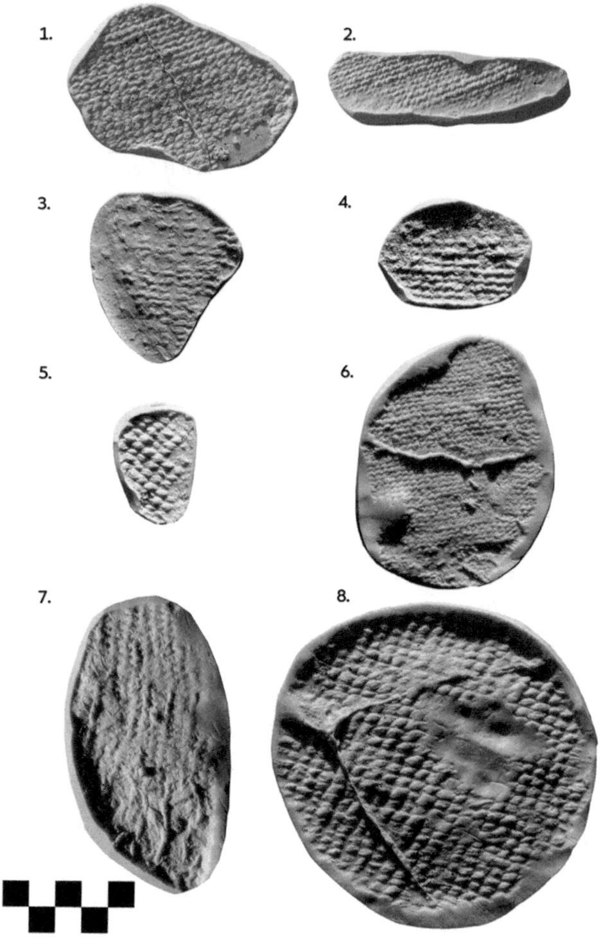

Fig. 5.3 Examples of textile imprints (positive casts): [1–2] Vrbjanska Čuka; [3] Pešterica; [4] Ġumušica; [5] Čuka Topolčani; [6] Čakovec; [7–8] Kanli Čair. (Photo by Z. Blažeska)

characteristic for the splicing method of yarn-making (Gleba and Harris 2018).[1] The varying weft diameters could also be affected by the uneven surfaces of the vessel bases, as their diameters are always larger on the protruded parts of the base (Fig. 5.3 [2]), while on the parts where the fabrics are frayed, the yarns untwist and look thicker.

The warp, or the passive vertical element of the fabric, is usually not visible due to the densely packed weft rows. The warp threads are visible only when the fabrics are damaged due to wear and fraying, or the weft rows are separated. We identified the warps only on two imprints from Kanli Čair. On one of them, it is an s-plied warp with a twist angle of 55° and a diameter of 1.80–2 mm (Fig. 5.3 [7]). The other has warps plied so loosely that at some points it looks as it is a double warp. The

[1] Gleba and Harris (2018: 2329) suggest that in the prehistoric archaeological contexts the fibres are spliced: "In splicing the strips of fibres are joined in individually, after having been stripped from the plant stalk directly and without or with only minimal retting."

single threads of the plied warp are 1–2 mm thick. Thread counts range from seven to 18 yarn elements per square centimeter (t/p/cm). In the samples where the weft diameters vary, several sections of the imprint were measured. Thread count of less than ten elements per square centimeter were measured at seven imprints, at 23 imprints t/p/cm was between 10–14 elements and eight imprints have 15–18 t/p/cm (Table 5.1). Due to the poor quality, we could not measure the thread count of seven imprints. Some of the imprints display densely packed wefts, with even tension and uniform twining twist angle of the weft, giving the fabric fine, smooth and even structure (Figs. 5.3 [4–6 and 8]), while the others are uneven, with thinner and thicker parts and bumps giving the rougher appearance of the fabric (Figs. 5.3 [1–3 and 7]). The analyzed textile imprints lack any marks of borders and selvedges.

We may only speculate on what kind of plant fibres was used for the textiles, as sheep wool was not appropriate for textile production before the Early Bronze Age (Wild 2003; Harris 2009). The yarns do not look as they were made of tree bast fibres. Flax, either cultivated or wild, or some other decorticated plants with bast (Adovasio and Maslowski 1988) could have been used. In personal communication with Goce Naumov, the project leader of the excavations at Vrbjanska Čuka in 2019, we were informed that nettle seeds were identified inside one of the houses, making nettle fibres a probable option.

The range of fabric properties exhibited in those textiles may imply that they could have been used for various purposes. We know from medieval Church-Slavonic texts (Makarijoska 2003), the Ottoman census registers in the archives and from the ethnographic data from the twentieth century that flax was cultivated and linen fabrics were used for clothing, towels, baby wraps, pillow cases and blankets (Nazim 2008; Nazim et al. 2011). Finer fabrics were used for clothing, while the coarser and thicker ones for blankets, bags and sacks or for winter clothes, such as shepherds' capes.

Life in those Neolithic communities can be compared, to some extent, to the subsistence way of life in the Macedonian villages up to the mid-twentieth century. For instance, the Neolithic house plans and constructions (Bailey 2000; Perlès 2001; Tolevski 2009) and basic simple house types from the twentieth century in North Macedonia (Tomovski 1996; Namičeva and Namičev 2020) have some common features: rectangular or square surface level buildings with one or more rooms, wattle and daub post-framed walls, central posts for supporting the roof, hard beaten, plastered clay floors, central open hearth, etc.

The Macedonian peasants from the twentieth century would sleep together on mats spread on the earthen floor, on slightly elevated clay platforms, next to the open fireplace, while the smoke was going out through a hole on the roof. The family would often share their only room with their livestock, separated only by a low fence made of twined branches (Nazim 2005). Several types of mats and blankets were used depending on the local traditions, the availability of raw materials and the social status. Mats were made of bundles of rye straw, joined by twined ropes, or straw was plaited in long braids joined together by sewing; rush mats were woven on a vertical loom; mats of maize leaves were twined in S/Z direction on a treadle loom (Nazim 2009). Thick and heavy textile mats, woven a on a treadle loom in plain or twill weaves, were made of hemp, flax, wool and goat hair. Very often they

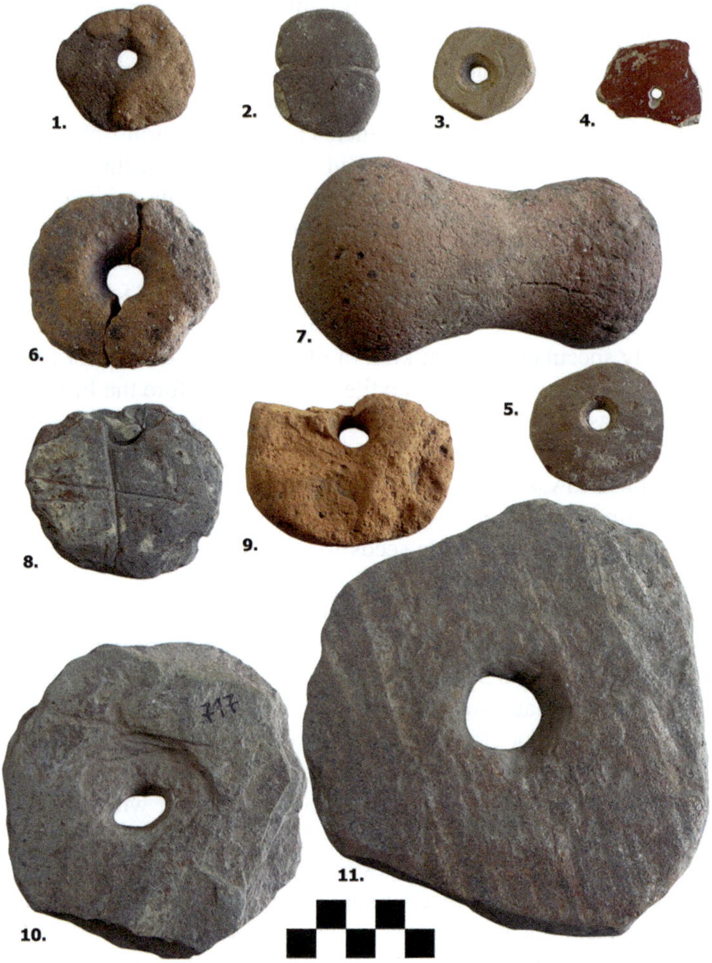

Fig. 5.4 Potential textile tools. Perforated pot sherds: [1] Pešterica; [2] Loom weights; [3–5] Vrbjanska Čuka; [6–7] Vrbjanska Čuka; [8–10] Čuka Topolčani. (Photo by Z. Blažeska)

were not dyed, instead using the natural hues of the fibres to produce striped or checkered patterns (Nazim 2008). Several people would be covered with a single blanket, similar to the textile mats, but larger. Untanned, joined sheep skins and heavy winter clothes were also used as sleeping covers (Nazim 2005).

Tools that could be related to the textile production were documented only in Vrbjanska Čuka, Čuka Tolopčani and Pešterica, most often in unknown contexts. Twenty one perforated, rounded sherds that could have been spindle whorls were found in Vrbjanska Čuka (Figs. 5.4 [3–5]) and only one in Pešterica (Fig. 5.4 [1]). Most of the fragments have circular or elliptical shape, but some of them have irregular shapes. The diameter of the fragments is between 2.5 and 6 cm and their thickness between 0.2 and 1 cm. The perforations on the fragments are often not placed in the center and usually are very small, between 0.2 and 0.5 cm. Only three sherds have perforations between 0.6 and 0.9 cm. On four of the sherds the perforations have not been finished. The weight of the sherds is between 20 and 63 g, and only one sample weighs about 160 g.

When we compare our measurements with those of the pierced pottery sherds from some other regions, only the small diameters of the perforations stand out (Good 2012; Orrelle et al. 2012; Rooijakkers 2012). They also have unbalanced shapes, some are too small and the perforations are not always centrally positioned. However, the experiment made by replicas of Late Neolithic pierced sherds from Jordan showed that they can be easily manufactured and that it was possible to produce yarns of each kind of fibre, even with the most asymmetrical sherds (Gibbs 2008). Pottery sherds were always readily available and easily manufactured, which was probably essential for their continuous and widespread use as spindle whorls, in parallel with various types of manufactured whorls. Perforated sherds have also been found at the other settlements in North Macedonia throughout the Neolithic (Kitanoski et al. 1987; Fidanoski 2009a). Discoid and biconical spindle whorls made of fired clay started to appear during the Middle and Late Neolithic. However, as we have previously noted that the yarns on the imprints from North Macedonia were produced by splicing, we agree with the suggestion of Gleba and Harris (2018) that the presence of spindle whorls is not evidence for draft spinning, and that spindles with pierced sherd whorls could be used just to stabilize the weft yarns with an additional twist at the splice points and to strengthen the warp by plying.

Twining may be performed without using any kind of loom, while warp-weighed looms are usually connected with the true weaving. Nevertheless, several objects from unknown contexts of a weight sufficient for potential loom weights, were found at Vrbjanska Čuka and Čuka Topolčani (Table 5.2). Two discoid shaped weights made of clay with central perforation were documented; one from Vrbjanska Čuka (Fig. 4.6) and one from Čuka Topolčani (Fig. 4.9), while another three discoid

Table 5.2 Perforated weights from Vrbjanska Čuka and Čuka Topolčani, now with the Institute for Protection of Cultural Monuments and Natural Rarities and Museum – Prilep (MP)

Sample no.	Museum no.	Settlement	Material	Diameter (cm)	Thickness (cm)	Perforation (cm)	Weight (gr)	Length (cm)
1	MP 15	Vrbjanska Čuka	Clay	6	2	1	95	/
2	MP 658	Čuka Topolčani	Clay	7	2	1	120	/
3	MP 717	Čuka Topolčani	Stone	10 × 11	1	1 × 1.5	212	/
4	MP 801	Čuka Topolčani	Stone	14 × 14.5	1.5–2	1.5	680	/
5	MP 660	Čuka Topolčani	Stone	6 × 6.5	0.7	/	70	/
6	MP 283	Vrbjanska Čuka	Clay	4	/	/	600	11.5
7	MP 307	Vrbjanska Čuka	Clay	4.7 × 4	0.5	/	20	/

stone weights were discovered at Čuka Topolčani. Two of them have irregular shapes and are centrally perforated (Fig. 4.10 and 4.11), while the third has a circular shape and an unfinished perforation close to its rim (Fig. 4.8). The last one could be an unfinished pendant, like the ones depicted on the Neolithic female figurines found at the sites close to Prilep and Veles (Kolištrkovska-Nasteva 2005: Fig. 72–73, 78). The waisted weight from Vrbjanska Čuka, made of fired clay (Fig. 4.2) and weighing only 20 g, is too light to be a loom weight and was most probably used as fishing net sinker. Larger weights of this shape were discovered in the Middle Neolithic horizon II at Tumba Madžari. A single dumbbell weight, weighing 600 g (Fig. 4.7) was found at Vrbjanska Čuka in an unknown context. Similar weights, usually referred to as spools, were also found at some other Neolithic settlements: Veluška Tumba (Simoska and Sanev 1975) and Porodin (Grbić et al. 1960) from Pelagonia, and Madžari, Govrlevo, Vrbnik and Stenče at the northwest of North Macedonia. A group of longer and thinner spools, discovered at Stenče placed around the oven, were described as net sinkers (Zdravkovski 2005).

Although a relatively small number of loom weights was found at the Early Neolithic settlements, from the end of the Early and during the Middle and Late Neolithic in Macedonia discoid, oval, semi oval, ovoid, spherical, conical, pyramidal and donut shaped loom weights started to appear in greater numbers (Simoska and Sanev 1975; Georgiev and Bilbija 1984; Temelkoski and Mitkoski 2006; Fidanoski 2009a, 2012). In the Middle Neolithic layer at Tumba Madžari, 56 loom weights were found on the floor, next to the altar (Sanev 1988). This is the only group with a large number of weights found so far. Although the loom weights indicate that the warp-weighted loom was familiar to the Neolithic population in North Macedonia, the earliest evidence for plain weaving is found on the textile imprints on two clay female figurines from the Chalcolithic settlements Crnobuki (Kolištrkovska-Nasteva 2005: 94, Fig. 78) and Bakarno Gumno in Pelagonia (Archaeological Museum of North Macedonia, Inv. no. 15). It is quite probable that during some period twining and weaving on a warp-weighted loom were practiced simultaneously.

5.4 Discussion

The technology of producing flexible textiles by twining plant fibres was already known to the people from the Upper Paleolithic sites in Central Europe, at least from 27,000 BP (Soffer et al. 2000; Good 2001; Adovasio et al. 2005). It is quite probable that twined textiles were also produced by the other Upper Paleolithic and Mesolithic cultures in the vast Eurasian area. About 13,000 years old imprints of twined textiles on ceramic sherds were found in the Amur River Basin at the Far East of Russia (Hyland et al. 2002). Starting from the eighth millennium BC, twined textiles and imprints on clay with various structural properties were identified at the Early Neolithic sites from the Middle East and Anatolia (Barber 1991; Gleba and

Nikolova 2009; Harris 2009; Alfaro Giner 2012). In the Balkans, the earliest imprints of twined textiles were discovered at the Early Neolithic settlement Nea Nikomedeia in Northern Greece, which was situated at the shores of the Thermaic Gulf during the last two centuries of the seventh millennium BC. Only close simple S-twist twining structures were preserved on the bases of 28 clay pots (Carington Smith 1977). The exclusive use of a close simple S-twist twining structure, common to the Early Neolithic sites in Northern Macedonia, should not be surprising considering the proximity of both regions and the northward movements of the Neolithic colonists along the river valleys. Although only this type of twining is present on the textile imprints from the Early Neolithic settlements in North Macedonia, a female steatopygic figurine, found at the Late Neolithic site Mramor is depicted wearing a skirt-like garment, which looks as it was made of an open twined fabric (Kolištrkovska-Nasteva 2005, Fig. 83).

Textile and mat imprints on pottery bases are, according to Özdoğan, one of the cultural markers of the Neolithic package (Özdoğan 2013). In the other parts of the Balkans they generally started to appear from the sixth millennia BC. Imprints of twined structures, mostly close diagonal, were found in Layer 7 (5950–5750 BC) at Aşağı Pınar in East Thrace (Özdoğan 2013), while at several Early Neolithic settlements in southwestern Bulgaria imprints of probably open simple S-twist twining with transposed weft were documented (Petkov 1926). Imprints of twined textiles were discovered at several Late Neolithic Vinča settlements in Serbia (c. 4800–4200 BC), where the closed diagonal twining with a Z-twist weft appeared most frequently (Adovasio and Maslowski 1988; Ninčić 2016). The largest number of imprints originates from Mali Borak and Vinča Belo Brdo (Ninčić 2016). During this period imprints of tabby weaves started to appear. At Divostin, about 75% of the textiles on the imprints found in five houses were twined and 25% were woven (Adovasio and Maslowski 1988). The close diagonal twining was also preferred in the Late Neolithic settlements in Romania, but there, the twist of the weft is in S direction (Mazăre 2012). This diversity of twining types at the Neolithic sites in the Balkans could be due initially to the different waves and routes of colonization and later as a result of local diversifications and contacts.

Carbonized flax seeds (*Linum usitatissimum* L.) were found in the Neolithic settlements in Greece, Bulgaria and Serbia (Kreuz et al. 2005; Valamoti 2011; Harris 2014; Ninčić2016), while in Croatia carbonized nettle seeds were identified (Jurić et al. 2001) at the Early Neolithic settlements along the Sava river (c. 6000 BC). At Gomolava, situated on the banks of Sava downstream in Serbia, during the transition from the Middle to the Late Neolithic, the use of wild flax (*Camelina sativa*) according to the seed analysis, had already declined, while the production of the cultivated flax (*Linum usitatissimum* L.) had grown (Ninčić 2016).

Pierced pottery sherds were found throughout Anatolia and the Middle East during the Neolithic and the Chalcolithic period (Barber 1991). The scarce finds of pot sherd discoid spindle whorls were typical for the Early Neolithic strata in Northern Greece (Carington Smith 1977; Perlès 2001). Sherd spindle whorls continued to exist during the Middle Neolithic in Greece, but in Thessaly they were gradually outnumbered by the flat discoid whorls made of clay, while conical and biconical whorls started to appear in the north (Carington Smith 1977; Tzachili 1996). Carington Smith suggested that pierced pottery sherds could have been prototypes for the later flat discoid clay whorls (1977), as most proper for spinning flax fibres.

The oldest finds of loom weights come from Ulucak, a settlement near Izmir in Aegean Western Anatolia, where clay loom weights and a mineralized piece of tabby weave on the shoulders of a human figurine were found in Level Vb (6400–6080 BC). Loom weights were also found in the Level Va (6230–5885 BC). In the level IVb (5990–5660 BC) 11 donut shaped, centrally perforated clay loom weights, were found deposited as a pile on the floor (Çilingiroğlu 2009; Çilingiroğlu et al. 2012). Until 5500 BC Ulucak is the only site in Western Anatolia where loom weights have been found. In the eastern and northern parts of Western Anatolia weights were absent until the fifth millennium BC (Schoop 2014). Between 5050 and 4250 BC, they mainly occurred as isolated finds. Nonetheless, from 4250–3000 BC almost all sites in West Anatolia produced loom weights and spindle whorls in large quantities (Schoop 2014).

Loom weights were also missing from the Early Neolithic sites in Greece. The first loom weights from Franchthi Cave in Argolid at the south and Sitagroi at the north of Greece, are dated around the mid-fifth millennium BC (Carington Smith 1977; Rasson 1988; Perlès 2001; Drašković 2004). However, in the other parts of the Balkans a great number of weights appeared much earlier, during the first half of the sixth millennium. Clusters of loom weights, in various shapes, were very often found at the Early Neolithic settlements. The chronological data from those settlements (Özdoğan 2011; Blagojević et al. 2017; Porčić et al. 2020), as well as the appearance of specific types of loom weights, indicate that the warp-weighted loom was travelling from the east to the west, generally following the river routes and valleys. The loom first appeared in eastern Thrace in Turkey and Bulgaria (Özdoğan 2013; Petrova 2015) and in a relatively short period spread throughout Bulgaria (Chohadzhiev 2004; Petrova 2015; Nikolov and Takorova 2018). Following the river routes to the south, loom weights started to appear in the territory of North Macedonia, while heading to the north they arrived to the early Neolithic settlements along the Middle Danube Valley in Serbia (Ninčić 2016) and Croatia (Minichreiler 2007) and in the Tisza Valley in Hungary (Barber 1991) and Romania (Mazăre 2014).

Objects similar to the waisted weight made of fired clay from Vrbjanska Čuka, but made of oval pebbles, were found at the Middle and Late Neolithic settlements in Greece: Servia, Saliagos and Rakhmani (Carington 1977; Rahmstorf 2015), at the Late Neolithic lake dwellings in Germany and in the Early Bronze Age contexts in Anatolia (Rahmstorf 2015) but they were considered to be net sinkers.

Although various interpretations for spools have been suggested, their function is not yet established. A large number of spools, the so called 'cocoon like' weights, from the Early Neolithic settlements in southwestern Bulgaria, sometimes found in piles or rows, clearly indicate that they were used as loom weights (Chohadzhiev 2004; Petrova 2015). Some of them are quite similar to the so-called spools found at the Neolithic settlements in North Macedonia. This is not surprising, knowing that the two regions are quite close to each other and their interactions are already verified (Mitrevski 2013). However, much more evidence should be gathered throughout Balkans and Anatolia in order to establish a reliable theory for the area of origin and the distribution patterns of the warp-weighted loom.

5.5 Conclusions

A textile production with characteristic local features was thriving at some of the Early Neolithic settlements on the territory of the Republic of North Macedonia, during the first centuries of the sixth millennium BC. Vrbjanska Čuka, Pešterica, Čuka Topolčani and Ġumušica, belonging to the Velušina-Porodin cultural group, make a compact group of settlements in the northern part of Pelagonia. Although Čakovec and Kanli Čair are positioned in the south-eastern fringes of the Anzabegovo-Vršnik cultural group, they share the same textile traditions.

We analyzed 45 textile imprints on clay, 22 perforated rounded pottery sherds and several objects that could be interpreted as loom weights. The majority of the imprints were preserved on the flat bases of ceramic vessels, one on the leg of an anthropomorphic house model and two on burnt daub fragments from a house wall. Most of the vessels were made of coarse clay. Pottery sherds with negative imprints of the fabrics were pressed upon dentist paste in order to obtain a positive mold. We visually identified the structure of the textiles and measured the yarn diameters, their twist directions and the thread count. All of the textiles were produced by close double twining with an S-twist of the weft. They are all weft faced and flexible, but most of them were worn out and frayed, which points to their secondary use during the modeling and drying of the pots. The standardized quality of the fabrics reveals a previous knowledge of the twining technology. Several other twining types were preferred at the Neolithic settlements in the other parts of the Balkans although the simple S twining was also used. The yarns were made from spliced, decorticated plant stalks, most probably flax. The thickness of the weft ranges from less than 1 mm to 5 mm. Warps could be identified only on two imprints from Kanli Čair; on one of them we could see the 1.80–2 mm thick S-plied warp with a twist angle of 55° while the warps of the other one are so loosely plied that their individual 1–2 mm thick threads could be easily observed. Thread counts range from seven to 18 yarn elements per square centimeter.

Tools that could be related to the textile production were documented only in Vrbjanska Čuka, Čuka Tolopčani and Pešterica. The 22 perforated rounded pottery sherds from Vrbjanska Čuka, Čuka Tolopčani and Pešterica were most probably used as spindle whorls for stabilizing the yarns with an additional twist at the splice points, and also for plying the wefts. Several objects from unknown contexts could have been loom weights: three perforated discoid stone weights from Čuka Topolčani, two discoid perforated weights made of clay from Vrbjanska Čuka and Čuka Tolopčani, one waisted weight of fired clay and one dumbbell shaped weight from Vrbjanska Čuka.

The loom weights from Ulucak, dating from 6400–6080 BC, may indicate that the origins of the warp-weighted loom should be searched in the Aegean coastal area at the south-west of Anatolia. Starting from the first quarter of the sixth millennium BC, loom weights were already present in the Balkans. Weaving on a warp-weighted loom was most probably gradually replacing the older twining technology. Scattered finds of loom weights during the Early Neolithic and the increase in their numbers during the Middle and Late Neolithic, as well as the parallel existence of both twined and plain weave imprints at some Late Neolithic sites in the Balkans, attest this process. The first imprints of plain weave found in North Macedonia were discovered on two clay figurines from a Chalcolithic settlement in Pelagonia.

Acknowledgments Special thanks to the curators Aleksandar Mitkoski from the Museum of Prilep, Elena Kanzurova and Irena Kolištrkovska-Nasteva from the Archeological Museum of North Macedonia, Goce Naumov from Museum of North Macedonia, Ljubo Fidanoski from the Museum of the City of Skopje, Dejan Ǵorǵievski from the Museum of Kumanovo and Slobodan Stojkovski from the Museum of Bitola for the useful information and for allowing us the access to the material.

References

Adovasio JM, Maslowski RF (1988) Textile impressions on ceramic vessels at Divostin. In: McPherron A, Srejović D (eds) *Divostin and the Neolithic of central Serbia.* Ethnology Monographs 10. University of Pittsburgh, Department of Anthropology, Pittsburgh, p 345–354.

Adovasio JM, Soffer O, Hyland D (2005) Textiles and cordage. In: Svoboda A (ed.) *Pavlov I southeast – a window into the Gravettian lifestyles.* Dolnověstonické studie 14. Archeologický ústav AV ČR, Brno, p 432–443.

Alfaro Giner C (2012) Textile from the Pre-Pottery Neolithic site of Tell Halula (Euphrates Valley, Syria). *Paléorient* 38:41–54.

Bailey DW (2000) *Balkan prehistory. Exclusion, incorporation and identity.* Routledge, London and New York.

Barber EJW (1991) *Prehistoric textiles. The development of cloth in the Neolithic and Bronze Ages with special reference to the Aegean.* Princeton University Press, Princeton.

Blagojević T, Porčić M, Penezić K, Stefanović S (2017) Early Neolithic population dynamics in the Eastern Balkans and the great Hungarian plain. *Documenta Praehistorica* XLIV. https://doi.org/10.2312/dp.44.2.

Carington Smith J (1977) *Spinning, weaving and textile manufacture in Prehistoric Greece: From the beginning of the Neolithic to the end of the Mycenaean ages; with particular reference to the evidence found on archaeological excavations.* Unpublished dissertation, University of Tasmania. Open Access Repository, https://eprints.utas.edu.au/11442/, accessed 19 Jan 2021.

Chohadzhiev A (2004) Weights and/or spools: Distribution and interpretation of the Neolithic "cocoon-like loom-weights". In: Nikolov V, Bačvarov K, Kalčev P (eds) *Prehistoric Thrace. Proceedings of the international symposium in Stara Zagora, 30.09–4.10.2003.* Institute of Archaeology with Museum – BAS, Regional Museum of History–Stara Zagora, Sofia, p 231–238.

Çilingiroğlu Ç (2009) Of stamps, loom weights and spindle whorls: Contextual evidence on the function(s) of Neolithic stamps from Ulucak, Izmir, Turkey. *Journal of Mediterranean Archaeology* 22.13–27. https://doi.org/10.1558/jma.v22il.3.

Çilingiroğlu A, Çilingiroğlu Ç, Çevik Ö (2012) Ulucak Höyük. Towards understanding the early farming communities of middle west Anatolia: Contribution of Ulucak. In: Özdoğan M, Başgelen N, Kuniholm P (eds) *The Neolithic in Turkey: New excavations and new research, Vol. 4 – Western Turkey.* Archaeology and Art Publications, Istanbul, p 139–135.

Drašković R (2004) Minijaturno posuđe, tegovi od pečene zemlje, kružne pločice, muzički instrumenti, predmeti kultne namene (Miniature vessels, weights made of fired clay, circular sherds, musical instruments, objects of cult purpose). In: Bogdanović M (ed.) *Grivac: Naselja Protostarčevacke i Vinčanske kulture (Grivac: settlements of Proto-Starčevo and Vinča culture).* Centar za naučna istraživanja Srpske akademije nauka i umetnosti Univerziteta u Kragujevcu i Narodni muzej u Kragujevcu, Kragujevac, p 367–380.

Fidanoski Lj (2009a) Predmeti od keramika, kamen, koska, rog i školka (Ceramic, stone, bone, antler and shell objects). In: Fidanoski Lj, Naumov G, Tolevski I, Ivkovska A (eds) *Neolitskite zaednici vo Republika Makedonija (The Neolithic communities in the Republic of Macedonia).* Dante, Skopje, p 159–168.

Fidanoski Lj (2009b) Periodizacija i hronologija na neolitskite kulturi (Periodization and chronology of the Neolilthic cultures). In: Fidanoski Lj, Naumov G, Tolevski I, Ivkovska A (eds) *Neolitskite zaednici vo Republika Makedonija (The Neolithic communities in the Republic of Macedonia).* Dante, Skopje, p 45–51.

Fidanoski Lj (2012) *Cerje-Govrlevo i Miloš Bilbija (Cerje-Govrlevo and Miloš Bilbija).* Muzej na Grad Skopje, Skopje.

Georgiev Z, Bilbija M (1984) Neolitska naselba kaj Govrlevo (Neolithic settlement at Govrlevo). *Kulturno nasledstvo* 9:39–48.

Gibbs KT (2008) Pierced clay disks and Late Neolithic textile production. In: Córdoba J, Molist M, Pérez C, Rubio I, Martínez S (eds) *Proceedings of the 5th international congress on the archaeology of the Ancient East, Madrid, April 3–8 2006,* vol II. Centro Superior de Estudios sobre el Oriente Próximo y Egypto, Madrid, p 91–93.

Gleba M, Nikolova A (2009) Early twined textiles from Sugokleya (Ukraine). *Archaeological Textiles Newsletter* 48:7–9.

Gleba M, Harris S (2018) The first plant fibre technology: Identifying splicing in archaeological textiles. *Archaeological and Anthropological Sciences* 11:2329–2346. https://doi.org/10.1007/s12520-018-0677-8.

Good I (2001) Archeological textiles: A review of current research. *Annual Review of Anthropology* 30:209–226.

Good I (2012) Changes in fibre use and spinning technologies on the Iranian plateau: A comparative and diachronic study of spindle whorls ca 4500–2500 BCE. *Paléorient* 38:111–126.

Grbić M, Mačkić M, Nadjiš P, Simoska D, Stalio B (1960) *Porodin, kasno-neolitsko naselje na Tumbi kod Bitolja (Porodin, Late Neolithic settlement at Tumba Near Bitola).* Narodni muzej Bitolj, Bitola.

Harris S (2009) Smooth and cool, or warm and soft: Investigating the properties of cloth in prehistory. In: Andersson S, Gleba M, Mannering U, Munkholt C, Ringgaard M (eds) *North European symposium for archaeological textiles X.* NESAT X. Ancient Textiles Series 5. Oxbow Books, Oxford, p 104–112.

Harris S (2014) Flax fibre: Innovation and change in the Early Neolithic – a technological and material perspective. *Textile Society of America 2014 biennial symposium proceedings: new directions: Examining the past, creating the future, Los Angeles, California, September 10–14, 2014.* http://digitalcommons.unl.edu/tsaconf/913, accessed 18 Jan 2021.

Hyland DC, Zhushchikhovskaya IS, Medvedev VE, Derevianko AP, Tabarev AV (2002) Pleistocene textiles in the Russian Far East: Impressions from some of the world's oldest pottery. *Anthropologie* XL:1–10.

Jurić I, Bogunović M, Dikić M, Balen, J (2001) Značajke poljoprivredne proizvodnje u naseljima starčevačke culture na prostoru između Vinkovaca i Slavonskog Broda u Hrvatskoj (Characteristics of agricultural production in the Starčevo Culture settlements at the area between Vinkovci and Slavonski Brod in Croatia). *Časopis za opća društvena pitanja* 56:1131–1158.

Kanzurova-Stojanova E (2017) The Neolithic settlement Kanli Čair-Bojkovci in Radoviš (Through archaeological excavations in 1987/88 and 2015). In: Stojanova Kanzurova E (ed.) *Dragiša Zdravkovski. In memoriam, cultural manifestations during the Neolithic period on the territory of the Republic of Macedonia and neighboring regions.* Archaeological Museum of Macedonia, Skopje, p 183–205.

Kitanoski B, Simoska D, Todorovik J (1987) Radin Dol –Neolitska naselba kaj Prilep (Radin Dol – Neolithic settlement near Prilep). *Macedonia Acta Archeologica* 7–8:7–29.

Kolištrkovska-Nasteva I (2005) *Praistoriskite dami od Makedonija – katalog (Prehistoric ladies from Macedonia – exhibition calatogue*). Muzej na Makedonija, Skopje.

Kreuz A, Marinova E, Shafer E, Wiethold J (2005) A comparison of early Neolithic crop and weed assemblages from the Linearbandkeramik and the Bulgarian Neolithic cultures: Differences and similarities. *Vegetation History and Archaeobotany* 14:237–258.

Makarijoska Lj (2003) *Leksikata na materijalnata kultura vo makedonskite crkovnoslovenski tekstovi (The lexicon of material culture in the Macedonian Church Slavonic texts).* Institut za makedonski jazik "Krste Misirkov", Skopje.

Mazăre NP (2012) *The craft of textile production at the Neolithic and Eneolithic communities in Transylvania.* Unpublished dissertation, University of Alba Iulia.

Mazăre NP (2014) Investigating Neolithic and Copper Age textile production in Transylvania (Romania). In: Nosch M-L, Michel C, Harlow M (eds) *Prehistoric, ancient Near Eastern and Aegean textiles and dress: An interdisciplinary anthology*. Ancient Textiles Series 18. Oxbow Books, Oxford, p 1–42.

Minichreiler K (2007) *Slavonski Brod Galovo. Monographiae Instituti Archaeologici,* Knjiga 1. Institut za arheologiju, Zagreb.

Mitrevski D (2013) Praistorija vo Republika Makedonija (Prehistory in the Republic of Macedonia). In: Kuzman P, Dimitrova E, Donev J (eds) *Makedonija: Mileniumski kulturno-istoriski fakti.* Media Print and Univerzitet Evro-Balkan, Skopje, p 83–266.

Namičeva E, Namičev P (2020) Tradicionalnata arhitektura vo prilepsko od 19 i početokot na 20 vek (The traditional architecture in the Prilep region from the 19th and the early 20th century). *PATRIMONIUM.MK* 13:673–699.

Naumov G, Mitkoski A, Talevski H, Anvari J, Przybila M, Stojanovski D, Antonin F, Sabanov A, Živajlević I, Dimitrijević V, Gibaja F G., Mazzucco N, Milevski Gj, Dumurđanov N, Pendić J, Blažeska Z and Stafanović S (2021) The Early Neolithic tell of Vrbjanska Čuka in Pelagonia. *Prähistorische Zeitschrift 96* (2):345–381. https://doi.org/10.1515/pz-2021-0007.

Nazim J (2005) Navikite pri spienje vo Makedonija vo tekot na 20 vek (Sleeping habits in Macedonia during the 20thcentury). *Zbornik Etnologija* 3:79–93.

Nazim J (2008) Za najstarite tipovi na postilači i pokrivači od Makedonija (About the oldest types of sleeping mats and covers in Macedonia). *Makedonski folklor* 65:63–74.

Nazim J (2009) Rogozini (Grass Mats). In: Svetieva A, Ashtalkovska A (eds) *Socijalni i duhovni aspekti na materijalnata kultura (Social and spiritual aspects of material culture).* Faculty of Natural Sciences and Mathematics, Institute of Ethnology and Anthropology, Skopje, p 44–60.

Nazim J (2011) Poslednite tkajački (The last weavers). *Zbornik Etnologija* 4:23–42.

Nazim J, Dimovska S, Ġorġiovska T, Hristova S (2011) Techniques and materials of production. In: Sumberg B (ed.) *Young brides, old treasures: Macedonian embroidered dress.* Exhibition catalogue, Museum of International Folk Art and The Macedonian Arts Council, Santa Fe, p 27–39.

Nikolov V, Takorova D (2018) *The dawn of European civilization: Slatina – Sofia 8 000 Years Ago.* Exhibition catalogue. Faber, Sofia.

Ninčić O (2016) *Tekstil u praistoriji na tlu Srbije (Textile in the prehistory in Serbia).* Unpublished dissertation, University of Arts, Beograd.

Orrelle E, Eyal R, Gopher A (2012) Spindle whorls and their blanks. In: Gopher A (ed.) *Village communities of the Pottery Neolithic Period in the Menashe Hills, Israel; Archaeological investigations at the sites of Nahal Zehora.* Emery and Claire Yass Publications in Archaeology, Institute of Archaeology, Tel Aviv University, Tel Aviv, p 632–656.

Özdemir A (2007) An experimental study of mat impressions on pot bases from Chalcolithic Gülpınar (Smintheion). In: Takaoğlu T (ed.) *Ethnoarchaeological investigations in rural Anatolia* 4. Ege Yayınları, Istanbul, p 73–85.

Özdoğan M (2011) An Anatolian perspective on the neolithization process in the Balkans. New question, new prospects. In: Krauβ R (ed.) *Beginnings – new research in the appearance of the Neolithic between Northwest Anatolia and the Carpathian Basin. Papers of the international workshop 8th–9th April 2009 Istanbul.* Rahden, Leidorf, p 23–33.

Özdoğan M (2013) Neolithic sites in the Marmara Region. In: Özdoğan M, Başgelen N, Kuniholm P (eds) *The Neolithic in Turkey: New excavations and new research, Vol. 4 – Western Turkey.* Archaeology and Art Publications, Istanbul, p 167–269.

Perlès C (2001) *The Early Neolithic in Greece. The first farming communities in Europe.* Cambridge University Press, Cambridge.

Petkov N (1926) Pletki of predistoricheskoto selishche Obreshcha pri Chelopechene (Textile structures from the prehistoric settlement Obreshta, v. Chelopechene). *Godishnik na Narodnia Muzej* IV:49–56.

Petrova V (2015) History of the warp-weighted loom from the Neolithic till the end of Antiquity (according to archaeological data from the present-day Bulgarian lands). *Studia Archaeologica Universitatis Serdicensis* 5:115–218.

Porčić M, Blagojević T, Pendić J, Stefanović S (2020) The timing and tempo of the Neolithic expansion across the Central Balkans in the light of the new radiocarbon evidence. *Journal of Archaeological Science: Reports* 33:1–12.

Rahmstorf L (2015) An introduction to the investigation of archaeological textile tools. In: Andersson Strand E, Noch M-L (eds) *Tools, textiles and contexts. Investigating textile production in the Aegean and Eastern Mediterranean Bronze Age.* Ancient Textile Series 21. Oxbow Books, Oxford and Philadelphia, p 1–23.

Rasson J (1988) Loom weights. In: McPherron A, Srejović D (eds) *Divostin and the Neolithic of central Serbia.* Ethnology Monographs 10. University of Pittsburgh, Department of Anthropology, Pittsburgh, p 337–343.

Rooijakkers CT (2012) Spinning animal fibres at Late Neolithic Tell Sabi Abyad, Syria. *Paléorient* 36: 93–19.

Sanev V (1987) Čakovec, naselba od raniot i sredniot neolit (Čakovec, settlement from the early and middle Neolithic). *Arheološki Pregled* 27:39–40.

Sanev V (1988) Neolitsko svetilište od Tumba vo Madžari, Skopsko, Preliminarno soopštenie od iskopuvanjata vo 1981 (Neolithic sanctuary fromTumba at Madžari, Skopje). *Macedoniae Acta Arheologica* 9:9–30.

Sanev V (1989) Bojkovci-Kanli Čair, neolitska naselba (Neolithic settlement Bojkovci-Kanli Čair). *Arheološki Pregled* 1987:36–37.

Schoop, U-D (2014) Weaving society in Late Chalcolithic Anatolia: Textile production and social strategies in the 4thmillennium BC. In: Horejs B, Mehofer M (eds) *Western Anatolia before Troy. Proto-urbanisation in the 4th millennium BC?: Proceedings of the international symposium held at the Kunsthistorisches Museum Wien, Vienna, Austria, 21–24 November, 2012.* Oriental and European Archaeology Vol. 1. Austrian Academy of Sciences Press, Vienna, p 421–446. https://doi.org/10.26530/oapen_625897.

Simoska D, Sanev V (1975) Neolitskata naselba Veluška Tumba kaj Bitola (Neolithic settlement Veluška Tumba near Bitola), *Macedoniae Acta Arhaeologica* 1:25–88.

Soffer O, Adovasio IM, Hyland DC (2000) The 'Venus' figurines, textiles, basketry, gender and status in Upper Paleolithic. *Current Anthropology* 41(4):511–537.

Temelkoski D, Mitkoski A (2006) Praistoriska naselba Mogila Senokos (Prehistoric settlement Mogila Senokos). *Macedoniae Acta Archaeologica* 17:55–80.

Tolevski I (2009) Arhitektura (Architecture). In: Fidanoski Lj, Naumov G, Tolevski I, Ivkovska A (eds) *Neolitskite zaednici vo Republika Makedonija (The Neolithic communities in the Republic of Macedonia).* Dante, Skopje, p 53–71.

Tomovski K (1996) Naselbi i narodna arhitektura (Settlements and traditional architecture). In: Tomovski K, Palikruševa G, Krsteva A (eds) *Etnologija na Makedoncite (The ethnology of Macedonians)*. Macedonian Academy of Sciences and Art, Skopje, p 81–128.
Tzachili I (1996) Weaving (catalogue). In: Papathanassopoulos GA (ed.) *Neolithic culture in Greece*. Nicholas P. Goulandris Foundation – Museum of Cycladic Art, Athens, p 137–139.
Valamoti SM (2011) Flax in Neolithic and Bronze Age Greece: Archaeobotanical evidence. *Vegetation History and Archaeobotany* 20(6). https://doi.org/10.1007/s00334-011-0304-4.
Wild J P (2003) Anatolia and Levant in the Neolithic and Chalcolithic Periods c. 8000–3500/3300 BC. In: Jenkins D (ed.) *The Cambridge history of western textiles*. Cambridge University Press, Cambridge, p 3–43.
Zdravkovski D (2005) Neolitska naselba Pod Selo – Tumba, selo Stenče (The Neolithic settlement Pod Selo – Tumba, Stenče). *Zbornik Arheologija* 2:25–30.

Zlata Blažeska is an archaeologist from Skopje, North Macedonia. She holds an MA degree in archaeology from the Cyril and Methodius University in Skopje. She has taken part in archaeological fieldwork working on documentation of finds, as well as in several conservation projects in North Macedonia and abroad. Her special area of interest is the archaeology of textile production, textile tools from the prehistoric settlements, as well as from later periods.

Jasemin Nazim graduated from the Cyril and Methodius University in Skopje, North Macedonia, in 1979. For 36 years, she worked as a curator for folk textiles in the Museum of North Macedonia in Skopje; she retired in 2018. She has participated in local and international research projects, international conferences, published a number of articles on traditional textiles, food history and identity in Macedonia and abroad. She has also curated several exhibitions on textiles in Skopje, Sofia, Zagreb, Istanbul, New York and Toronto. Her current area of interest is archaeological textiles.

Chapter 6
Tradition versus Innovation? Horizontal (Ground-) and Vertical (Warp-Weighted) Looms at Koukonisi, Lemnos

Tina Boloti

Abstract The densely occupied Bronze Age settlement on Koukonisi (an elliptical islet located in the innermost part of Moudros bay on Lemnos) was excavated under the direction of Chr. Boulotis and the auspices of the Academy of Athens from 1994 to 2016, with intermissions. Located in the north part of the islet, it provides an undisturbed sequence from the Early Bronze Age to the early Late Bronze Age, with sporadic finds of the early Mycenaean period (LH II–LH III A1) and a substantial phase of occupation corresponding to the developed Late Bronze Age (LH III A2–LH III B periods). Given the equation 'presence of loom weights = use of the warp-weighted loom', this paper discusses the introduction of the vertical (warp-weighted) loom in Koukonisi in the advanced or at the close of the Middle Bronze Age, when the earliest loom weights are detected in the settlement. Nearly 40 loom weights, discoid in their vast majority, either with a grooved or flattened top, can be assigned to the prominent South Aegean, and particularly Cretan type of loom weights, which indicates the adoption of Cretan weaving technology and probably reflects increased specialization in the textile industry. In any case, the suggested technological transmission presupposes a degree of human mobility within the Aegean during the Bronze Age. In search of a predecessor type of loom in Koukonisi, I argue that the most probable option would have been the horizontal (ground-) loom, although no direct evidence for its existence has so far been found in the settlement. Nevertheless, finds from the Middle Bronze Age Greek mainland and the wider Bronze Age Eastern Mediterranean makes this suggestion reasonable, while the practice of other weaving techniques cannot be excluded.

Keywords Aegean Bronze Age · North-Eastern Aegean · Weaving technology · Textile implements · Looms · Loom weights

---

T. Boloti (✉)
Koukonisi Archaeological Project, Lemnos, Greece

A. Ulanowska et al. (eds.), *Ancient Textile Production from an Interdisciplinary Perspective*, Interdisciplinary Contributions to Archaeology,
https://doi.org/10.1007/978-3-030-92170-5_6

6.1 The Prehistoric Settlement on Koukonisi (Lemnos)

The prehistoric settlement on Koukonisi, an elliptical (c. 500 m × 400 m) low (max. ht. 10.6 m) islet located in the innermost part of Moudros bay on Lemnos in the north-eastern Aegean, was excavated under the direction of Chr. Boulotis and the auspices of the Academy of Athens from 1994 to 2016. The excavations brought to light the remains of a densely occupied Bronze Age settlement in the northern part of the islet (Fig. 6.1), whose material culture shows strong links to the North East Aegean and the West Anatolian regions. The southward orientation of the bay would have ensured safe anchorage for ships heading towards the northern Aegean, Macedonia, Thrace, Troy or the Dardanelles, providing, along with harbour facilities, replenishment of victuals and water supplies. The privileged geographical location of Koukonisi is surely one of the decisive reasons for its longevity "as a major *emporium* for almost two millennia" (Boulotis 2009: 179). Having followed the developments of the two other major settlements in the Early Bronze Age[1] on Lemnos (i.e. Poliochni and Myrina), Koukonisi continued to flourish in the Middle Bronze Age (Koukonisi IV) and early Late Bronze Age (Koukonisi III), as the most important settlement of the island, providing an undisturbed sequence from the Early Bronze Age (Poliochni yellow, Troy IIg) to the early Late Bronze Age (Fig. 6.2; Poliochni violet, Troy VI b/c, Boulotis 1994; 1997: 230; Boulotis 2009:

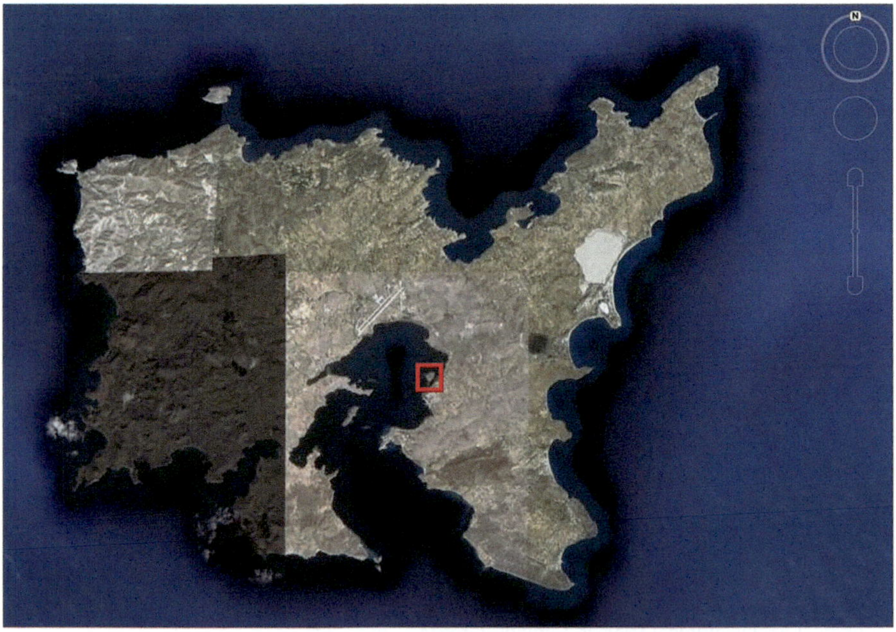

Fig. 6.1 Lemnos island. Koukonisi islet is framed in red. (Image drawn from Google Maps)

[1] For a chronological framework of the Aegean Bronze Age see Tartaron 2008: 84, Tab. 1.

<table>
<tr><th>Ayia Irini</th><th>Crete</th><th>Mikro Vouni</th><th>Troy</th><th>Poliochni</th><th>Koukonisi</th></tr>
<tr><td>VIA</td><td>LM IA</td><td>I</td><td rowspan="2">VI b/c</td><td rowspan="2">Violet</td><td rowspan="2">III</td></tr>
<tr><td rowspan="2">V</td><td rowspan="2">MM III</td><td>II</td></tr>
<tr><td>III</td><td rowspan="2">VI a</td><td rowspan="5">Brown</td><td rowspan="5">IV</td></tr>
<tr><td rowspan="2">IV</td><td rowspan="2">MM II</td><td>IV</td></tr>
<tr><td>V</td><td rowspan="3">V</td></tr>
<tr><td rowspan="3">Hiatus</td><td rowspan="4">(MM I)
(EM III)
(EM IIB)</td><td>VI</td></tr>
<tr><td>VII</td></tr>
<tr><td>VIII</td><td>IV</td><td>Hiatus?</td><td rowspan="2">V</td></tr>
<tr><td>III</td><td>IX</td><td>IIg-III</td><td>Yellow</td></tr>
</table>

Fig. 6.2 Comparative chronological table for the crucial sites and areas mentioned in the paper. (After Girella and Pavúk 2015: 394, Fig. 2)

181; Girella and Pavúk 2015: 394, Fig. 2). Finds of the early Mycenaean period (LH II–LH III A1) were also identified but were far less abundant. The next substantial phase of occupation corresponds to the developed Late Bronze Age, namely the LH III A2 and LH III B periods (Koukonisi II). The presence of Mycenaean pottery, as well as other typical Mycenaean objects, such as figurines of canonical Phi and Psi types, is a clear indication of a direct connection between Koukonisi and the Greek mainland during the fourteenth and the thriteenth centuries BCE (Boulotis 1997: 265, Fig. 28; Boulotis et al. 2017). This period, however, was not the first time that Koukonisi was in contact with people from the south Aegean. Minoan or Minoanizing pottery and other small finds, including textile tools, discovered in fair quantities in the settlement, indicate contacts with Crete and the south Aegean as early as the late Middle Bronze Age.

6.2 Loom Weights in the Koukonisi Settlement

Among the textile-related finds in Koukonisi, the most interesting appear to be *circa* 40 clay loom weights so far brought to light. These artefacts, which constitute the focal point of this paper, begin to appear in the settlement in the advanced Middle Bronze Age or at its close, unlike approximately 140 spindle whorls of various shapes and sizes, which are found in almost every strata of the excavation from the mid-third millennium BCE onwards (Boulotis 2009: 186). Illustrating household activities in the early Late Bronze Age settlement (Fig. 6.3), the loom weights are concentrated in a specific area (Trench 7 and, particularly, in Trenches 8 and 9), where other Minoan and 'Minoanizing' elements, such as specific household and

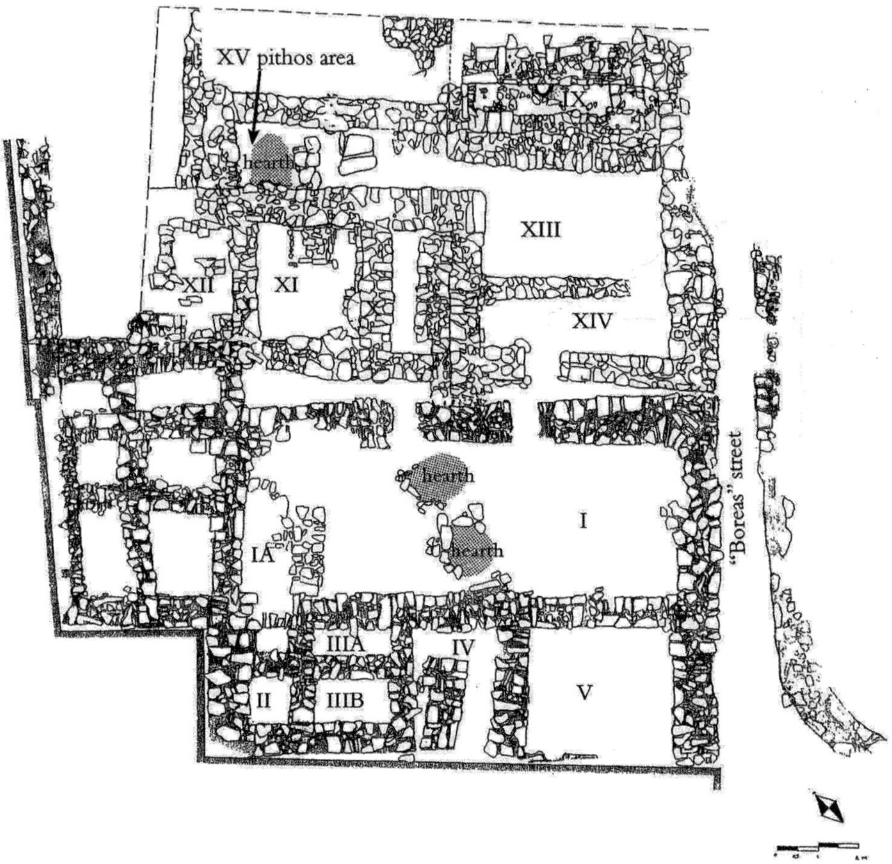

Fig. 6.3 Plan of the Minoanizing sector of the settlement (Trenches 8 and 9). Image drawn from Boulotis 2009: 180, Fig. 5. Courtesy of Chr. Boulotis

industrial activities (e.g., use of the tubular drill, use of terracotta *tuyères* of the straight cylindrical type) and prestige artefacts (e.g. painted pottery, small plaster fragments, stone vases) are also prominent (Boulotis 2009: 180–181, Figs. 5, 7). Except for one dubious example, all loom weights belong to the discoid type (rounded or elliptical) with one (occasionally two) suspension holes. The most important feature, however, is the grooved or the flattened top on some of them (18 and 11 examples respectively, Fig. 6.4a–d), which belong to the distinctive type of Cretan-style loom weights (Boulotis 2009: 186). The latter are considered, along with conical cups, as the hallmarks of strong Minoan influence throughout the Aegean by the mid-second millennium BCE (for the role of loom weights within this process see, Davis 1984: 161–163; Cutler 2012: 146–148; 2019: 81–82). One of them features an impressed finger mark beneath the suspension hole (Fig. 6.4a), closely comparable to a specimen from Ayia Irini V (Davis 1986: 38 [S-30], Pl. 66).

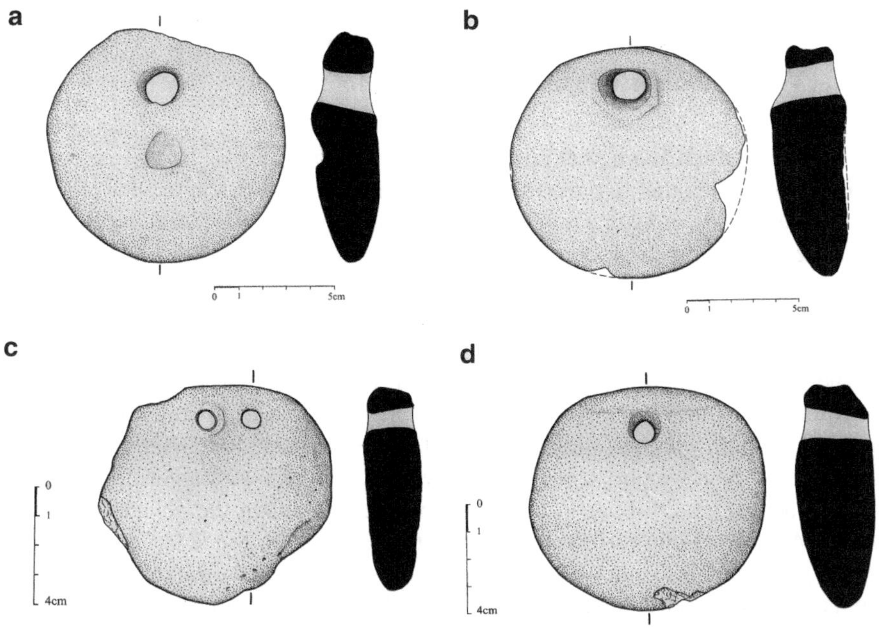

Fig. 6.4 (**a**–**d**)Selection of loom weights from Koukonisi by find location: (**a**–**b**) Trench 8, Room XIV, (**c**) Trench 8, Room XV, (**d**) Trench 9, Room IIIA. Courtesy of Chr. Boulotis

Similar in shape to these clay loom weights is a stone example (with incomplete suspension hole), which was found in 1996 in the ploughing surface.

With regard to the functional attributes on which the technological weaving optimum depends, intact loom weights from Koukonisi vary in weight from 54 to 305 g, while their thickness ranges from 1.6 to 3.2 cm. Their diameter lies between 5.6 and 10.5 cm and the diameter of the suspension hole ranges from 0.4 to 2.2 cm.

Given that the weight of the loom weights affects the type of fabric they are best suited to produce, the loom weights from Koukonisi appear to be suitable for fine rather than coarser fabrics, since they are classified as small ($50 \leq w < 250$ g) and medium ($250 \leq w < 750$ g).

Although the identification of perforated clay objects as loom weights can be highly controversial (some are called 'net sinkers', others 'spool-shaped weights' or just 'perforated objects', see Rahmstorf 2015: 7–9) and a thorough study of the textile implements (e.g. pierced sherds) from Koukonisi is pending (given that perforated discs in the Aegean are generally agreed to be loom weights), the present evidence allows us to correlate the appearance of the aforementioned discoid loom weights to the adoption of the warp-weighted loom in the settlement in the advanced Middle Bronze Age.

6.3 Weaving in the Koukonisi Settlement: The Horizontal (Ground-) Loom versus the Vertical (Warp-Weighted) Loom?

The warp-weighted loom was one of the three main types of looms used in the ancient world; the other two were the horizontal ground loom and the upright loom, i.e. the vertical two-beam loom (Carington Smith 1992: 675; Andersson Strand 2015: 52–54). Since the looms themselves were made of perishable material they are rarely preserved. Nevertheless, during the Bronze Age there is evidence for the presence of all of them in the Eastern Mediterranean region (Andersson Strand 2015: 50).

The warp-weighted loom, an upright loom-type on which weights are used to apply tension to the vertically hanging warp threads, is attested mainly thanks to the presence of loom weights (Barber 1991: 100; Cutler 2012: 146). Given the equation 'presence of loom weights = use of warp-weighted loom', we suggest the adoption of the warp-weighted loom in the Koukonisi settlement in the advanced Middle Bronze Age, when the earliest loom weights are detected. Since large numbers of spinning implements from the earliest phases of habitation confirm extensive textile production in the settlement (Boulotis 1997: 260), it is essential to investigate the method of weaving practiced on Koukonisi during the Early Bronze Age and the early phases of the Middle Bronze Age, prior to the adoption of the warp-weighted loom.

Surprisingly, although the warp-weighted loom is present in the North Eastern Aegean already from the Early Bronze Age, as indicated by the loom weights found in a number of settlements such as Thermi on Lesbos (Lamb 1936: 163), Emporio on Chios (Hood 1981–1982: 632–633), Heraion on Samos (Isler 1973: 174; Milojčić 1961: 39, Fig. 37.22) as well as at Troy (Guzowska and Becks 2005: 280), we lack similar evidence from the island of Lemnos. This is true not only of Koukonisi but also of the two other main, more or less contemporary, excavated settlements of Myrina and Poliochni (Bernabò-Brea 1964: 588–590, 655–658; Bernabò-Brea 1976: 280–284), although some terracotta weights from Early Bronze Age Myrina, identified by the excavator as possible loom weights, need to be examined (Panatsi 2004: 84–86).

Since the absence of evidence cannot always be taken as evidence of absence, the apparent dearth of loom weights in Lemnos prior to the appearance of the discoid variety in Koukonisi obliges us to examine the possibility that the loom weights either have not survived or have not been recognized as such. The use of loom weights made of perishable material seems possible, although unpractical; for instance, for loom weights made of wood, it would be difficult to achieve a practical weight to size ratio (Cutler 2012: 147–148, n. 31). Furthermore, the use of stones as loom weights, a practice attested both ethnographically and archaeologically (see Hoffmann 1974: 20–21), necessitates perforation or shaping in order to fasten the warp threads securely (Carington Smith 1977: 99). By contrast, clay, a readily obtainable resource, seems to allow easier control of the shape and weight of the produced loom weights than stone (see Hoffmann 1974: 21). In this case, "weights of poorly baked clay may not have survived: unbaked weights have been recovered from sites in the southern Aegean, although they are harder to detect" (Cutler 2012: 147–148, n. 31).

Leaving aside the scenario of loom weights made of perishable material, stone or unfired clay, undetected in Koukonisi during the Early Bronze Age and the early phases of the Middle Bronze Age, we have to turn to the two other types of looms known in the Eastern Mediterranean during that era. The oldest loom type among them is considered to be the horizontal ground loom, used mainly for tabby and basket weaves. The earliest depiction is dated to the Late Neolithic and comes from Badari, Egypt, where there is also information about its function. Thanks to iconographic representations and models, we know that the warp threads were stretched over two beams fastened with four corner-pegs, while the heddle rod is supported with heddle jacks (Andersson Strand 2015: Fig. 2.20). In general, two weavers are depicted, sitting on either side of the loom, changing the shed, entering and beating the weft (Andersson Strand 2015: 52; 2018).

The vertical two-beam loom, which stands upright like the warp-weighted loom, is thought to have originated in Syria or Mesopotamia, although its earliest representation occurs in Egypt during the last part of the second millennium BCE, where "it could have been developed in connection with the introduction of wool" (Andersson Strand 2015: 54). Wool is quite easy to dye and this could have inspired tapestry weaving, for which the two-beam loom is considered most suitable (Barber 1991: 113). Nevertheless, tabby and twill fabrics can also be produced on this loom (Barber 1991: 116; Andersson Strand 2015: 54).

Given the apparently late appearance of the vertical two-beam loom in the wider Eastern Mediterranean, I focus on the horizontal ground loom. In this context, there is another implement which might be of interest: the spools (see Pavúk 2012: 123–126; for unpierced cylindrical or slightly 'waisted' spools of later Bronze Age date see Cutler 2016: 178–179). Longitudinally pierced spools are associated with the horizontal loom although their exact function remains uncertain (Cutler 2019: 82). Possibly "they were used for preparing the warp" (Guzowska et al. 2015: 317). So far longitudinally pierced spools are undetected on Koukonisi, unless they have escaped attention in the pottery bags, something that seems implausible. Nevertheless, "supposedly numerous pieces" of spools are reported from Lemnos, although "their date and affiliation are not always clear" (Pavúk 2012: 124, 130, n. 37). From nearby Lesbos, examples occur in the Early and Middle Bronze Age phases of the settlement at Thermi (Pavúk 2012: 124, 130, n. 36).

Occurrence of longitudinally pierced spools is widely noted on the Greek mainland, as well as in Western Anatolia, in the Middle Bronze Age. In Western Anatolia there is a surprisingly similar situation to that in the North Eastern Aegean: a period of use of the warp-weighted loom in the Early Bronze Age, followed by disappearance of loom weights from archaeological contexts in the Middle Bronze Age, when spools are detected (Pavúk 2012: 129–130, Pl. XXXIVa – distribution map and a short catalogue of spool finds in the Aegean).

Returning to Koukonisi, I propose as a working hypothesis that the horizontal ground loom was in use in the settlement during the Middle Bronze Age (at least prior to the presence of loom weights in the settlement) and probably during the Early Bronze Age, given the absence of loom weights and the occurrence, although vague, of spools on Lemnos during this period. In any case, we have to bear in mind that if the spools were associated either with the warping of a horizontal loom or with plaiting, this suggests that two different textile production methods may have co-existed. A parallel use of both types of loom (namely the horizontal and the warp-weighted loom) at Ayia Irini (on Keos) during the Middle Bronze Age has been suggested by Jack Davis, who argued that the horizontal one was used essentially for domestic production while the warp-weighted was used by specialist weavers. The horizontal loom probably went out of use gradually in the Late Bronze Age "as more and more weaving was done by specialists" (Davis 1984: 163). A similar situation probably occurs at Troy (VI), where "the horizontal loom was used alongside the vertical loom, although no direct evidence for the existence of the horizontal loom has been found" (Guzowska et al. 2015: 317). It seems that the horizontal loom went gradually out of use by the end of Troy VI (LH IIIA2), when "textile production became dominated by specialists working with the warp-weighted loom" (Pavúk 2012: 127).

Evidently, all the above-mentioned loom types have advantages and disadvantages. Since the main aim remains always to achieve the best yield and the most desirable result, different looms may have been used concurrently for producing different types of textiles (Andersson Strand 2015: 55). Undoubtedly the Bronze Age textile craftspeople would have been aware of other techniques as well (e.g. making different types of bands, using different braiding techniques or other textile production techniques not recognizable in the archaeological record, Barber 1991: 311–382), "which is partly due to the fact that these techniques either do not require any tools at all, or the tools are made of perishable material" (Andersson Strand 2015: 57).

6.4 The Adoption of the Warp-Weighed Loom at Koukonisi. Specialists versus Domestic Weavers?

The adoption of the warp-weighted loom at Koukonisi in the advanced Middle Bronze Age and especially of the grooved discoid loom weights, i.e. loom weights of Minoan type, can be seen in a wider context of cultural innovation. In the North Eastern Aegean, grooved discoid loom weights have also been found at Mikro Vouni on Samothrace (Becks and Guzowska 2004: 103; Girella and Pavúk 2016: 19, contemporary with MM II on Crete), Thermi (Pavúk 2012: 130), Emborio (Hood 1981–1982: 633) and Troy (Becks and Guzowska 2004), as well as at a number of sites on the West Anatolian coast. Nevertheless, their main diffusion area is the wider Southern Aegean, where, apart from Crete, they are detected in the

Peloponnese and on the islands of Aigina, Kythera, Keos, Melos, Thera, Kos and Rhodes (for a distribution map of Aegean sites with grooved loom weights see Pavúk 2012: 130, Pl. XXXIVb). This process, which must have happened already in the Middle Bronze Age, "seems to have reached its peak in LM IA" (Pavúk 2012: 122).

Elisabeth Barber, as well as Jill Carington Smith, demonstrated that grooved loom weights[2] have a certain technological advantage over the non-grooved ones: instead of hanging directly from the warp, the grooved loom weights were attached to a horizontally mounted rod, a simple solution which "prevented warp from twisting, held weights in one position and in the unchanging intervals, and largely reduced noise at work" (Guzowska and Becks 2005: 281). New explanations on the use of grooved loom-weights have been proposed more recently by Brendan Burke (2010: 47), Carol Cheval (2008: 21) and Agata Ulanowska (2019). Based on iconographic analysis of Minoan glyptic, Burke proposed that the rod "would provide an equal tension of the warp threads and, possibly, would help in keeping the even spacing of the warp, while Cheval suggested that "the grooves were merely post-production marks without any specific function" (Ulanowska 2019: 742–743). Ulanowska herself also comes to this conclusion on the basis of archaeological experiments (Ulanowska 2019: 752). If the outcome of this technological novelty was the production of finer fabrics (Tzachili 1990: 385), it is justifiable to consider not only the purely technical aspect of the new technology, but also its eventual social role. How was this new weaving technology introduced in the settlement? Who were its users and who were the recipients of its better 'products'?

When considering how this new weaving technology may have been introduced in advanced Middle Bronze Age or early Late Bronze Age Koukonisi, we need to take into consideration the diffusion pattern of the loom weights in the excavation. Given their strong concentration in the southwest part of the settlement, i.e. Trench 7 and, particularly in Trenches 8 and 9 (Fig. 6.3), where other Minoan and Minoanizing elements have been detected as well, I cannot avoid leaning towards Boulotis' conclusion that the Minoan population dwelt in this specific quarter of the settlement (Boulotis 2009: 180–181, Figs. 5, 7; 2011: 18). This possibility, which suggests a degree of mobility within the Aegean during the Bronze Age, indicates one way in which Cretan weaving technology may have been transmitted (Cutler 2012: 150). These immigrant weavers from within the Minoan or Minoanizing world, either men or women, may have carried with them either complete sets of loom weights or examples to act as templates for making further weights. Since loom weights are not objects that are likely to have any perceived intrinsic value, it is probable that the non-local weights detected in a settlement would have travelled with their users, rather than been imported independently. Concerning the origin of the loom weights from Koukonisi we are obliged to wait for the results of chemical/

[2]An overview of the explanations proposed so far concerning the functionality of grooves on the upper edge of some discoid loom weights is provided in Ulanowska 2019: 741–743.

petrographic analysis to confirm the observations of preliminary macroscopic examination.

Whether or not this is how the new weaving technology was introduced, once it was established on Koukonisi, it was probably part of a fundamental change in the industry of the entire settlement since there is no reason to believe that the Minoan loom weights were restricted in their use to the Minoan immigrants alone (for a comparable situation on Keos, see Davis 1984: 162–163). Since technical expertise in weaving can take several years to achieve (Cutler 2012: 149), we would suppose that there was a sustained interaction between the immigrant and local weavers. During this phase the experienced individuals would intervene as necessary to assist the learners, providing a progressively diminishing scaffold to support and extend the novices' more limited capabilities, until full expertise was achieved.

As the warp-weighted loom was adopted, grooved loom weights would probably have been used with caution. Only after having passed this stage, would they have become fully 'domesticated' and incorporated into the local weaving practices. The grooved loom weights could be used either according to their designed role, in separate sets, attached to a rod, or together with non-grooved weights on one loom depending on their position in relation to heddle, as has been postulated in some cases for Minoan weaving (Evely 2000: 502), although the technical justification for this needs to be explained.

Alongside local weavers using – as I hypothesize – the horizontal ground loom, immigrant weavers and any other users of the warp-weighted loom would have produced Cretan-style fabrics with complex patterns "since [on the warp-weighted loom] individual sections of warp threads can easily be lifted independently of each other" (Cutler 2012: 151). Considering the fame and high status of the Minoan textiles, the grooved loom weights could designate their owners and/or users as possessing higher status, prestige or skill in the community of weavers. At whom or what was their textile production aimed? The Koukonisian elite? Export or exchange within the wider North Eastern Aegean? These rhetorical questions remain open.

The introduction of the warp-weighted loom on Koukonisi probably reflects increased specialization in the textile industry, and is an innovation which becomes more striking when evidence from earlier habitation levels is considered. Since, so far, there are no preserved textiles from the settlement and no looms have been detected, the possibility that other weaving techniques were also being practiced cannot be excluded. The few fragments of Bronze Age textiles recovered in the Aegean to date are tabby weaves (also known as plain weaves) (for an overview of archaeological textiles from Bronze Age Greece see Spantidaki and Moulherat 2012: 187–194). However, since the existing sample of textiles is so small, only the loom weights recovered from sites across the Aegean can provide significant insight into the range of fabrics that could have been made.

Iconography of the second millennium BCE in the Aegean – and especially Late Bronze Age iconography – is abundant concerning sartorial types and practices of both sexes (e.g. Jones 2015; Shaw and Chapin 2016). However, it offers little evidence in relation to fibres and weaving techniques, since the latter prove to be hardly recognizable. Although it seems plausible to assume that Aegean painters have some technical knowledge of textile production through cross-craft interactions (Brysbaert 2007), contemporary iconography seems to represent products which are technically far more advanced than any remains of the actual archaeological textiles. As a consequence, any attempt to reconstruct textiles based on iconography must be treated with caution – experimental recreation of repetitive patterns and complex textile designs presupposes the use of fibres and techniques which we consider appropriate for the time, as well as replicas of actual tools. For this reason, analysis of archaeological textile tools remains the most reliable means to understand contemporary textile production.

6.5 Conclusions

The appearance of loom weights in the Bronze Age settlement on Koukonisi islet in the advanced Middle Bronze Age indicates the use of the warp-weighted loom, an upright type of loom on which weights apply tension to the vertically hanging warp threads. Given the lack of evidence for other materials for clothing (e.g. leather) and the ample presence of spindle whorls in the settlement from the Early Bronze Age onwards, which points to established textile production on the site prior to the introduction of the warp-weighted loom, I suggest that its most probable predecessor would have been the horizontal ground loom. Although no direct evidence for the existence of the latter has been found so far (e.g. longitudinally pierced spools), evidence from the Middle Bronze Age Greek mainland and the wider Bronze Age Eastern Mediterranean makes this suggestion reasonable (Pavúk 2012: 123–126). It is true that the presence and practice of other weaving techniques on Koukonisi cannot be excluded. In any case, only the forthcoming systematic study of the corpus of textile implements on Koukonisi can shed further light on the *chaîne opératoire* of textile production in the settlement.

However, if we accept that specialist weavers, perhaps from Crete or elsewhere in the southern Aegean (Cutler 2019: 86–88; Landenius Enegren 2020), were settled at Koukonisi in the advanced Middle Bronze Age, it seems reasonable to suppose that they would have transmitted the Cretan technology through sustained interaction with the local weavers, since technical expertise in weaving can take several years to achieve. The occurrence of *ra-mi-ni-ja* ('Lāmniai/'Lēmnian women', cf. Aura Jorro 1993: s.v. *ra-mi-ni-ja*) among others (from east Aegean islands – and from different localities on the south-west coast of Anatolia) in the thirteenth century Linear B tablets from Pylos (see recently Olsen 2014: 95–100) indicates indirectly the efficiency of the supposed immigrants trainers: the abilities of Lemnian women to weave must have been exceptional, since they were employed at palatial workshops on the Greek Mainland.

Whether or not this is how the new (Cretan) weaving technology was introduced in Koukonisi, once established, it led evidently to a fundamental change in the industry of the entire settlement.

Acknowledgements I would like to thank the British School at Athens for the award of a grant from the Vronwy Hankey Memorial Fund for Aegean Studies (2019), which allowed me to carry out work in the archaeological Museum of Lemnos during the summer of this year. Special thanks are also due to Christos Boulotis for the opportunity to study the corpus of the textile implements on Koukonisi and for fruitful discussions on the site. I am very grateful to Robin Barber for his comments on the text and the solutions he offered in order to avoid language pitfalls. Last, but not least, the anonymous reviewer and the editors for identifying points for clarification.

References

Andersson Strand E (2015) The basics of textile tools and textile technology – from fibre to fabric. In: Andersson Strand E and Nosch M-L (eds) *Tools, textiles and contexts. Investigating textile production in the Aegean and Eastern Mediterranean Bronze Age*. Ancient Textile Series 21. Oxbow Books, Oxford and Philadelphia, p 39–60.

Andersson Strand E (2018) Early loom types in ancient societies. In: Siennicka M, Rahmstorf L, Ulanowska A (eds) *First textiles: Beginnings of textile manufacture in Europe and the Mediterranean*. Ancient Textile Series 32. Oxbow Books, Oxford and Philadelphia, p 17–30.

Aura Jorro F (1993) *Diccionario Micénico*, Vol. II. Consejo Superior de Investigaciones Científicas, Madrid.

Barber EJW (1991) *Prehistoric textiles. The development of cloth in the Neolithic and Bronze Ages with special reference to the Aegean*. Princeton University Press, Princeton.

Becks R, Guzowska M (2004) On the Aegean-type weaving at Troia. *Studia Troica* 14:101–115.

Bernabò-Brea L (1964) *Poliochni, città preistorica nell'isola di Lemnos*. Monografie della Scuola Archeologica di Atene e delle Missioni italiane in Oriente, volume I: testo e tavole. L'Erma di Bretschneider, Roma.

Bernabò-Brea L (1976) *Poliochni, città preistorica nell'isola di Lemnos*. Monografie della Scuola Archeologica di Atene e delle Missioni italiane in Oriente, volume II: testo e tavole. L'Erma di Bretschneider, Roma.

Boulotis C (1994) Énas néos proïstorikós oikismós sto Koukonísi Límnou). In: *Límnos filtáti. Praktiká tou 1ou Synedríou gia to Aigaío, Mýrina 21–24 Avgoústou 1992*. Dímos Myrinaíon, Athína, p 19–36.

Boulotis C (1997) Koukonísi Límnou: Téssera chrónia anaskafikís érevnas. Théseis kai ypothéseis. In: Doumas Ch, La Rosa V (eds) *I poliόchni kai i próimi epochí tou Chalkoú sto vóreio Aigaío, diethnés synédrio, Athína 22–25 Aprilíou 1996.* Scuola Archeologica Italiana di Atene, Panepistímio Athinón, Athína, p 230–272.

Boulotis C (2009) Koukonisi on Lemnos: Reflections on the Minoan and Minoanising evidence. In: Macdonald CF, Hallager E, Niemeier W-D (eds) *The Minoans in the Central, Eastern and Northern Aegean – new evidence. Acts of a Minoan Seminar 22–23 January 2005 in collaboration with the Danish Institute at Athens and the German Archaeological Institute at Athens.* Monographs of the Danish Institute at Athens 8. The Danish Institute at Athens, Athens, p 175–218.

Boulotis C (2011) PSIFOS IMITELIS í ti mas léei éna mikró téchnergo apó ton oikismó tis Chalkokratías sto Koukonísi tis Límnou. *Anáskamma* 5:13–31.

Boulotis C et al. (2017) Koukonísi Límnou. Prótes paratiríseis apó ti meléti tis mykinaïkís keramikís. In: *To archaiologikó érgo sta nisiá tou Aigaíou, diethnés epistimonikó synédrio, Ródos, 27 Noemvríou – 1 Dekemvríou 2013,* tom. *B*. Ypourgeío Politismoú kai Athlitismoú, Mytilíni, p 45–62.

Brysbaert A (2007) Cross-craft and cross-cultural interactions during the Aegean and Eastern Mediterranean Late Bronze Age. In: Antoniadou S, Pace A (eds) *Mediterranean crossroads*. Pierides Foundation Publications, Athens, p 325–359.

Burke B (2010) *From Minos to Midas: Ancient cloth production in the Aegean and in Anatolia.* Ancient Textile Series 7. Oxbow Books, Oxford and Oakville.

Carington Smith J (1977) *Spinning, weaving and textile manufacture in Prehistoric Greece. From the beginning of the Neolithic to the end of the Mycenaean Ages; With particular reference to the evidence found on archaeological excavations.* Unpublished dissertation, University of Tasmania. Open Access Repository, https://eprints.utas.edu.au/11442/, accessed 19 Jan 2021.

Carington Smith J (1992) Spinning and weaving equipment. In: Macdonald WA, Wilkie NC (eds) *Excavations at Nichoria*, Vol. 2. University of Minnesota Press, Minneapolis, p 674–711.

Cheval C (2008) Protohistoric weaving, the Minoan loom-weights: A first approach. In: Alfaro Giner C, Karali L (eds) *Vestidos, textiles y tintes: studios sobre la producción de bienes de consumo en la Antigüedad. Actas del II Symposium Internacional sobre Textiles y Tintes de Mediterráneo en el mundo antiguo (Atenas, 24 al 26 de noviembre, 2005).* PURPUREAE VESTES II. Universitat de València, València, p 19–24.

Cutler J (2012) Ariadne's thread: The adoption of Cretan weaving technology in the wider southern Aegean in the mid-second millennium B.C. In: Nosch M-L, Laffineur R (eds) *KOSMOS. Jewellery, adornment, and textiles in the Aegean Bronze Age. Proceedings of the 13th International Aegean Conference, University of Copenhagen, Danish National Research Foundation's Centre for Textile Research, 21–26 April 2010.* Aegaeum 33. Peeters, Leuven and Liège, p 145–154.

Cutler J (2016) Fashioning identity: Weaving technology, dress and cultural change in the Middle and Late Bronze Age southern Aegean. In Gorogianni E, Pavúk P, Girella L (eds) *Beyond thalassocracies: Understanding processes of Minoanisation and Mycenaeanisation in the Aegean.* Oxbow Books, Oxford, p 172–185.

Cutler J † (2019) Arachne's web: Women, weaving and networks of knowledge in the Bronze Age Southern Aegean. *The Annual of the British School at Athens* 114:79–92.

Davis JL (1984) Cultural innovation and the Minoan thalassocracy at Ayia Irini, Keos. In: Hägg R, Marinatos N (eds) *The Minoan thalassocracy. Myth and reality. Proceedings of the third international symposium at the Swedish Institute in Athens, 31 May–5 June, 1982.* Svenska Institutet i Athen and Paul Åström, Stockholm, p 159–165.

Davis JL (1986) *Ayia Irini: Period V. Keos. Results of the excavations conducted by the University of Cincinnati under the auspices of the American School of Classical Studies at Athens,* vol. V. Philipp von Zabern, Mainz on Rhine.

Evely D (2000) *Minoan crafts: Tools and techniques. An introduction.* Studies in Mediterranean Archaeology XCII:2. Paul Åström, Jonsered.

Girella L, Pavúk P (2015) Minoanisation, acculturation, hybridisation: The evidence of the Minoan presence in the North East Aegean between the Middle and Late Bronze Age. In: Stampolides N, Çiğdem M, Kopanias K (eds) *NOSTOI. Indigenous culture, migration and integration in the Aegean islands and Western Anatolia during the Late Bronze and Early Iron Age. Proceedings of the international conference held in Istanbul 31/3/11–3/4/11.* Koç University Press, Istanbul, p 387–420.

Girella L, Pavúk P (2016) The nature of Minoan and Mycenaean involvement in the northeastern Aegean. In: Gorogianni E, Pavúk P, Girella L (eds) *Beyond thalassocracies. Understanding processes of Minoanisation and Mycenaeanisation in the Aegean.* Oxbow Books, Oxford and Philadelphia, p 15–42.

Guzowska M, Becks R (2005) Who was weaving at Troia? On the Aegean style loomweights in Troia VI and VIIa. In: Laffineur R, Greco E (eds) *EMPORIA. Aegeans in the Central and Eastern Mediterranean: Proceedings of the 10th international Aegean conference, Athens, Italian School of Archaeology, 14–18 April 2004.* Aegaeum 25. Université de Liège, University of Texas at Austin, Program in Aegean Scripts and Prehistory, Liège and Austin, p 279–286.

Guzowska M, Becks R, Andersson Strand E, Cutler J, Nosch M-L (2015) Textile tools from Troia, western Anatolia. In: Andersson Strand E, Noch M-L (eds) *Tools, textiles and contexts. Investigating textile production in the Aegean and Eastern Mediterranean Bronze Age.* Ancient Textile Series 21. Oxbow Books, Oxford and Philadephia, p 309–328.

Hoffmann M (1974 [1964]) *The warp-weighted loom: Studies in the history and technology of an ancient instrument.* Robin and Russ Handweavers, McMinnville Oregon, USA (later edition of Hoffmann's book in the USA).
Hood S (1981–1982) *Excavations in Chios 1938–1955. Prehistoric Emporio and Ayio Gala.* Thames and Hudson, London.
Isler HP (1973) An Early Bronze Age settlement on Samos. *Archaeology 26*(3): 170–175.
Jones BR (2015) *Ariadne's threads: The construction and significance of clothes in the Aegean Bronze Age.* Aegaeum 38. Peeters, Leuven and Liège.
Lamb W (1936) *Excavations at Thermi in Lesbos.* Cambridge University Press, Cambridge.
Landenius Enegren H (2020) Women, men, girls and boys: Gendered textile work at Late Bronze Age Knossos. In: Harlow M, Michel C, Quillen L (eds), *Textiles and gender in Antiquity: From the Orient to the Mediterranean.* Bloomsbury Publishing, London, p 95–110.
Milojčić V (1961) *Samos. Die prähistorische Siedlung unter dem Heraion, Grabung 1953 und 1955.* Rudolf Habelt, Bonn.
Olsen B (2014) *Women in Mycenaean Greece: The Linear B tablets from Pylos and Knossos.* Routledge, London.
Panatsi A (2004) Evrímata. In: Archontídou A, Kokkinofórou M (eds) *I Mýrina tis próimis epochís tou Chalkoú.* Ypourgeío Politismoú, K′ Eforeía Proïstorikón and Klasikón Archaiotíton, Límnos, p 71–87.
Pavúk P (2012) Of spools and discoid loom weights: Aegean type weaving at Troy revisited. In: Nosch M-L, Laffineur R (eds) *KOSMOS. Jewellery, adornment, and textiles in the Aegean Bronze Age. Proceedings of the 13th International Aegean Conference, University of Copenhagen, Danish National Research Foundation's Centre for Textile Research, 21–26 April 2010.* Aegaeum 33. Peeters, Leuven and Liège, p 121–130.
Rahmstorf L (2015) An introduction to the investigation of archaeological textile tools. In: Andersson Strand E, Nosch M-L (eds) *Tools, textiles and contexts. Investigating textile production in the Aegean and Eastern Mediterranean Bronze Age.* Ancient Textile Series 21. Oxbow Books, Oxford and Philadelphia, p 1–23.
Shaw M, Chapin A (2016) *Woven threads: Patterned textiles of the Aegean Bronze Age.* Ancient Textile Series 22. Oxbow Books, Oxford and Philadelphia.
Spantidaki Y, Moulherat C (2012) Greece. In: Gleba M, Mannering U (eds) *Textiles and textile production in Europe from Prehistory to AD 400.* Ancient Textile Series 11. Oxbow Books, Oxford and Oakville, p 185–202.
Tartaron Th (2008) Aegean Prehistory as world archaeology: Recent trends in the archaeology of Bronze Age Greece. *Journal of Archaeological Research* 16:83–161.
Tzachili I (1990) All important yet elusive: Looking for evidence of cloth-making at Akrotiri. In: Hardy DA, Doumas CG, Sakellarakis JA, Warren PM (eds) *Thera and the Aegean world III. Proceedings of the third international congress, Santorini, Greece, 3–9 September 1989, Vol. I. Archaeology.* The Thera Foundation, London, p 380–389.
Ulanowska A (2019) Why are some discoid loom weights grooved? Answers from experimental archaeology on the functionality of weaving tools in the Bronze Age Aegean. In: Pieńkowska A, Szeląg D, Zych I (eds) *Stories told around the fountain. Papers offered to Piotr Bieliński on the occasion of his 70th birthday.* University of Warsaw Press, Warsaw, p 733–758.

Tina Boloti is a postdoctoral researcher, holding a PhD from the University of Crete, Greece (2016). Her thesis, which examines the functional and symbolic role of cloth and clothing in rituals in the Aegean Late Bronze Age, constitutes a combined study of the related iconography and the Linear B archives. She has participated in excavations and research programs of Bronze and early Iron Age sites on Crete, Mainland Greece and NE Aegean, including the ongoing Koukonisi Archaeological Project, a combined field and study research program of the prehistoric settlement on the islet of Koukonisi (Lemnos) under the auspices of the Academy of Athens.

Chapter 7
Weaving Traditions in Archaic Sicily: The Case Study of Portella Sant'Anna at Monte Polizo

Gabriella Longhitano

Abstract Investigation of textile traditions in Archaic Sicily has traditionally focused on loom weight types and how Greek and Punic cultures affected indigenous weaving traditions. The available archaeological evidence seems to show that some indigenous communities were reluctant to adopt foreign cultural elements and maintained distinctive and peculiar textile traditions.

This paper offers the opportunity to extend the discussion on weaving traditions to the methodological approach and enhances our understanding of ancient textile traditions through the case study of Portella Sant'Anna. This area, located to the north of the Acropolis at the site of Monte Polizo, yielded a group of loom weights of different weights and sizes fallen down from a potential working loom. In this paper, the evidence has been analysed and an attempt has been made to understand whether and how these loom weights might have worked together in a loom set-up. This assemblage has then been compared to other loom weight clusters found at the site. The result might be suggestive of a particular way to set up a loom, which could create particular appearances in the warp threads.

Keywords Textile tradition · Technology · Loom weights · Loom set-up · Archaic Sicily · Portella Sant'Anna

7.1 Weaving Traditions in Archaic Sicily

Textile culture, namely the way in which people use clothing tradition as a form of self-description (Gleba 2017: 1206), is the result of a variety of factors ranging from technological and technical aspects to the way in which textiles are wrapped, folded and combined to cover the body. All these factors are dependent on either practical

G. Longhitano (✉)
Department of Human Sciences, University of Catania, Catania, Italy
e-mail: gabriella.longhitano@studium.unict.it

A. Ulanowska et al. (eds.), *Ancient Textile Production from an Interdisciplinary Perspective*, Interdisciplinary Contributions to Archaeology,
https://doi.org/10.1007/978-3-030-92170-5_7

or cultural reasons. Technical aspects of setting up a loom (i.e. the tension applied to the warp threads, the number and size of the loom weights etc.) are primarily connected to the product being created. However, techniques, patterns and colours can also be determined by the specific traditions and aesthetic values of an individual community. Every society favours specific textiles, techniques, raw materials, colours, patterns and decorations because they are perceived as more appropriate for conveying social messages. Thus, the value of textiles and raw material can be different across societies. Our understanding of these factors and choices is often limited by the nature of the archaeological record and investigation often reveals a very limited selection of the variables of each textile culture.

In Sicily, archaeological research on textile culture has gained momentum (Quercia and Foxhall 2014; Landenius Enegren 2015, 2017; Militello et al. 2016; Quercia 2018), although it is almost exclusively based on textile tools, which are the most common evidence related to textile production on the island. In Archaic contexts, loom weights are found in large numbers, while spindle whorls and other textile implements are much less numerous. Greek and some western Greek communities seem to share a similar tradition of setting up looms with sets of light, standardised and well-fired loom weights (Longhitano 2020, 2021). The latter are mainly truncated pyramids with either square or rectangular bases. From the fifth century BC, some communities started using disc, discoid and lenticular loom weights.[1] Textile production was characterised by tabby weave and, in particular, weft-faced tabby. This evidence is consistent with what has emerged in other parts of southern Italy, where fabrics were characterised by yarn requiring 5–10 g tension and a low thread count of 4–7 threads per cm.[2] In contrast, textile traditions at indigenous[3] sites have received little attention from scholars but in the last decade loom weights from a few sites have been preliminarily analysed (Balco and Kolb 2009; Quercia and Foxhall 2014; Landenius Enegren 2015, 2017). In western Sicily, loom weight tradition is particularly variegated. Beyond truncated pyramids, parallelpiped[4] and cubic loom weights are commonly attested (e.g. Segesta,[5] Monte

[1] Discoid loom weights are attested in Ramacca (Albanese and Procelli 1988–1989), Sabucina (Longhitano 2021), Himera (Allegro et al. 1976: 83; 218; 462; 558), Camarina (Pisani 2008: 143) during the fifth century BC. However, this loom weight shape is more common from the fourth century BC onwards. For an overview of the evidence, see Meo 2015: 44–47.

[2] A few fragments of mineralised textiles were recovered from a number of sites in southern Italy: Fondo Artico tomb 104 at Cumae, Ripacandida, Chiaromonte, Ascoli Satriano (Gleba et al. 2018), Tomba Spina 418 at Paestum (Gleba and Meo 2017).

[3] Here, this term is used to indicate local peoples, who show a material culture different from that of Greeks and Punics.

[4] This type of loom weights has been referred to as 'Punic block' in Balco and Kolb (2009).

[5] Landenius Enegren 2017.

Maranfusa,[6] Montagnola di Marineo,[7] Entella,[8] Monte Castellazzo di Poggioreale[9]), while discoid and lenticular shapes became more common from the end of fifth – beginning of fourth centuries BC.[10] These loom weight shapes coexist in the Punic site of Motya, where types of loom weights and decorations belonging to culturally different craft traditions were adopted (Rossoni 2002; Nigro 2004, 2005, 2007; Landenius Enegren 2015: 144–147).

This paper aims to enhance our understanding of textile traditions in Archaic Sicily by considering fascinating evidence from the indigenous site of Monte Polizo, which provides us with information on the weaving tradition at the site.

7.2 Monte Polizo and the Area of Portella Sant'Anna

Monte Polizo is a settlement located in the hinterland of western Sicily, between the two competing settlements of Selinous and Motya, a Greek and a Punic centre respectively. The main settlement at Monte Polizo was established during the second half of the seventh century BC and abandoned around 550 BC, possibly as a result of conflicts. At the time of abandonment, the site covered approximately 20–25 hectares and housed a population of up to 3000 individuals (Morris and Tusa 2004: 79). The settlement was constructed at the highest point of a hill and along its slopes, most extensively on the eastern and western ends. The town plan was irregular and the layout of the houses was often adapted to the undulating terrain, preventing any kind of unitary grid. Excavations carried out intermittently between 1970 and 2006[11] uncovered the highest area (the so-called Acropolis), the areas to the east (areas A and B), those to the south and to the north (Portella Sant'Anna) of the Acropolis[12] (Fig. 7.1). Monte Polizo has been traditionally investigated with the main focus of shedding light on the ethnic group. Based on its location and material culture, the site has been connected with the Elymians (e.g. Tusa 1988–89, 2000,

[6] De Simone 2003: 356, Fig. 288, no. 66.

[7] Spatafora and De Simone 2007: 38.

[8] Nenci 1992.

[9] Giglio Cerniglia et al. 2012: Fig. 427.

[10] Segesta (Biagini 2008), Entella (Bove 1995), Montagnola near Marineo (Spatafora and De Simone 2007).

[11] The settlement was investigated by an international team: The Scandinavian Sicilian Archaeological project (SSAP) uncovered four houses to the south of the highest point (reports 1998–2001, reports 2002–2003, Prescott et al. 1998). The University of Stanford explored the Acropolis, revealing other houses and a cult area. Preliminary reports have been published in https://web.stanford.edu/group/mountpolizzo/index.htm; Morris et al. 2001, 2003. The Soprintendenza of Trapani dug the area to the south and to the north of the Acropolis (Tusa 1972; Lentini 2004).

[12] A comprehensive publication of the excavations is still lacking and only a few partial works have been published so far. The major contribution has been made by Mühlenbock (2008), who has presented preliminarily findings about some contexts and material.

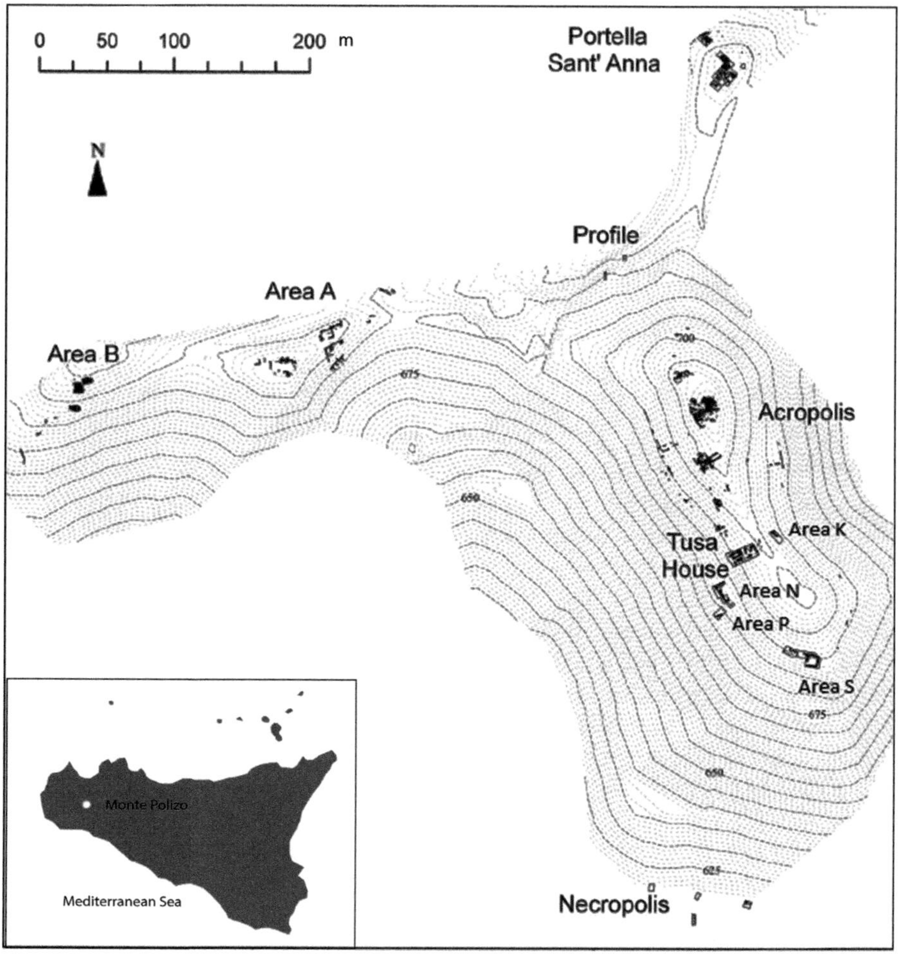

Fig. 7.1 The main excavation areas at Monte Polizo mentioned in the text. (After Mühlenbock 2008)

2004), although this has been challenged (e.g. Albanese Procelli 2003: 18–25). As a whole, the pottery, architecture and cult practices were particularly conservative and appear reluctant to adopt foreign traditions (Mühlenbock 2008, 2013). Similarly, its weaving tradition shows distinctive features with loom weights that differ substantially from those of other nearby sites and, so far, have no Sicilian parallels. More than 100 loom weights were found at the site, although many of them were recovered in a poor and friable condition, likely due to poor manufacturing and firing processes. The remarkably high number of fragmentary loom weights preserved as simple clay lumps points to the possibility that many more loom weights once existed.

In 2004 and 2005, some limited trenches were opened up in the area of Portella Sant'Anna to estimate the extent of the town and the urban plan. The area was not fully investigated and a preliminary excavation brought to light several multi-room buildings, which might have been part of a workshop district. In one of these buildings, a room yielded a group of sixteen loom weights. From the scatter pattern, the excavator concluded that the loom weights might have fallen from a working loom at the time the site was destroyed.[13] However, it should be observed that no post-holes in the floor or traces of carbonized loom were found. Stratigraphic information has been obtained from excavation documentation housed in the storerooms of the museum of Salemi (where the loom weights are housed) and through personal communication with the excavator.

7.3 Loom Weights from Portella Sant'Anna

The sixteen loom weights are of different shapes and weight (Fig. 7.2). Ten (of which two fragmentary) are similar in shape, size, weight and manufacture and are comparable to the majority of the loom weights found at the site of Monte Polizo. They are truncated pyramidal in shape with rectangular bases and a single hole drilled down through the thinner side. The sides are usually tapered, although sometimes the taper is hardly evident or absent resulting in more rectangular or actual rectangular faces. These variations are likely due to the manufacturing process and would hardly have affected the weaving process. The perforated faces are notably thinner, approximately half the thickness of the non-perforated ones. Such loom weights might have been used either with their perforated side or with the non-perforated sides facing weavers, resulting in a notable difference in fabric density. However, it seems very unlikely that they were used with the larger sides facing weavers, as this usually reaches 4.5–5.5 cm and would produce a very open fabric. Therefore, here the perforated sides are considered as thickness. These loom weights weigh between 256 and 280 g, with only one exceeding 300 g. One (psa3683; Fig. 7.3) bears an incised sign on the top made before firing, resembling a Σ, which might have been a manufacturing or potter's mark. This would imply either a Greek presence at the site or a non-local loom weight.[14] However, both hypotheses are purely speculative as not supported by other archaeological evidence. On the other hand, this sign could be interpreted as a decorative motif like those commonly attested on indigenous pottery of western Sicily (see Vassallo 1999: 131, Fig. 144; Spatafora 2003: 132, Fig. 143, no. 87; 150–151; 147, Fig. 155, no. 24; Mühlenbock 2008: 105, Fig. 47).

There are two more truncated pyramidal loom weights (psa3671 and psa3676), which are different in weight and size (369 g and 279 g respectively) and thinner

[13] Recovered in trench 20005, layer 142 (from the information on the box where the loom weights are stored). A pair of tweezers was also found close to the loom weights. I warmly thank the excavator Dr. Assia Ingoglia for sharing with me the details of the context and the discovery of the loom weights.

[14] Loom weights bearing a Σ are attested at Himera (Allegro et al. 1976: 143, 676 cat. no. 90).

Fig. 7.2 The loom weights from Portella Sant'Anna. (Photo by G. Longhitano)

Fig. 7.3 Psa 3683 with an incised sign on the top. (Photo by G. Longhitano)

than the first group of loom weights. Psa3671 bears an incised hourglass shaped decoration (Fig. 7.4), which might have been derived from the motif of triangles commonly attested in the indigenous pottery (Spatafora 2003: 149–150; 223 Fig. 204, no. 21). These two loom weights are made of grey clay. Additionally, three heavier truncated pyramidal loom weights with no tapered sides (rectangular) were also recovered. They weigh c. 840–870 g, except for a slightly lighter one (psa3668),

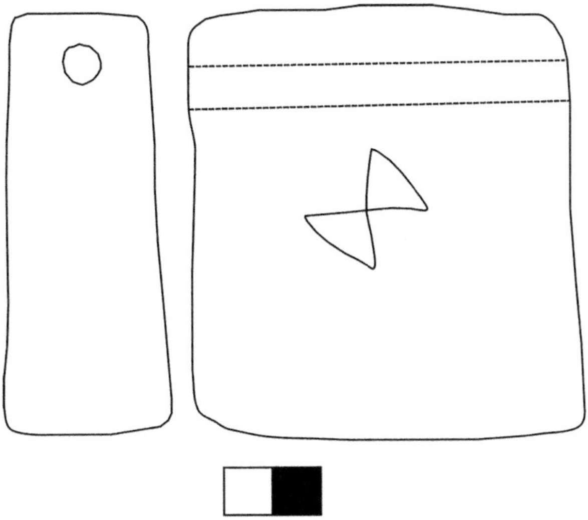

Fig. 7.4 Psa 3671 with an incised hourglass shaped decoration. (Photo by G. Longhitano)

which has been holed in the middle of one of the largest sides after firing. This might originally have weighed as much as the other two. All of them are 6.4 cm thick. Except for psa3671 and psa3676, the remaining loom weights are made of reddish clay in a coarse fabric rich in coarse grained inclusions.

Finally, there is a rectangular prism-shaped loom weight with two holes through the long side (psa3677) weighing 160 g and made of grey clay. Only a few examples have been recovered so far in Sicily and Italy (e.g. Gleba 2008: 132; Quercia 2017: 246, Fig. 7.3b). The function of these artefacts is not quite clear, but they are usually associated with weaving.

7.4 Method and Results

The sixteen loom weights have been analysed by applying the methodology developed at the Centre for Textile Research of Copenhagen (CTR) (Mårtensson et al. 2009; Andersson Strand and Nosch 2015), which enables the reconstruction of the range of textiles that could have been produced using these loom weights. Thus, an attempt was made to calculate the optimal loom set-up based on the hypothesis that all of the loom weights were being used together on the same loom. However, if the hypothesis were correct, this would raise the question of how such a varied selection of loom weights would work together. How would the heavy loom weights work along the others? What was the function of the heaviest loom weights?

According to the CTR methodology, it is possible to estimate the tension range that can be applied to a specific loom weight. As can be seen from the Table 7.1, calculations of potential loom set-ups with loom weights of c. 260–280 g and c. 4.5–5 cm thick show that they would work better with a warp tension of c. 10–20 g

Table 7.1 Calculations of possible loom set-ups in a tabby weave for 260–280 g loom weights from Portella Sant'Anna

Truncated pyramidal, weight 268 g, width 4.6 cm				
Warp thread tension	No. of warp threads per loom weight	No. of warp threads per two loom weights	Threads/cm	Evaluation
10 g	27	54	11	Optimal
20 g	13	26	6	Possible
30 g	9	18	4	Possible?

per thread, producing fabrics of 6–11 threads per cm in a balanced tabby, or even more in a weft-faced tabby. But would they be suitable for working with 30 g tension? Four warp threads per cm is a limit value in the CTR guidelines (Mårtensson et al. 2009: 392), thus it cannot be taken as a sharp division and assessment can only be made through experimental archaeology. If these loom weights were used with a 30 g tension yarn, they would produce a very open tabby, unless weft-faced was preferred.

The calculations for tabby weave with loom weights of over 800 g demonstrate that they would work with a tension of 30 g and even more, producing a fabric of 6–10 threads per cm in a balanced tabby or more in a weft-faced tabby (Table 7.2). Some observations can be made by comparing the calculations of the lightest loom weights with those of the heaviest loom weights. It seems that the overlapping 30 g tension might possibly have been a preferred option, if all the loom weights worked in the same set-up.

Skilled weavers might have worked out the warp in order to get a balanced distribution of threads. Also, since the tension applied to the warp threads does not always need to be the highest possible, weavers might have even varied the level of tension in order to create textiles with different appearances. Differences in tension could affect the appearance of the threads, making the warp yarns loose or hard.[15] Alternatively, might the difference in tension point to the use of different twist or different yarn diameters in warp? In the case of Portella Sant'Anna, would the difference of warp diameter imply the use of a different raw material (i.e. animal and plant fibres)? So far, however, the example of simultaneous use of animal and plant fibres is attested only in weft (e.g. goat hair and supplementary threads of nettle fibre), while warp is made using only one type of fibre (e.g. linen).[16] The insertion of different thread diameter to create decorative patterns is sometimes attested. There is an interesting case of a textile fragment from Cape Zoster,[17] in which decorative pattern is created by inserting thicker threads in regular intervals in both

[15] Examples of textiles made of yarns variously tensioned are attested in Classical Attica (Spantidaki 2016: 180).

[16] This is the case of a carbonised band from Bronze Age Chania, Crete (Spantidaki and Moulherat 2012: 189). The use of different raw material in weft and warp is also attested from the Kerameikos in Classical Attica (Spantidaki 2016: 111–112 cat. nos 8, 10, 12).

[17] Cape Zoster 2 (Spantidaki 2016: 110 cat. no. 4).

Table 7.2 Calculations of possible loom set-ups in a tabby weave for over 800 g loom weights from Portella Sant'Anna

Truncated pyramidal (rectangular), weight 840 g, width 6.6 cm				
Warp thread tension	No. of warp threads per loom weight	No. of warp threads per two loom weights	Threads/ cm	Evaluation
30 g	28	56	10	Optimal
40 g	21	42	7	Possible
50 g	17	34	6	Possible

directions. However, these hypotheses remain speculative and need to be verified by experimental tests.

In light of the lack of the exact find spots, it is not possible to reconstruct the combination of loom weights in the set. However, it is possible that a certain combination of heavier and lighter loom weights could create particular decorative effects on the fabric through the natural colour of the plant and animal fibres. In this scenario, the incisions on the set might have functioned as markers for a given pattern in a weave, thus facilitating the work for weavers. Also, psa3671 and psa3676 might have worked with the other loom weights providing a similar tension, although it is not possible to say why they differ in shape from the others. Hypothetically, placing all the truncated pyramidal loom weights in two rows and measuring their average width would produce a cloth of around 40 cm wide, which might be consistent with a narrow band, although it is possible that other loom weights have not survived.

In the light of what has emerged from Portella Sant'Anna, it is worth comparing these results with the evidence from the other part of Monte Polizo. In Area A, three structures, referred to as House 1, 2 and 3, were uncovered. Small clusters of loom weights were found in each house. Usually, concentrations of no more than 25–30 loom weights were found in rooms,[18] probably stored in bags. Particularly the cluster of 25 loom weights found in Room I in House 1 was likely stored there, and might thus have belonged to different sets for different loom set-ups (Fig. 7.5).

They are similar to those from Portella Sant'Anna, with a few displaying a rectangular shape as they have no tapered sides. Their weights predominantly range between 100 and 300 g. Despite the wide range, it seems possible to distinguish different groups of similar loom weights within the cluster on the basis of distinguishable decorations, weight classes and size. It is notable that an identical stamp of concentric circles on the top is found on a group of six similar loom weights, (Fig. 7.6), which are 100–200 g and c. 3.5 cm thick. Three other weights that are 250–300 g and 7–8 cm thick, display no tapered sides (rectangular) and bear a motif of a dot inside a circle (Fig. 7.6). It is impossible to say whether these impressions are maker or owner marks. They might have been made by using a personal object,

[18] In House 2, room III yielded 18 loom weights (Mühlenbock 2008: 150). Approximately 20 loom weights were recovered in room I and II in House 3 (Mühlenbock 2008: 157).

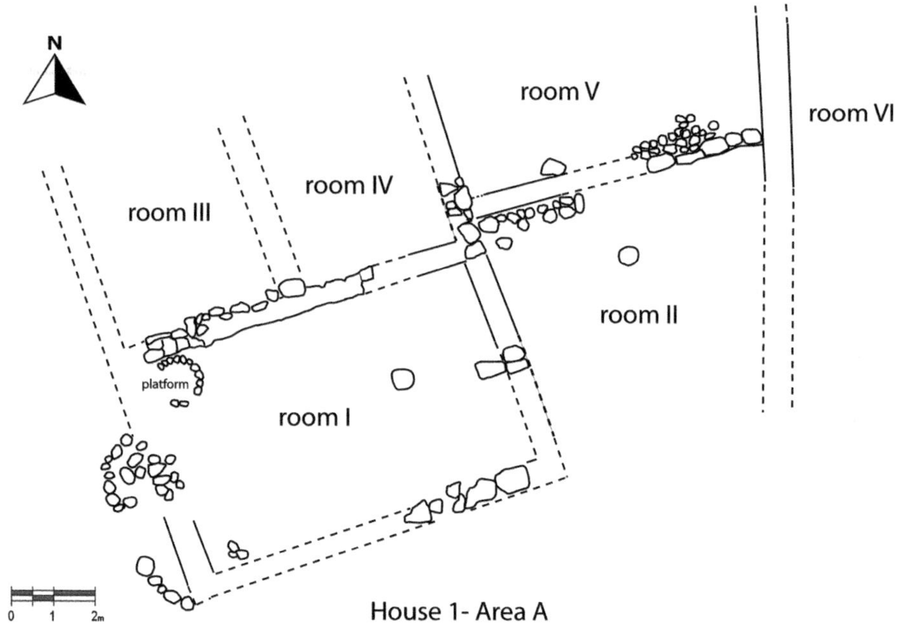

Fig. 7.5 House 1 with indication of the platform where the loom weights were found. (After Mühlenbock 2008)

although they closely resemble common geometric motifs on indigenous pottery (Vassallo 1999: 131, Fig. 144; Spatafora 2003: 147, Fig. 155, no. 37; 15, Fig. 159, no. 84; Mühlenbock 2008: 105, Fig. 47, no. 18) and on metal objects (Vassallo 1999: 96, Fig. 106). Interestingly, the remaining loom weights of the cluster belong to both ranges of weight and size of these two groups but have no stamps. While it is doubtful that the loom weights with stamps constituted a complete set as it would have been particularly small, one might wonder whether there was a specific reason for using distinctive groups of weight and marks.

Loom weights of 100–200 g might have worked well with 5–10 g tension, while loom weights of 250–300 g with 10–20 g tension (Tables 7.3 and 7.4). Thus, it cannot be completely ruled out that all the loom weights worked together in the same set-up. Indeed, the overlapping 10 g tension might have been a common tension. It is notable that the combination of lighter truncated pyramidal and heavier rectangular weights and the presence of some decorated loom weights resemble that found at Portella Sant'Anna. Thus, is it possible that the group of weights from House 1 worked together in the same set? Like that from Portella Sant'Anna, would this set work using different types of yarn diameter, twist or different raw materials?

Fig. 7.6 Truncated pyramidal loom weights with two different motifs of circles on the top. (Photo by G. Longhitano)

Table 7.3 Calculations of potential loom set-up for tabby weave for 100–200 g loom weights from House 1, Room I

Truncated pyramidal, weight 121 g, thickness 3.7 cm				
Warp thread tension	No. of warp threads per loom weight	No. of warp threads per two loom weights	Threads/cm	Evaluation
5 g	24	48	13	Optimal
10 g	12	24	6	Possible

Table 7.4 Calculations of potential loom set-up for tabby weave for 250–300 g loom weights from House 1, Room I

Truncated pyramidal, weight 288 g, thickness 4.4 cm				
Warp thread tension	No. of warp threads per loom weight	No. of warp threads per two loom weights	Threads/cm	Evaluation
10 g	29	58	13	Possible
20 g	14	28	6	Optimal

7.5 Conclusion

Monte Polizo had a distinctive loom weight tradition in terms of shape and decoration. The peculiar shape might be due to the need to ensure that heavy loom weights would dry out thoroughly and fire quickly. However, the functional parameters of the loom weights affect weaving: calculations point to the use of warp yarns requiring 10–30 g or even more tension, which differs from what has emerged elsewhere in Sicily, where the most common tension is about 5–10 g.[19] It is possible that weft-faced tabby was produced, creating relatively dense fabrics. The weave tradition at the site appears to fit with what was common in Sicily, although a different range of tension seems to point to the use of both thin and medium-thick threads. It is notable that a peculiar combination of light and heavy, decorated and non-decorated loom weights seems to be common at Monte Polizo, which might imply the use of different yarn diameter or raw material and/or a different technique for tying loom weights to the loom.

The discovery of the loom weights at Portella Sant'Anna and the comparison with the other evidence from Monte Polizo represent a crucial contribution to the study of weaving tradition at the site. The loom weight clusters found in many rooms at the site might indicate the use of rather small looms and the possible production of narrow bands. The inhabitants at Monte Polizo might have produced narrow cloth strips to be sewn onto a broader cloth to create particular decorative patterns, belts perhaps folded two or more times, or other kinds of bands. Unfortunately, there is no iconographic evidence to compare to this data. Some late written sources in Sicily attest the use of tunics with belts, corsets and garments decorated with purple bands and colourful cloaks among other fine quality clothing (Brugnone 2008: 53–54). Moreover, beyond garments, there is evidence on the island for the use of cloth as domestic furnishings, such as bedcovers, cushions and wall decorations, which might have been consistent with narrow cloth bands (Brugnone 2008: 58). However, given the lack of other evidence, this remains purely speculative and further research and experimental tests are needed to answer all the above questions.

Acknowledgements I warmly thank the Soprintendenza of Trapani for giving me the permission of studying and publishing this material (authorisations prot. no. 1397 dated 16 March 2018 and prot. no. 3262 dated 28 June 2019). Particularly, I thank Dr R. Giglio, Dr R. Guazzelli and Dr G. Mammina for kindly allowing me to study the textile implements at the site. This research was part of my PhD project, which was partially funded by the Pasold Research Fund, Peet Travel Award and Andrew Sherratt Grant. I am grateful to my supervisors, Prof. L. Foxhall and Dr M. Fitzjohn, to Dr M. Gleba and Prof. P. Militello for their support. I sincerely thank Alistair Dickey for his comments and suggestions on the final draft of this paper.

[19] Selinous (Quercia 2018), Segesta, Monte Iato, Motya (Landenius Enegren 2015), Himera, The Cittadella hill, Sabucina, Monte Maranfusa (Longhitano 2020, 2021).

References

Albanese R-M, Procelli E (1988–1989) Saggi di scavo nelle contrade Castellito e Montagna negli anni 978, 1981 e 1982. *Notizie degli scavi di Antichità* 42–43:7–148.

Albanese Procelli R-M (2003) *Sicani, Siculi, Elimi: forme di identità, modi di contatto e processi di trasformazione.* Longanesi, Milano.

Allegro N, Belvedere O, Bonacasa N, Bonacasa-Carra R-M, Epifanio E, Di Stefano C, Joly Elda A, Manni-Piraino M-T, Schmiedt G, Tullio A, Tusa-Cutroni A (1976) *Himera, 2. Campagne di scavo* 1966–1973. L'Erma di Bretschneider, Roma.

Andersson Strand E, Nosch M-L (eds) (2015) *Tools, textiles and contexts. Investigating textile production in the Aegean and Eastern Mediterranean Bronze Age*. Ancient Textile Series 21. Oxbow Books, Oxford and Philadelphia.

Balco W, Kolb M (2009) Loomweights as material culture indicators: A western Sicilian case study. In: Oniz H (ed.) *Soma* 2008. *Proceedings of the XII symposium on Mediterranean Archaeology, Eastern Mediterranean, University Famagusta, North Cyprus, 5–8 March.* Archaeopress, Oxford, p 177–182.

Biagini C (2008) Pesi da telaio e oscilla. In: Camerata-Scovazzo R (ed.) *Segesta, 3. Il sistema difensivo di Porta di Valle. Scavi 1990–1993.* SAP, Mantova, p 645–670.

Bove A (1995) Rocca d'Entella: gli oscilla (1985–1989). In: Nenci G (ed.) *Entella I, Stamperia della Scuola normale superiore.* Stamperia della Scuola normale superiore, Pisa, p 163–168.

Brugnone A (2008) Tessuti, costumi e mode nella Sicilia antica. *Kokalos* 49:51–85.

De Simone, R. (2003) Oggetti fittili, terrecotte, metalli, oggetti in pietra, astragali in osso. In: Spatafora F (ed.) *Monte Maranfusa. Un insediamento nella media valle del Belice. L'abitato indigeno.* Assessorato regionale dei beni culturali ambientali e della pubblica istruzione, Palermo, p 347–378.

Giglio Cerniglia R, Falsone G, Sconzo P (2012) Nuove ricerche a Castellazzo di Poggioreale. Campagne 2008–2009. In: Ampolo C (ed.) *Sicilia occidentale. Studi, rassegne, ricerche. Atti delle settime giornate internazionali di studi sull'area elima e la Sicilia occidentale nel contesto mediterraneo Erice, 12–15 ottobre 2009 Workshop,* II. Edizioni della Normale, Pisa, p 239–250.

Gleba M (2008) *Textile production in pre-Roman Italy.* Ancient Textiles Series 4. Oxbow Books, Oxford.

Gleba M (2017) Tracing textile cultures of Italy and Greece in the early first millennium BC. *Antiquity* 144:1205–1222.

Gleba M, Meo F (2017) L'abbigliamento funerario. In: Niola M, Zuchtriegel G (eds) *Action painting. Rito & arte nelle tombe di Paestum.* Editrice politecnica, Napoli, p 141–191.

Gleba M, Heitz C, Landenius Enegren H, Meo F (2018) At the crossroads of textile cultures: Textile production and use at the South Italian Archaic site of Ripacandida. *Journal of Mediterranean Archaeology* 31(1):27–51.

Landenius Enegren H (2015) Loom weights in Archaic South Italy and Sicily: Five case studies. *Opuscula: Annual of the Swedish Institutes at Athens and Rome* 8:123–155.

Landenius Enegren H (2017) The loom weights from the Scarico di Grotta Vanella. Evidence for a sanctuary on the North Acropolis of Segesta? In: Brøns C, Nosch M-L (eds) *Textiles and cult in the Mediterranean area in the first millennium B.C., Copenhagen, 21–22 November 2013.* Ancient Textiles Series 31. Oxbow Books, Oxford and Philadelphia, p 104–111.

Lentini F (2004) *Saggi di scavo archeologico presso l'acropoli di Monte Polizzo. Giornale di scavo.* Sovrintendenza di Trapani (unpublished excavation diary).

Longhitano G (2020) Textile production in Archaic Sicily: Spinning and weaving at the indigenous sites of the Cittadella hill and Monte Maranfusa. In: Bustamante-Álvarez M, Sánchez López EH, Jiménez Ávila J (eds) *Redefining textile handcraft, structures, tools and production processes.* PURPUREAE VESTES VII. University of Granda, Granada, p. 277–286.

Longhitano G (2021) *Textile activity and cultural identity in Sicily between the Late Bronze Age and the Archaic period.* Ancient Textiles Series 37. Oxbow Books, Oxford.

Mårtensson L, Nosch M-L, Andersson Strand E (2009). Shape of things: Understanding a loom weight. *Oxford Journal of Archaeology* 28(4):373–398.
Meo F (2015) *L'attività tessile a Herakleia di Lucania tra III e I secolo a.C. Fecit Te 7*. Scienze e Lettere, Rome.
Militello P, Longhitano G, Messina T (2016) Tra indigeni e Greci: la filatura e la tessitura in area nissena. In: Congiu M, Panvini R (eds) *Atti Indigeni e Greci tra le valli dell'Himera e dell'Halykos, Caltanissetta, giugno 2012*. Paruzzo, Caltanissetta, p 64–85.
Morris I, Jackman T, Blake E, Tusa S (2001) Stanford University excavations on the Acropolis of Monte Polizzo, Sicily, I: Preliminary report on the 2000 Season. *Memoirs of the American Academy in Rome* 46:253–271.
Morris I, Jackman T, Blake E, Garnand B, Tusa S, Hnatiuk T, Matthews W, Stika H-P (2003) Stanford University excavations on the Acropolis of Monte Polizzo, Sicily, III: Preliminary report on the 2002 season. *Memoirs of the American Academy in Rome* 48:243–315.
Morris I, Tusa S (2004) Scavi sull'acropoli di Monte Polizzo, 2000–2003. *Sicilia Archeologica* 37:35–84.
Mühlenbock C (2008) *Fragments from a mountain society. Tradition, innovation and interaction at Archaic Monte Polizzo, Sicily.* GOTARC Serie B. Gothenburg Archaeological Thesis 50, Gothenburg.
Mühlenbock C (2013) Having an axe to grind: An examination of tradition in the Sicilian Iron Age. In: Bergerbrant S, Sabatini S (eds) *Counterpoint: Essays in archaeology and heritage studies in honour of Professor Kristian Kristiansen.* BAR International Series 2508. Archaeopress, Oxford, p 401–409.
Nenci G (1992) Entella. Relazione preliminare della campagna di scavo 1989. *Annali della Scuola normale superiore di Pisa* 22:617–759.
Nigro L (ed.) (2004) *Mozia–X. Zona C: Il Kothon, Zona D: Le pendici occidentali dell'Acropoli. Zona F: La Porta Ovest. Rapporto preliminare della XXII campagna di scavi – 2002 condotta congiuntamente con il Servizio Beni Archeologici della Soprintendenza Regionale per i Beni Culturali e Ambientali di Trapani.* Quaderni di Archeologia Fenicio-Punica 1, Roma.
Nigro L (ed.) (2005) *Mozia-XI. Zona C. Il Tempio di Kothon. Rapporto preliminare delle campagne di scavi XXIII-XXIV (2003–2004) condotte congiuntamente con il Servizio Beni Archeologici della Soprintendenza Regionale per i Beni Culturali e Ambientali di Trapani.* Quaderni di Archeologia Fenicio-Punica 2, Roma.
Nigro L (ed.) (2007) *Mozia–XII. Zona D. La 'Casa del Sacello domestico', il 'Basamento meridionale' e il Sondaggio stratigrafico I. Rapporto preliminare delle campagne di scavi XXIII e XXIV (2003–2004) condotte congiuntamente con il Servizio Beni Archeologici della Soprintendenza Regionale per i Beni Culturali e Ambientali di Trapani.* Quaderni di Archeologia Fenicio-Punica, Roma.
Pisani M (2008) *Camarina. Le terrecotte figurate e la ceramica da una fornace di V e IV secolo a.C.* L'Erma di Bretschneider, Roma.
Prescott C, Mühlenbock C, Englund E (1998) *Sicilian-Scandinavian archaeological project.* Unpublished annual report 1998.
Quercia A, Foxhall L (2014) Weaving relationships in areas of cultural contacts: Production, use and consumption of loom weights in pre-Roman Sicily. In: Lipkin S, Vaianto K (eds) *Focus on archaeological textiles: multidisciplinary approaches*. Monographs of the Archaeological Society of Finland 3. Archaeological Society of Finland, p 88–101.
Quercia A (2017) Textile production and technological changes in the Archaic societies in Magna Graecia: the case of Torre di Satrano, (Lucania, Italy). In: Gleba M, Laurito R (eds) *Contextualising textile production in Italy in the 1st millennium BC.* ORIGINI 40. Gangemi Editore, Roma, p 243–258.
Quercia A (2018) Weaving during the Archaic period in south Italy: two key studies. In: Busana MS, Gleba M, Meo F (eds) *Textiles and dyes in the Mediterranean economy and society. Proceedings of the VIth international symposium on textiles and dyes in the Ancient Mediterranean world*

(Padova–Este–Altino, Italy 17–20 October 2016). PURPUREAE VESTES VI. Libros Pórtico, Valencia, p 145–156.
Rossoni G (2002) Pesi da telaio. In: Famà M-L (ed.) *Mozia, gli scavi nella "Zona A" dell'abitato.* Edipuglia, Bari, p 315–320.
Spantidaki S (2016) *Textile production in classical Athens*. Ancient Textiles Series 27. Oxbow Books, Oxford and Philadelphia.
Spantidaki Y, Moulherat Ch (2012) Greece. In: Gleba M, Mannering U (eds) *Textiles and textile production in Europe: from Prehistory to AD 400.* Ancient Textile Series 11. Oxbow Books, Oxford and Oakville, p 185–200.
Spatafora F, De Simone R (2007) *Makella, La montagnola di Marineo.* Guida breve, Palermo.
Spatafora F (ed.) (2003) *Monte Maranfusa. Un insediamento nella media Valle del Belice.* Beni Culturali Palermo, Palermo.
Tusa V (1972) Monte Polizzo-Scavi 1970. *Sicilia Archeologica* 5:119–121.
Tusa V (1988–89) Sicani ed Elimi. *Kokalos* 34–35:47–70.
Tusa S (2000) Ethnic dynamics during pre- and proto-history of Sicily. *Journal of Cult Heritage* 1 Sup.17–Sup.28:17–28.
Tusa S (2004) Historical reference frame in western Sicily since the end of II to the beginning of I millennium BC. In: Mühlenbock C, Prescott C (eds) *The Scandinavian Sicilian archaeological project: Archaeological excavations at Monte Polizzo Sicily, reports 1998–2001*. GOTARC serie C 56. Göteborg Universitet, Institutionen för Historiska Studier, Göteborg, p 17–26.
Vassallo S (1999) *Colle Madore: Un caso di ellenizzazione in terra sicana, Regione siciliana.* Assessorato dei beni culturali e ambientali e della pubblica istruzione, Palermo.

Gabriella Longhitano is an archaeologist specialized in textile tools from Bronze Age-Archaic Italy and Greece. She received her PhD from the University of Liverpool in 2019. In her PhD, she investigated the range of indigenous traditions of making and using textile tools in Sicily between the thirteenth and the fifth centuries BCE. Her research interests included the study of the impact of cultural interactions between indigenous and incomers on their textile traditions. She is currently carrying out a research project at the University of Catania, Italy, on textile tools from the Archaic site of Prinias in Crete, Greece.

Chapter 8
Visible Tools, Invisible Craft: An Analysis of Textile Tools across the Iron Age South Coast of Britain

Lewis Ferrero

Abstract Textile tools are vital pieces of evidence in areas where few or no ancient textiles survive; while a lot of work has focused on the analysis of these tools, such as spindle whorls and loom weights, few studies have compared textile tools across large areas or various settlements. The prehistoric textile tools of Britain in particular have received very little attention, which has stymied our understanding of the textile industries prior to the social and cultural upheavals of the Roman conquest. This article presents the preliminary results of the author's doctoral research analysing the Iron Age spindle whorls and loom weights from various settlements across the South coast of Britain and comparing the data with other sites within the same county and with sites from other counties. The initial results indicate a variety of textile traditions and organisation of textile production and craft specialisation.

Keywords Textile tools · Textile archaeology · British archaeology · Iron age · Craft organisation

8.1 Introduction

Textile production supplied communities with one of the fundamental materials of society used in every context, from furnishings and clothing to cordage and netting (Barber 1992). Spinners and weavers contributed important elements to almost every section of society. However, textiles rarely survive in Britain due to the damp climate, leaving textile tools as the best available evidence for what was produced. Even so, poor preservation takes its toll on organic textile tools like bone or antler needles and weaving combs, effectively destroying a large proportion of evidence (Poole 1984).

L. Ferrero (✉)
University of Cambridge, Cambridge, UK
e-mail: lferrero@live.co.uk

A. Ulanowska et al. (eds.), *Ancient Textile Production from an Interdisciplinary Perspective*, Interdisciplinary Contributions to Archaeology,
https://doi.org/10.1007/978-3-030-92170-5_8

Fortunately, there is already a methodological framework in place to analyse spindle whorls and loom weights to determine the sort of yarn or fabric the tools are best suited to making, developed by Mårtensson et al. (2009) at the Centre of Textile Research (CTR), Andersson Strand and Nosch (2015) at "Textiles, Tools, Texts and Contexts Research Program" (TTTC) and by Karina Grömer (2005) on the Hallstatt textiles (cf. Andersson et al. in this volume). Despite the preservation issues with organic tools, the study of textile tools and their context remains the best method of studying ancient textiles and the organisation of textile workers in areas where fabrics do not survive. By studying the materials, make, and standardisation of tools, it is possible to gain culturally specific insights into how textile work and workers were organised, how such organisation varied from settlement to settlement or region to region, as well as indicators into the importance of textile production in the larger economy of a particular society.

This paper presents a preliminary report of a larger study of textile tools in various settlements across the south coast of Britain. The methodology and the initial data from these sites will be discussed here, along with a brief discussion of the early results.

8.2 Previous Work

It is well known that spindle whorls and looms weights are better able to survive deposition and deterioration in the archaeological record, which makes it easier to identify the presence of textile production at any site. The importance of textile tool analysis is well documented from the various experiments performed at the CTR (Mårtensson et al. 2009) and TTTC (Andersson Strand and Nosch 2015). These experiments determined which properties of various tools, such as spindle whorls and loom weights, directly affected the nature of the final textile (Mårtensson et al. 2009: 378). The weight and diameter of spindle whorls were most important since the weight directly influenced the diameter of the thread produced (Grömer 2005; Kania 2013), which in turn affected what tension was required for weaving (Mårtensson et al. 2007, 2009: 378, 392). The lighter the spindle whorl, the thinner the thread produced; the heavier the whorl, the thicker the thread. Likewise, the thinner the thread, the less tension needed during weaving and the more threads that could be tied onto a loom weight in order to distribute the weight between them. On the other hand, the thicker the thread, the heavier the tension required on the threads and the fewer threads that could be attached to each loom weight. The most efficient loom setup in these experiments proved to be between 5 and 50 warp threads attached to each loom weight (Mårtensson et al. 2009: 392). Fabrics could still be woven loosely or tightly depending on what the weaver wanted, no matter the thickness of the yarns used. The diameter of a spindle whorl is also important to note, as it will affect the speed at which the whorl turns and how long it can turn for. For example, a spindle whorl with a faster twist is more suitable for spinning flax, the fibres of which are longer and with more overlap, thicker wool, or plying threads together; those with slower twists are best suited to spinning thin wool threads (Grömer 2016: 88–89).

Moreover, it is possible to judge the degree of standardisation of textile tools, which can indicate specialisation when combined with other analytical methods (DeRoche 1995: 27–28). The presence of such standardisation in textile tools has been noted in other areas, particularly in Bronze and Iron Age textile workshops around the Mediterranean where they are connected to the production of homogenous fabrics (Burke 2005; Gleba 2007; Cutler et al. 2013; Dimova 2016; Marín-Aguilera 2019). Analysis of textile tools and their standardisation (or lack thereof) can also demonstrate the range of possible fabrics produced in a settlement or region, which itself can indicate the degree of craft specialisation at these sites. However, these methods have not been applied to textile tools from British Iron Age sites on a large scale and rarely to individual sites, such as Danebury (Cunliffe and Poole 2000) and Maiden Castle. Combining such methods will provide new information comparable to contemporary European sites and textiles, improving our understanding of Britain's craft capabilities compared to the continent.

Discussing textile production as a specialist craft comes with its own challenges; simply finding evidence of any production, like a spindle whorl or loom weight, does not mean there was a dedicated craftworker on site, only that the household or settlement had some form of textile work present at it. Traditionally, archaeologists rely on evidence of surplus production in the form of more craft debris or tools than would be needed to supply a household (Costin 1991; Gleba 2007; Marín-Aguilera 2019: 229–230). This excess production allows craftworkers to supply items for local and foreign exchange and control how their products are distributed (Brumfiel and Earle 1987; Costin 1991), which is the most widely accepted separation between household production and specialist, often workshop-based, craft activities (van der Leeuw 1977; Costin 1991; Marín-Aguilera 2019).

Usually, for textile production to be deemed an *economic* rather than a *subsistence* activity, a site must produce large amounts of textile tools. The early Iron Age site of Smolenice-Molpír in West Slovakia produced 2200 spindle whorls and almost 200 loom weights (Belanová Štolcová and Grömer 2010: 12–13); similarly, Bronze Age Malia produced over 600 loom weights (Cutler et al. 2013), and Gordion in Turkey had over a thousand spindle whorls and loom weights from several buildings (Burke 2005).

Textile work in Iron Age Central Europe demonstrates a high degree of standardisation and specialisation, similar to the evidence found in Western Europe and the Mediterranean (Gleba 2008; Grömer 2016).

These large numbers of textile tools in Bronze and Iron Age Mediterranean societies have been used as an important factor in determining the location of production centres and the organisation of workers and production. The use of specialist textiles in contexts such as religious sanctuaries shows the importance of this craft and the status of its workers in their societies (Andersson Strand and Nosch 2015; Barber 1992; Burke 2005; Marín-Aguilera et al. 2019). No similar work using textile production has been conducted on British contexts of the Bronze or Iron Age for a number of reasons. British Iron Age sites often lack very high numbers of tools for a variety of reasons, especially due to poor preservation of organic material like bone or antler needles and weaving combs (Poole 1984). Furthermore, early excavations (particularly those before the first half of the twentieth century) rarely record the precise number of textile tools found (for examples of this, see Fox et al. 1949–1950, Brailsford and Richmond 1952, and Dudley 1956). Finally, textile production in the British Iron Age often took place in domestic spaces – roundhouses – rather than a separate, dedicated weaving or spinning area (DeRoche 1995; Cunliffe 2005). However, this does not mean that specialist production of textiles could not take place in these domestic areas, but there is a long-standing bias against the perceived importance of domestic (often female dominated) spaces and the activities performed in these places among archaeological literature that has only recently been challenged. The study of textile tools is an excellent way to determine both whether specialist craft activity occurred in these domestic areas and whether such activity produced items for household consumption or for trading purposes.

8.3 Method

This study chose four to five sites per county with five or more textile tools in order to create a suitably large dataset. These sites included a range of settlement types, including farmsteads, open settlements, regional types like rounds or banjo enclosures, and fortified sites (Fig. 8.1). It was more important to have a range of site types than an even spread of across each region, so that an assessment could be made on the visible differences (if any) in textile production between sites.

The methods of tool analysis were based on the methodologies developed by Mårtensson, Andersson Strand, and Nosch (2009). Measurements were taken of the weight, height, and diameter of spindle whorls and the weight, width, and depth of loom weights. Analysis of textile tools was limited to spindle whorls and loom weights, despite the presence of weaving combs and needle at some sites. These organic tools rarely survived outside of areas with chalk-rich soil and no such tools

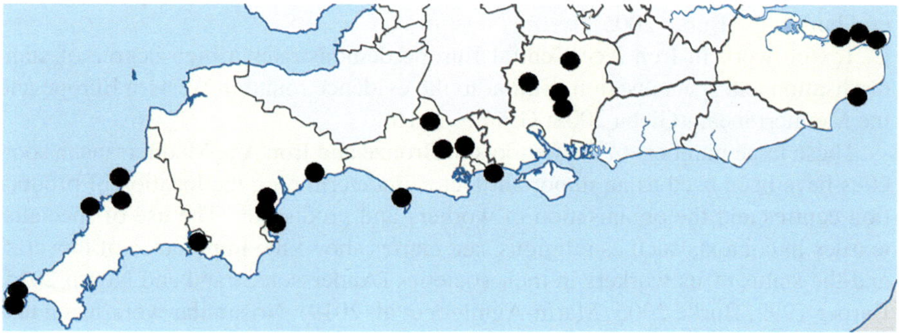

Fig. 8.1 The locations of all sites included in this study (by L. Ferrero)

survived in the more acidic clay of Devon and Cornwall. Bone and antler combs in particular are a very interesting aspect of British textile archaeology and the function of these items remains highly debated (Tuohy 1999; Beamer 2018; and cf. Beamer in this volume). Many combs were also frequently decorated, suggesting a personal connection with the object that other tools lacked (such as loom weights, or spindle whorls made from reused pottery). Heavy wearing around the teeth of these combs, as well as the recutting of teeth to extend the comb's use-life, clearly demonstrates that these were practical items (Tuohy 1999). The placement and direction of these wear marks has called into question the use of these tools on the wider warp-weighted loom; Tina Tuohy suggested that the combs were instead used as beaters on narrow strap looms or with tablet weaving (1995, 1999). There is also no set standard in recording or analysing weaving combs (Beamer 2018) and researchers are forced to track down the artefacts in question to gather the necessary data, which may not be possible with older antiquarian collections that may have lost or given away some combs (personal experience).

I felt that the inclusion of these tools risked skewing the results, given the issues of preservation across the study area, and thus these were omitted from the study. Time restrictions made it impossible to expand the study area further than Cornwall, Devon, Dorset, Hampshire, and Kent, limiting this study to the southern coast of Britain. Unfortunately, East and West Sussex were excluded from this study due to the lack of physical data; the settlements that produced five or more textile tools were excavated before the Second World War and those tools could not be located in any museum collections.

8.4 Results

8.4.1 South West

The data collected from the South West clearly demonstrates that spinning was conducted across this region and at all types of sites. In comparison, loom weights are not common at all sites, although these tools are found at several types of settlement, such as:

- Fortified settlements – The Rumps (Fig. 8.2, no. 5), Trevelgue Head (Fig. 8.2, no. 3), and Blackbury Castle (Fig. 8.2, no. 10);
- Enclosed settlements – Trevisker (Fig. 8.2, no. 4) and Milber Down (Fig. 8.2, no. 8);
- Open sites – Dainton (Fig. 8.2, no. 7).

There are far fewer textile tools available for research in Britain than are found in contemporary European sites; this may be reflective of a smaller population at the surveyed sites in this study, although it is certainly one of the issues encountered in British archaeology that has prevented any previous study of textile tools on the same scale as in Europe. Despite the small sample sizes, there is a huge difference

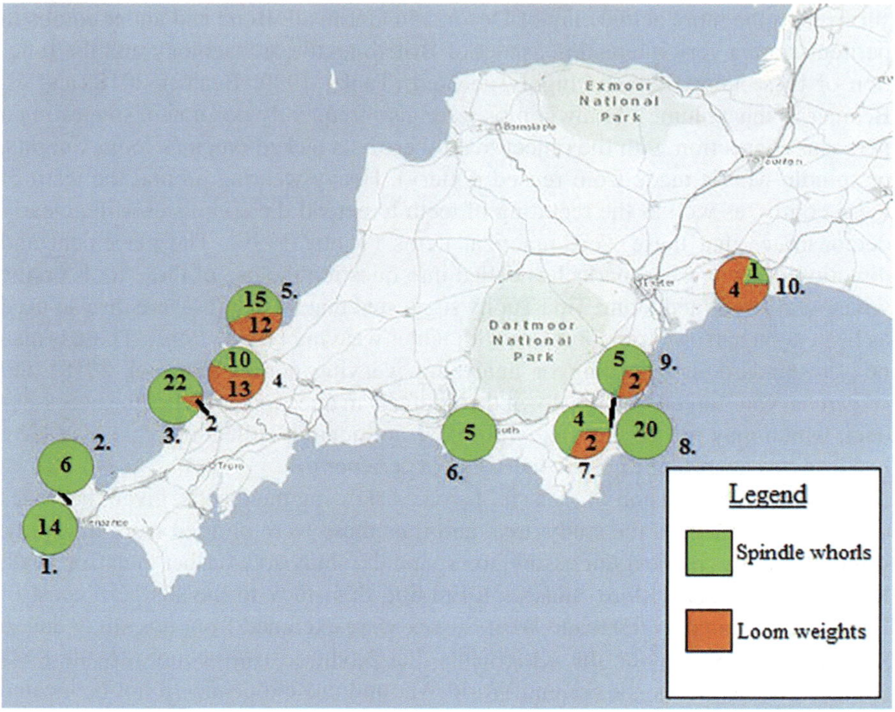

Fig. 8.2 Map of tool types found across all site types in the South West: 1) Carn Euny, 2) Bodrifty, 3) Trevelgue Head, 4) Trevisker, 5) The Rumps, 6) Mount Batten, 7) Dainton, 8) Kent's Cavern, 9) Milber Down, 10) Blackbury Castle (by L. Ferrero)

in the distribution of tools in the South West compared to the other areas of this study that I argue should not be ignored. The weaving tools of Devon and Cornwall appear restricted to certain areas rather than settlement types, which could suggest that different areas focused on different stages of textile production (Table 8.1). Indeed, the distribution of textile tools overall may suggest a form of craft specialisation, possibly focused on quantity rather than quality. However, further detailed studies and future excavations may reveal more information not yet available that could disprove this theory.

The tool data provides an interesting insight into the forms of fabric produced in these sites. Cornish spindle whorls weigh between 5 and 10 g and measure 13 and 50 mm in diameter. The wider range of diameters combined with such restricted weights is unusual and appears best suited to producing extremely fine yarn at a variety of speeds. This would allow the spinners to work with fine, short animal fibres and possibly fine plant fibres. Similarly, the loom weight data demonstrates that these tools are best suited to working with fine threads; these tools weigh from as little as 50 g to as much as 400 g. These weights could work with threads requiring a range of tensions, which could allow the weavers to produce textiles with a loose, open weave or a tight, warp-faced fabric. It is important to note that the loom

Table 8.1 The minimum, average, and maximum data from Cornwall and Devon

County	Site		SW weight (g)	SW diameter (mm)	LW diameter (mm)	LW diameter (mm)
Cornwall	Bodrifty	Minimum	5	12	N/A	N/A
		Average	24	19.6	N/A	N/A
		Maximum	42.2	26	N/A	N/A
	Carn Euny	Minimum	9	16.4	N/A	N/A
		Average	19.1	33.79	N/A	N/A
		Maximum	61	46	N/A	N/A
	The Rumps	Minimum	6	29	11	3
		Average	20.84	37.67	151.16	7.9
		Maximum	44.4	63.6	543.6	16.7
	Trevelgue Head	Minimum	9	14	57	7
		Average	19.18	32.04	N/A	N/A
		Maximum	46.8	45.7	657	36.2
	Trevisker	Minimum	6	22.7	6	1.8
		Average	19.65	32.86	380.35	20.78
		Maximum	46.25	42	1372.8	92
Devon	Blackbury Castle	Minimum	N/A	N/A	78	48
		Average	N/A	N/A	1174.5	137.25
		Maximum	N/A	N/A	2040	183
	Dainton	Minimum	18	34.7	166	20
		Average	39	44	N/A	N/A
		Maximum	53.8	50	272	44
	Kent's Cavern	Minimum	4.2	20	N/A	N/A
		Average	23.69	32.6	N/A	N/A
		Maximum	48.4	42	N/A	N/A
	Milber Down	Minimum	8	28	64.4	13
		Average	17.92	31.25	N/A	N/A
		Maximum	64.4	65	181.25	15
	Mount Batten	Minimum	26	13	N/A	N/A
		Average	38.2	16.5	N/A	N/A
		Maximum	68	19	N/A	N/A

weight data in Cornwall is unique compared to other counties due to the nature of its material. The Cornish loom weights are made from local slate, resulting in disc-shaped weights that are extremely thin compared to the triangular shape more typical of British Iron Age loom weights. Compared to other counties, these tools are aimed at only producing extremely fine textiles, which suggests highly specialised and skilled craft workers.

In comparison, Devon spindle whorls have a wider and heavier range of weights, between 15 and 42 g, and a smaller range of diameters, between 25 and 48 mm. These whorls would be able to quickly produce a range of yarns, from 0.4 to 1.5 mm in diameter. This would allow spinners to work with either animal and plant fibres, which could then be used to weave a variety of textiles. The loom weight widths varied a large amount, between 10 and 94 mm, while the weights were more restricted, between 50 and 180 g. This combination could allow weavers to work with yarns of fine to medium thicknesses to produce a range of open or balanced textiles. This evidence makes it clear that the textile workers of Devon produced a larger variety of textiles compared to the textile workers of Cornwall.

8.4.2 Central South

Dorset has remarkably different distribution compared to Devon and Cornwall; while weaving tools only appear at a few sites in the South West, there is a more even distribution of weaving and spinning tools across Dorset (Fig. 8.3, nos 1–5), suggesting that some amount of textile production occurred at each site. Unfortunately, the low number of textile tools overall makes it difficult to draw definite conclusions. There is one exception; Gussage All Saints (Fig. 8.3, no. 4) has a larger proportion of loom weights than all other sites, which was interpreted by excavators as evidence that this site had a possible weaving workshop.

The pattern in Hampshire is different again (Fig. 8.3, nos 6–10). Here, all sites except Kennel's Farm have an overwhelming number of loom weights. Kennel's Farm produced one spindle whorl and several weights that excavators presumed to be loom weights, although further analysis showed these were too heavy to weave with (Ferrero forthcoming). Kennel's Farm has therefore been excluded from further analysis as it no longer met the requirements of this study, which perfectly demonstrates how the study of textile tools can also identify artefacts not involved in textile production.

The only significant number of spindle whorls were found at Danebury hillfort, suggesting that spinning was a significant activity at this site compared to others. The high proportion of loom weights is a strong indication that this region produced a greater quantity of textiles than others. It is important to remember that this does not necessarily mean Hampshire sites produced textiles of better quality, simply that these produced more than other areas.

The spindle whorls in Dorset have the largest range of weights (5–55 g) and diameters (26–57 mm) compared to the other counties (Table 8.2). This would allow spinners to produce very fine to extremely thick yarns (0.2–2 mm in diameter) and would suit a range of animal and plant fibres. The need to produce a variety of textiles is also reflected in the loom weight data, with its wide range of weights and thicknesses. Compared to the South West (and Cornwall in particular), the textile producers of Dorset seem to prefer adaptability, which would allow them to produce textiles for any purpose. While this appears to suggest less specialist textile production, the number of tools found at fortified sites (Fig. 8.3, nos 1, 3, 5) are larger than necessary for domestic production and indicate some form of large-scale, lower-quality textile production for trade or exchange.

The Hampshire spindle whorls are very similar in weight and diameter to those from Dorset (6–40 g and 22–43 mm respectively and would be able to produce

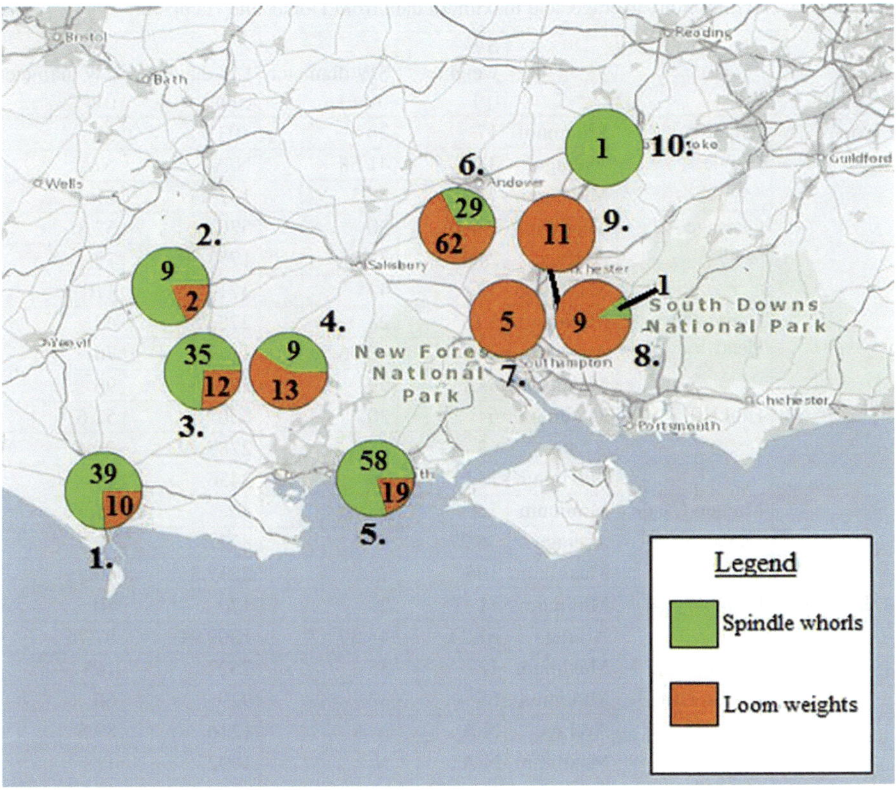

Fig. 8.3 Map of tool types found across all site types in the Central South: 1) Hengistbury Head, 2) Allard's Quarry, 3) Hod Hill, 4) Gussage All Saints, 5) Maiden Castle, 6) Danebury, 7) Zions Hill Farm, 8) Easton Lane, 9) Winnall Down, 10): Kennel Farm (by L. Ferrero)

similar yarns and textiles. The majority of loom weights from Hampshire display a similar range of weights and thicknesses, which would allow these to easily weave balanced fabrics with the variety of thread thicknesses produced at the Hampshire sites.

8.4.3 South East

Overall, few textile tools of any type were excavated from Kent (Table 8.3). The majority of spindle whorls came from the oppida site of East Wear Bay (Fig. 8.4, no. 1), although whorls were present at all sites. The large number of whorls found at this site is highly unusual for Kent, although it does fit the number commonly found in fortified sites elsewhere in the study area. Such a large disparity in spindle whorl numbers across Kent could be because conditions were particularly destructive and

Table 8.2 The minimum, average, and maximum data from Dorset and Hampshire

County	Site		SW weight (g)	SW diameter (mm)	LW diameter (mm)	LW diameter (mm)
Dorset	Allard's Quarry	Minimum	17	26	303.3	31
		Average	38.7	41.58	N/A	N/A
		Maximum	71	82.3	1313.3	61
	Gussage All Saints	Minimum	9.4	30	520	52.8
		Average	24.1	35.9	1753.79	72
		Maximum	36	44	3710	110
	Hengistbury Head	Minimum	5.8	28	278.75	40
		Average	21	42.76	1177.5	76
		Maximum	84.2	59	2574.6	96
	Hod Hill	Minimum	6	20	650	52.6
		Average	62	38.3	2287.99	98.4
		Maximum	134	89	3456	164
	Maiden Castle	Minimum	7.3	28	40	20
		Average	36.27	43.6	877.3	59.7
		Maximum	106	71	2237.3	102
Hampshire	Danebury	Minimum	11	29	170	40
		Average	61.24	48.59	1369.94	87.76
		Maximum	122	85	2622	165
	Easton Lane	Minimum	65	32	929	60
		Average	N/A	N/A	1216	89.5
		Maximum	N/A	N/A	1992	118
	Kennel Farm	Minimum	3	42	1100	120
		Average	N/A	N/A	2334.5	192
		Maximum	N/A	N/A	3183	418.25
	Winnall Down	Minimum	N/A	N/A	1216	42
		Average	N/A	N/A	1591	89.5
		Maximum	N/A	N/A	3331	104
	Zions Hill Farm	Minimum	N/A	N/A	522	77
		Average	N/A	N/A	1261.3	91.6
		Maximum	N/A	N/A	2760	106

too few survived, or because the majority of textiles were produced in these fortified sites and distributed to the smaller satellite settlements nearby. Loom weights follow a similar distribution pattern to those in the South West, with most of these tools restricted to the same area: Highstead and Thanet Earth (Fig. 8.3, nos 4–3 respectively). As in the South West, this suggests that weaving was restricted to specific areas rather than certain types of settlement. That said, the limited number of textile tools in Kent severely curtails the conclusions that can be drawn on the forms of textile production in different settlements, as well as how these were organised.

Given the restricted range of weights, the spindle whorls in Kent are primarily limited to the production of medium thickness yarns (approx. 0.6–1 mm in

Table 8.3 The minimum, average, and maximum data from Kent

County	Site		SW weight (g)	SW diameter (mm)	LW diameter (mm)	LW diameter (mm)
Kent	East Wear Bay	Minimum	7	7.3	1340	132
		Average	23	38.5	N/A	N/A
		Maximum	46	62	1423.75	158
	Highstead	Minimum	45	40	1400	106
		Average	N/A	N/A	N/A	N/A
		Maximum	52	40	1750	109
	Sunset Caravan Park	Minimum	20	21	N/A	N/A
		Average	31	32.2	N/A	N/A
		Maximum	44	39	N/A	N/A
	Thanet Earth	Minimum	14	33.5	291	54
		Average	N/A	N/A	847.2	59.7
		Maximum	N/A	N/A	1177.5	70

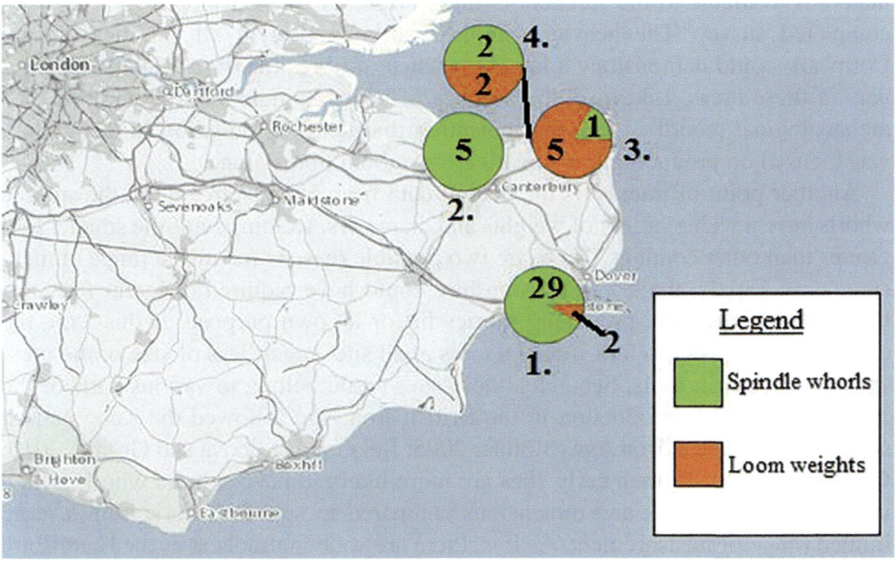

Fig. 8.4 Map of tool types found across all site types in the South East: 1) East Wear Bay, 2) Sunset Caravan Park, 3) Thanet Earth, 4) Highstead (by L. Ferrero)

diameter) and unable to spin yarns as thick as Hampshire. The spinners of Kent were able to spin animal and plant fibres equally well, as both the weights and diameters of these tools are able to be adapted to either.

As there are very few loom weights, identifying tool patterns for weaving is more difficult. However, two rough clusters stand out from the dataset: the first weighs 800–1080 g and measures 56–70 mm in thickness, which is best suited for medium to thick yarn in a balanced weave. The second weighs 1340–1750 g and measures

100–158 mm in thickness, which is best suited for weaving thicker yarns in a balanced weave.

8.5 Discussion

A closer look at the textile tool data across the study area quickly highlights several interesting trends across the study area. In most counties, the majority of spindle whorls are made to various sets of weights and dimensions; this data was used to determine the range of threads produced at each settlement, using the methods developed after experiments carried out by the TTTC project at the CTR (Mårtensson et al. 2006: 7). The tool groups discussed above can produce a range of thread thicknesses, from very fine (0.2–0.3 mm) to very thick (1.5–2 mm). However, Cornwall's spindle whorls are restricted to fine threads alone. This remarkable departure from the common dimensions of spindle whorls used elsewhere could be reflective of the materials available to the spinners in these areas, which is explored further in the completed survey. The heavier spindle whorls present in all counties (except Cornwall) could demonstrate a larger presence of fibres like flax or nettle by spinners in these areas. Likewise, the absence of heavy spindle whorls in Cornwall's archaeological record is a strong indication that textile production in this county was focused on producing fine threads from animal fibres alone.

Another point of interest is the overall data from Hampshire. Here, the spindle whorls have a wider variety of weights and diameters, forming multiple smaller tool groups than other counties. There are two possible reasons for such a range of measurements. Firstly, the sites in Hampshire could have required different forms of textiles, with each site producing fabrics fit for its own purpose. If this were the case, I would expect to find weaving tools at all sites regardless of size or the presence of other craft tools. Secondly, these sites could belong to various parts of the Iron Age; if textile production in the British Iron Age followed the same pattern seen in the European Iron Age (Stöllner 2005; Belanová Štolcová and Grömer 2010; Grömer et al. 2013), then early sites are more likely to have spindle whorls with a wide range of weights and dimensions compared to spindle whorls with a more limited range of measurement. As it is, there are weaving tools at every Hampshire settlement included in this study, which could indicate that each settlement in this area produced textiles for its own use. In sites able to support a community of workers, such as Danebury, textile workers could spend more time on spinning and weaving surplus textiles as household specialists. Unfortunately, there are too few textile tools remaining to determine whether individual spinners and weavers worked together as a workshop or were otherwise organised as a larger community of specialists that each focused on their own part of the textile production process.

Unlike spindle whorls, loom weights from different counties do not clearly share a range of weights and dimensions that could indicate the most common requirements for these tools. Instead, each county appears to make loom weights to suit its own regional needs, most likely based on the sort of textiles each area produced.

Most areas of overlap appear to be largely co-incidental, although Dorset and Hampshire have a majority of loom weights that share the same range of weights and dimensions. Most likely because hillforts produced most of the loom weights in these counties: Danebury in Hampshire and Maiden Castle, Hengistbury Head, and Hod Hill in Dorset. As regional centres, these were most likely also centres of craft production, able to produce extra textiles for further distribution or trade. When considered alongside the range of spindle whorl weights and dimensions in these counties, these sites most likely produced a range of textiles instead of a more standardised but limited form of textiles. This hypothesis will be examined further in the completed study, with a closer look at the data from fortified sites alone. The smaller settlements in the study area provide a greater challenge in determining the most common form of loom weight, although Dorset clearly has a wider variety of loom weights compared to Hampshire. This larger spread of data could represent smaller sites producing loom weights to suit individual needs, rather than making tools to weave a single form of textile. Kent is most difficult to discuss; although the loom weights have a similar range of weights and dimensions to Dorset and Hampshire, this region produced the fewest weaving tools and so provides the least data for weaving practices.

Cornwall and Devon also present intriguing data overlaps. As previously mentioned, Cornish loom weights (like its spindle whorls) are lighter and thinner than those from any other county, partly due to the slate that they are made from. This slate is the most common stone found across Cornwall and is easily cut to similar depths and weights, allowing the Iron Age tool makers to produce highly standardised textile tools. Devon's loom weights share similar weights, although these tools are deeper than those from Cornwall. The limited weights of these weaving tools suggest these counties either produced highly specialised textiles (e.g., very fine textiles compared to other areas) or were highly adaptable. Lighter loom weights would enable weavers to alter the tension placed on the warp threads more easily than with heavier weights, which would allow a weaver to make fine textiles, open weaves, warp-faced fabric, or thick fabrics, tight weaves, as well as working with any thread thickness. It is only possible to determine whether these are specialist or more general tools by comparing the loom weight data with the spindle whorl data from these regions.

The organisation of textile workers, and thus a sense of the larger modes of textile production, can be gained via an examination of tool distribution. The evidence from Cornwall indicates a division existed between where spinning was conducted and where weaving took place. Interestingly, it seems that this divide focuses more on the location of the settlements themselves than the type of settlement. While spinning occurred at all sites, weaving was limited to sites that were either in or close to trading centres. This pattern is similar to those of the Danebury Environs Project, where weaving tools were clustered in the hillfort while spindle whorls were more common in the smaller settlements surrounding it (Cunliffe and Poole 2000). Devon follows a similar pattern of textile tool distribution to Cornwall, with weaving tools clustered according to geographical areas rather than settlement type, especially in the Torbay region (see Fig. 8.3, nos 7–9). It should be noted that more work could be done in north Devon to find sites with a suitable amount of textile tools to establish if this pattern is repeated across the county. Nonetheless, both counties demonstrate weaving cultures made up of areas with specialised weavers, close to trade hubs, producing unusually fine yarns and fabrics.

This is partly due to the prevalence of slate in Cornwall, as previously discussed. While slate is rarer in Devon, Kimmeridge shale is a common material for spindle whorls in the Torbay region. This imported material was usually used for personal ornaments, although the cores are easily adapted into spinning tools. The numbers of Kimmeridge shale spindle whorls found in such a restricted area heavily suggests that these tools held value in Torbay and that spinning was an important task in this area. The decorations that cover most of the Torbay spindle whorls provide further evidence for this; this form of heavy decoration indicates a public declaration of individual ownership not seen in other whorls made of local materials. It is especially notable due to the fact that whorls are usually considered by archaeologists to be a common, utilitarian tool, easily made and easily discarded. Such decoration could make it easier to distinguish one craftperson's tools from another's or these could indicate that spinning was a public activity, where decorations could draw attention to the user's skill. It certainly suggests that spinning, spindle whorls, and (possibly) the craftspeople were considered important parts of this region's identity.

Milber Down (Fig. 8.3, no. 9) had little evidence for other craft activities, as well as no foreign or traded artefacts. This indicates that the occupants here were self-sufficient, although it is significant that textile tools outnumbered other craft tools, which suggests this site produced a larger quantity of textiles than necessary for such a small site. On the other hand, the excavations at Kent's Cavern (Fig. 8.3, no. 8) and Dainton (Fig. 8.3, no. 7) had very little evidence for occupation during the Iron Age, suggesting that these were not permanent settlements. Silvester (1986) proposed that Kent's Cavern was a seasonally-used site for preparing and spinning raw fibres before transporting the finished yarns elsewhere, which would be easier than moving raw wool or plants. Considering the poor preservation conditions that affect other sites in Devon, I would suggest that the Torbay region demonstrates that the pattern seen in Cornwall continued across this county as well, where separate geographical areas conducted different stages of textile production. However, this will remain hypothetical until further evidence can be found via excavations.

Compared to the South West region, Dorset produced spinning and weaving tools in all of its sites, which suggests that each site could produce the textiles that they needed, if in small quantities. On the other hand, Gussage All Saints has a larger proportion of loom weights and could have produced a larger quantity of textiles than any other Dorset site, possibly because Gussage All Saints had a specialist weaver. This pattern of self-sufficiency in textile production is not as clear in Hampshire, where settlements are characterised by a larger proportion of loom weights than elsewhere in the study area. The presence of these tools in so many types of site strongly suggests that each settlement made textiles for its own use, instead of a few sites weaving textile that were then distributed to other nearby settlements. Given the increased presence of loom weights compared to the other regions in the study area, it also indicates that weaving may have been a significant activity for the people of this region. Although this does not mean that the fabrics were of higher quality than other regions, only that more could be made. On the other hand, it is strange that Hampshire has fewer spindle whorls and other textile tools compared to the rest of the study area. Cunliffe and Poole (2000) noted that the smaller sites around Danebury had larger proportions of spindle whorls than loom weights and proposed that these small settlements supplied the larger fortified sites. However, the data presented here has been unable to confirm this, although the parameters under which this information was collected differed a great deal from those of the Danebury Environs Project. Nonetheless, this phenomenon is further explored and discussed in the finished study.

Finally, there is limited evidence from the South East, where most of the finds were collected from rescue excavations. Such excavations often led to a number of issues in determining the context of the textile tools found: small sites with a restricted understanding of the layout and character of the overall settlement, reports that are hard to access, methods of reporting and excavation that differ according to the archaeological unit responsible, and trouble in tracking the finds for analysis. Of the limited evidence gained thus far from Kent, it appears that this region loosely follows the pattern established in Cornwall: weaving restricted to sites within a small area while spinning remains a common activity.

8.6 Conclusions

The preliminary results of this study demonstrate a new way for British archaeologists to explore the organisation, production, choices, and status of textile workers in prehistory from a fresh perspective that is more nuanced and thorough than before. This study highlighted differences in textile production and traditions across the south coast of Britain by proving that the weaving techniques in Cornwall and Devon (specifically the Torbay area) had a number of differences compared to those in Kent, Hampshire, and Dorset. The close analysis of the dimensions and weights of various textile tools performed in this study was able to identify that Cornwall and Devon were producing far finer textiles – most likely only with wool – than elsewhere in the study area.

More work could be done elsewhere in Britain and Europe on the analysis of textile tools and the distribution of these artefacts, as well as significant changes in tool shape or materials across time, settlement types, and/or geographical areas. This will be able to prove whether the distinctive nature of British sites shown in this study is common throughout Britain and Ireland, and whether similar evidence exists in contemporary European settlement. So far, the common consensus holds that northwest Europe developed a system of producing large quantities of standardised textiles in the Iron Age. By applying the methods of analysis used in this study, it will be possible to gain further insights into the choices available to textile workers and a better indication of the contributions these workers made to their communities. Furthermore, it will be possible to not only identify areas, large or small, where the textile workers deviated from the common pattern of production or organisation, but also highlight the most likely reasons for such deviation. This would also allow for further study and discussion of group identities, social status of skilled individuals, and the textile traditions across large areas.

The next stage in processing the data from these sites will see this information separated by site type and compared to data from different forms of settlement, in order to determine whether different sites produced specific fabrics or a variety of textiles. The distribution of textile tools will also be investigated further by comparing the regional distributions to the Iron Age use of the local landscape. For example, noting areas of rough ground where grazing occurred compared to areas of arable land could indicate why certain sites only produced spindle whorls or loom weights. If a region had an abundance of land that was unsuitable for agriculture and used for grazing, sites located in or near these grazing areas would have direct access to the raw material needed for spinning yet unsuitable for producing food crops. These sites would then be able to trade or exchange spun wool for crops grown by sites surrounded by arable land, which would produce woven textiles. This is explored further in my doctoral thesis and the possible contribution that this method has to exploring craft organisation, regional economy, and land use.

References

Andersson Strand E, Nosch ML (2015) *Tools, textiles, and contexts: Investigating textile production in the Aegean and Eastern Mediterranean Bronze Age*. Ancient Textile Series 21. Oxbow Books, Oxford and Philadelphia.

Barber EJW (1992, third edition) *Prehistoric textiles: The development of cloth in the Neolithic and Bronze Age with special reference to the Aegean*. Princeton University Press, Princeton.

Beamer J (2018) Long-handled weaving combs – a rubric for recording. Paper presented at Exploring textiles and textile working from Prehistory to AD 500: Liverpool University Textile Symposium, 29 Oct 2018.

Belanová Štolcová T, Grömer K (2010) Loom-weights, spindles and textiles. Textile production in Central Europe from the Bronze Age to the Iron Age. In: Andersson Strand E, Gleba M, Mannering U, Munkholt C, Ringgaard M (eds) *North European symposium for archaeological textiles X*. NESAT X. Ancient Textiles Series 5. Oxbow Books, Oxford, p 9–20.

Brailsford J, Richmond I (1952) British Museum excavation at Hod Hill, Dorset. *The British Museum Quarterly* 17(3): 49–50.

Burke B (2005) Textile production at Gordion and the Phrygian economy. In: Kealhofer L (ed.) *The archaeology of Midas and the Phrygians. Recent work at Gordion*. University of Pennsylvania Press, Philadelphia, p 69–81.

Brumfiel EM, Earle T (1987) Specialization, exchange, and complex societies: An introduction. In: Brumfiel EM, Earle T (eds) *Specialization, exchange, and complex societies*. Cambridge University Press, Cambridge, p 1–9.

Costin CL (1991) Craft specialization: Issues in defining, documenting, and explaining the organization of production. *Archaeological Method and Theory* 3:1–56.

Cunliffe B (2005, fourth edition) *Iron Age communities in Britain*. Routledge, London.

Cunliffe B, Poole C (2000) *Introduction: The Danebury Environs Programme: The prehistory of a Wessex landscape,* Vol 1. English Heritage and Oxford University Committee for Archaeology Monograph 49. English Heritage and Oxford University Committee for Archaeology, Oxford.

Cutler J, Nosch M-L, Andersson Strand E (2013) Textile production in Quartier Mu. In: Poursat J-C (ed.) *Fouilles exécutées à Malia. Le Quartier Mu V. Vie quotidienne et techniques au Minoen Moyen* II. Études Crétoises 34. École française d'Athènes, Athènes, p 95–118.

DeRoche CD (1995) *Textile production in Britain during the first millennium B.C.* Unpublished dissertation, University of Cambridge. https://doi.org/10.17863/CAM.37739.

Dimova B (2016) Textile production in Iron Age Thrace. *European Journal of Archaeology* 19(4):652–680. https://doi.org/10.1080/14619571.2016.1164457.

Dudley D (1956) An excavation at Bodrifty, Mulfra, near Penzance. *Archaeological Journal* 113:1–32.

Ferrero, L.J. (Forthcoming) *Visible Tools, Invisible Craft.* Unpublished dissertation. Cambridge: University of Cambridge.

Fox A, Raleigh Radford CA, Rogers EH, Shorter, AH (1949–1950) Report on the excavations at Milber Down, 1937–8. *Proceedings of the Devon Archaeological Exploration Society* 4:27–65.

Gleba, M (2007) Textile production in Proto-Historic Italy: From specialists to workshops. In: Gillis C, Nosch M-LB (eds) *Ancient textiles: Production, craft and society.* Ancient Textiles Series 1. Oxbow Books, Oxford, p 71–76.

Gleba M. (2008) *Textile production in pre-Roman Italy.* Ancient Textiles Series 4. Oxbow Books, Oxford.

Grömer K (2005) Efficiency and technique – experiments with original spindle whorls. In: Bichler P, Grömer K, Hofmann-de Keijzer R, Kern A, Reschreiter H (eds) *Hallstatt textiles – technical analysis, scientific investigation and experiments on Iron Age textiles.* BAR S1351. Archaeopress, Oxford, p 107–116.

Grömer K (2016) *The art of prehistoric textile making.* Veröffentlichungen der Prähistorischen Abteilung 5. Natural History Museum Vienna, Vienna.

Grömer K, Kern K, Reschreiter A, Rösel-Mautendorfer H (eds) (2013) *Textiles from Hallstatt. Weaving culture in Bronze and Iron Age salt mines. Textilien aus Hallstatt. Gewebte Kultur aus dem bronze- und eisenzeitlichen Salzbergwerk.* Archaeolingua 29. Archaeolingua Publishers, Budapest.

Kania K (2013) Soft yarns, hard facts? Evaluating the results of a large scale hand-spinning experiment. *Archaeological and Anthropological Sciences* 7:113–130. https://doi.org/10.1007/s12520-013-0167-y

Marín-Aguilera B (2019) Weaving rural economies: Textile production and societal complexity in Iron Age Southwestern Iberia. *World Archaeology* 51(2):226–251.

Marín-Aguilera B, Rodríguez-González E, Celestino S, Gleba M (2019) Dressing the sacrifice: Textiles, textile production and the sacrificial economy at Casas del Turuñuelo in fifth-century BC Iberia. *Antiquity* 42:933–953. https://doi.org/10.15184/aqy.2019.42.

Mårtensson L, Andersson E, Nosch M–L, Batzer A (2006) *Technical report: Experimental archaeology. Part 2:2 Whorl or bead? Tools and Textiles – Texts and Contexts research programme.* Copenhagen: The Danish National Research Foundation's Centre for Textile Research https://ctr.hum.ku.dk/research-programmes-and-projects/previous-programmes-and-projects/tools/technical_report_2-2__experimental_arcaheology.pdf, accessed 22 Jan 2021.

Mårtensson L, Andersson E, Nosch M–L, Batzer A (2007) *Technical report: Experimental archaeology. Part 3 Loom weights. Tools and Textiles – Texts and Contexts research programme.* Copenhagen: The Danish National Research Foundation's Centre for Textile Research https://ctr.hum.ku.dk/research-programmes-and-projects/previous-programmes-and-projects/tools/technical_report_3__experimental_archaeology.pdf, accessed 22 Jan 2021.

Mårtensson L, Nosch M–L, Andersson Strand E (2009) The shape of things: Understanding a loom weight. *Oxford Journal of Archaeology* 28(4):373–389.

Poole C (1984) The structural use of daub, clay, and timber. In: Cunliffe B (ed.) *Danebury: An Iron Age hillfort in Hampshire. Vol. 1: The excavations 1969–1978: The site.* Council of British Archaeology Research Report 52:110–123.

Silvester RJ (1986) The later prehistoric and Roman material from Kent's Cavern, Torquay. *Devon Archaeological Society Proceedings* 44: 9–38

Stöllner T (2005) More than old rags – textiles from the Iron Age salt-mine at the Dürrnberg. In: Bichler P, Grömer K, Hofmann-de Keijzer R, Kern A, Reschreiter H. (eds) *Hallstatt textiles – technical analysis, scientific investigation and experiments on Iron Age Textiles*. BAR S1351, Oxford, p 161–174.
Tuohy T (1995) *Prehistoric combs of antler and bone*. Unpublished dissertation, University of Exeter.
Tuohy T (1999) *Prehistoric combs of antler and bone:* Vols 1, 2. BAR 285. Archaeopress, London.
van der Leeuw S (1977) Towards a study in the economics of pottery making. In: Beek BL, Brant RW, Gruenman van Watteringe W (eds) *Ex Horreo*. Cingvla 4. Albert Egges van Griffen Instituut voor Prae- en Protohistorie, University of Amsterdam, Amsterdam, p 68–76.

Lewis Ferrero is a final year doctoral research student at the University of Cambridge, UK, studying textile archaeology in the Iron Age and its importance and contribution to society. They are also interested in gender archaeology, third or other genders, and the status of spinners and weavers in prehistory.

Chapter 9
Combing the Data: Re-evaluating 'Weaving' Combs in the Textile Production Sequence during the British Iron Age

Jennifer Beamer

Abstract Bulleid and Gray cemented the association of long-handled combs from Iron Age contexts with weaving in the early twentieth century. Since then, the long-handled comb has been generally included as part of the weaving *chaîne opératoire*. Early attempts were made to assess use-wear; however, it was not until the 1970s and 1980s that wear patterns were critically examined. Building on a century of research, Tina Tuohy's doctoral project further examined wear patterns and offered an alternative proposal, citing use-wear as evidence. In archaeological research, combs are frequently associated with other known textile production tools, which has embedded their place in the prehistoric textile toolkit for Britain. Although this notion is generally agreed, their utility within the textile production sequence remains questionable. In illustrations, combs are usually depicted with the warp-weighted loom, despite Tuohy's proposition to the contrary. Analyzing the history of long-handled comb research revealed the necessity for re-evaluation This chapter presents preliminary data on a recording rubric and methodology for assessing utilitarian comb function from the archaeological evidence from Danebury hillfort (Hampshire, UK).

Keywords Iron age · Long-handled combs · *Chaîne opératoire* · Use-wear · Recording methodology · Functional analysis

---

J. Beamer (✉)
University of Leicester, Leicester, UK
e-mail: jkb32@leicester.ac.uk

A. Ulanowska et al. (eds.), *Ancient Textile Production from an Interdisciplinary Perspective*, Interdisciplinary Contributions to Archaeology,
https://doi.org/10.1007/978-3-030-92170-5_9

9.1 Introduction

Long-handled 'weaving' combs are pervasive textile tools in Iron Age Britain (800 BC–43 AD). Since they are so widespread, these combs have been used as a 'type fossil' by past scholars to define cultural horizons during this period (Hodson 1964: 103). The majority of recovered combs were located in southern Britain and typically range in length from 127–155 mm (Tuohy 1999: 21). The width at the dentate, the tines or prongs, is quite narrow, 18–36 mm as recorded from Danebury hillfort (Cunliffe 1984). Though the dentate usually occurs at one end of the long handle, some examples of double-ended dentate combs (18) do exist and mainly come from Glastonbury Lake Village and Meare Lake Village (Tuohy 1999: 35). The exact distribution of decorated and undecorated combs is tricky since identifying decoration on broken combs is easier than assuming broken combs were undecorated. In Tuohy's (1999: 24–25) analysis of southern Britain combs, 109 complete combs were undecorated and 313 complete and broken combs were decorated. Some combs have evidence of re-cuts to prolong their use-life. Tuohy (1999: 21) reviewed the frequency of re-cuts at six principal sites in southern Britain and showed that 43 whole combs exhibited these recuts.

Previous textile archaeologists have held differing opinions on the purpose of these slender-handled, multi-pronged bone tools, and there is little modern unified consensus. A review of research on combs revealed ongoing arguments of their utilitarian function, where some believed they were used as weft beaters, whereas others thought they were for fibre processing. In the late nineteenth century, Scottish combs were discussed and debated in terms of their utilitarian function and sources from the ancient world were often consulted by antiquarians to corroborate their function. My doctoral project featured a multi-scalar approach that was aimed at re-evaluating the basic tools of the toolkit of textile production—loom weights, spindle whorls, long-handled combs, and needles – and where they were placed in the operational sequence. As part of this research, it was relevant to examine these functional debates and pioneer a new method to analyse combs.

The social aspects of long-handled combs tend to be featured in broader discussions of Iron Age society due to their status as decorated, non-metal objects representing the subversion of a typically privileged position (Chittock 2014). Not all combs are decorated which makes the decision to decorate a very intriguing aspect. Therefore, the rarity of decorated objects during British Iron Age has brought long-handled combs into discussions related to motifs and materialities (e.g., Joy 2011). In addition, their connection with warp-weighted loom weaving now comes as standard (e.g., Crowfoot 1945; Cunliffe 1984; DeRoche 1995). To investigate the social functions of these combs, however, it is imperative to also consider their utilitarian use. Utilitarian function refers to the physical function of a tool which can be analysed through an examination of accumulated wear patterns. Few experiments using long-handled combs exist (see Ling Roth 1918; Reynolds 1972; Tuohy 1995), so assumptions about function is primarily based on theoretical considerations. Gaining an understanding of the social uses of long-handled combs can be best achieved through an in-depth study of their functional uses. Without this utilitarian consideration, a nuanced understanding of their social impact cannot be reached. Long-handled combs continue to be regarded as proxy type-specimens of textile production and a prerogative exists to question these long-held assumptions of use. The recently discovered anthropomorphic comb discussed by Fitzpatrick (2020) foregrounded the unresolved debate of functionality and demonstrated the need to devise a system to analyse combs in a functionalistic way so the rarity of this discovery could be better understood.

A biographical approach to long-handled comb study led to the development of a recording methodology that foregrounded the range of possible utilitarian functions. Connected with the object biography approach (Joy 2015; Jones et al. 2016), the historical literature was consulted to develop the categories for the recording methodology. This rubric of recording considered manufacturing trends and use-wear as well. The following sections of this chapter outline the geographical distribution of combs, as well as the history of the study of long-handled combs to demonstrate the need for a standardized recording methodology. The combs from Danebury hillfort served as a pilot case study to illustrate the efficacy of the rubric. The conclusions that can be drawn at this developmental stage confirm the utility of the rubric and reveals the increased interpretive value of the combs.

9.2 Geographical Distribution of Long-Handled Combs

Though previously regarded as a solely British textile tool, long-handled combs have been discovered in in France (Hodder and Hedges 1977: 28), Ireland (Hallén and O'Neill 1994), Frisia (Roes 1963)[1] and the Netherlands (Tuohy 1992). Most combs date to the Iron Age (800 BC–AD 43) period in Britain. Some combs have been recovered from Middle and Late Bronze Age contexts in Great Britain and may demonstrate continuity throughout later prehistory (Hodder and Hedges 1977: 17; Tuohy 1999, volume 2). Instances of long-handled combs in Romano-British contexts have been considered residual (Bulleid et al. 1917).

9.3 History of Research on Long-Handled Combs

Many scholars have been interested in long-handled comb typology, though some have also focused on their function. Mostly, the discussion has centered on comb decoration and the geographic distribution of these styles. The substantial contributions of John W. Hedges (1973) and Tina Tuohy (1995) on long-handled combs was

---

[1]Though MacGregor (1985: 189) mentioned that these combs might be residual or else Anglo-Saxon or Viking in date.

built on the seminal work of antiquarian researchers (e.g., Coughtrey 1871; Bulleid and Gray 1911; Henshall 1950). Other relatively recent studies (e.g., Hodder and Hedges 1977; Sellwood 1984) have spotlighted these combs, though still predominantly in terms of typology and style rather than function. Michael Ryder (1997) and Ro Bailey (1999) have critically considered comb function and Helen Chittock (2014) evaluated how the presence of decoration on combs may have impacted the social sphere through a discussion of material choice. For how long-handled combs came to reside in the *chaîne opératoire* of textile production, and why, the following subsections provide context.

9.3.1 The Antiquarian Era of Study

Significant attention was accorded to these artefacts when Millen Coughtrey (1871) called them long-handled combs (Fig. 9.1) to distinguish their characteristic long handle. This distinction implied a different function than for personal grooming (Roes 1963: 27; MacGregor 1985: 73–99) according to visible wear patterns. The patterns he referred to were 'striations'. Joseph Anderson (1883) assessed similar combs recovered from the Broch of Burrian in Orkney. Previously, Anderson (1871: 551) had determined that the wear patterns suggested a non-combing purpose and proposed their implementation in weaving based on this transverse grooving—probably 'striations'. He linked these combs to the *pecten textoris*[2] of the ancient world (Anderson 1871: 552–7). Citing the research of Keller (1866: Pl.28, Fig. 8) at the Swiss Lake Villages, Anderson (1883: 253) concluded that long-handled combs were probably weft beaters.

Augustus Lane Fox (1881) became involved with this discussion during the post excavation period at Mount Caburn; one comb had two lateral grooves near the base of the teeth which he thought was for combing flax or weaving. Fox (1881: 432) also considered alternatives from ethnography, such as for scraping fat from hides. He noted their resemblance to the Broch of Burrian combs and described the blunted apices of the combs as use-wear from packing weft into cloth (Fox 1881: 433), citing the eighth century AD composite comb from Birka (Stolpe 1874: 625, Fig. 12) as evidence. The morphological similarities of both combs likely led to this conclusion. Though Fox explored various functions, he mainly considered them as optimal for flax combing or for weaving.

[2] Anderson (1871: 553–557) offers several sources that cite a description which matches the long-handled comb. He notes from Rous, Archaeologica Attica that a "χερχίδ [sic], or *pecten* or the sley, *like a comb*" is operating as a way to beat weft into place (emphasis in original). A note from Juvenal's *Satires* (9.30) illustrates a disgruntled man who complains of an improperly woven garment: "Et male percussas textoris pectine Galli…and insufficiently struck *with a comb* of a Gaulish weaver" (emphasis in original). These explanations, and the others he cites which are not printed here, are enough for Anderson to draw the conclusion that a comb was used as a beater on Greek and Roman looms, and that if Iron Age weavers in Gaul are using the same warp-weighted loom technology, then the long-handled combs from Iron Age Britain were the *pecten textoris*.

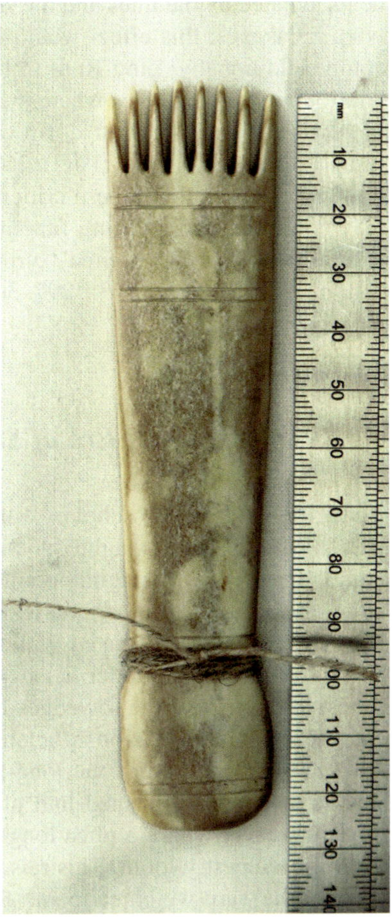

Fig. 9.1 A replica comb based on a nearly complete long-handled comb (SF628) from Danebury hillfort, UK. (Photo by J. Beamer)

Discovered in 1892, Glastonbury Lake Village (BC 250–50, Coles and Minnitt 1995: 199–206) has had a profound impact on Iron Age comb research. The few combs that were previously recovered were isolated finds (Fox 1881: 431–434); the 89 well-preserved combs discovered at Glastonbury increased the relevance of their study. In the 1911 publication, Bulleid and Gray (1911: 266) immediately called them long-handled *weaving* combs. Along with spindle whorls and loom weights, they considered combs as indisputable evidence for warp-weighted loom weaving based on the *pecten textoris* reference in Roman weaving practices.[3] By the time the Glastonbury Lake Village excavations were published, the weaving theory was definitively established.

Ling Roth (1918: 129–140) criticized the weaving theory put forward by Bulleid and Gray (1911) and cited a few problems with this notion. The varying number of

[3] They also reference modern Scandinavian weavers who use the *spatha* or 'weaving sword'.

dents, created by the tines and the spacing between them, could cause abrasion and warp breakages; this effect would be exacerbated by the natural curvature of the comb. Additionally, Ling Roth (1918: 133) noted that the illustration used in the Glastonbury publications, where the fell of the cloth grew upwards and the function of beating forced the weft downwards, did not represent a warp-weighted loom setup (Bulleid and Gray 1911: 269, Fig. 57). Though Bulleid and Gray (1948: 61–62) followed up on these criticisms with the Meare Lake Village publications, they confirmed the weaving function based on the association with other known textile tools. They also cited Cyril Fox (cited in Smith 1925: 152–153) and Sir Mortimer Wheeler (1943: 297–303) as additional proponents of the 'weaving' theory.

9.3.2 The Modern Era of Study

By the 1970s, Iron Age scholars were fully aware of weaving combs and their association with the warp-weighted loom (Hodder and Hedges 1977). However, Hodder and Hedges (1977: 17) rightly pointed out that long-handled combs were recovered alongside a variety of other non-textile related artefacts, suggesting that combs may not be as strongly associated with textile production as previously thought. They also raised two important criticisms. First, Crowfoot (1945: 158) had deemed them necessary to prevent the selvedges from waisting on a warp-weighted loom, which can occur when there is insufficient slack in the weft. Crowfoot suggested this is done by the curvature of the long-handled comb thrusting the warps apart as the weft is inserted and pushed into place. However, Hedges (1973: 140) provided a modern weaver with a replica long-handled comb who confirmed that it would be ideal for a tapestry loom. This raised doubt that long-handled combs were strictly used for the warp-weighted loom.[4] Second, Hodder and Hedges (1977: 19) cited the experimental work conducted by Ling Roth (1918), who stated that the curvature of the combs created distortions in the warp making it ineffectual, and possibly causing unnecessary abrasion on the warp yarns. Having questioned the association with warp-weighted weaving, Hodder and Hedges (1977) concluded with no alternative interpretation.

A typological study of decoration and morphology was standard practice in the study of combs by the 1980s. Working on Danebury, Lyn Sellwood (1984) analyzed a geographically discrete comb assemblage and added a new parameter to the study of long-handled combs: use-wear.[5] Despite the disparity of comb sizes, the observed wear patterns were both distinctive and consistent (Sellwood 1984: 375). Her analysis confirmed the functional implications of use-wear. Sellwood established that

[4] Cunnington (1923) also doubted their suitability for weaving.

[5] Use-wear is the accumulation of physical friction on the surface of a material that may be from manufacture or use. See Marreiros et al. (2015) for an historical account of use-wear studies.

weaving was a function of long-handled combs based on this use-wear, despite being aware of contemporary research (Hodder and Hedges 1977: 19; *contra* Reynolds 1972).

Tuohy (1995) expanded upon use-wear patterns and developed a systematized recording scheme. Tuohy observed wear which was commonly on the inside of outer tines only, an unexpected deviation of warp-weighted loom usage.[6] Tuohy (1999: 56–9) believed that use on a band loom would be a more appropriate explanation, including for why combs were occasionally found with other textile production equipment. However, this project still kept combs within a narrow interpretive scope.

This is the scholarly foundation on which subsequent research on long-handled combs has been based. Despite previous scholars emphasizing the need clarify function through use-wear analysis, typological approaches focusing on morphological character and decoration continued to dominate research trends. However, a typological distinction of style is limited and there has been little progress in resolving the debate between the functional attributes. Bailey (1999) presented a series of counter-arguments to the weft-beating function, and cited their inadequacy based on use-wear. Combs that were broken and recut had signs of continued use, which is not in accordance with weaving. Instead, Bailey (1999) suggested they were ideal for flax combing which is supported by the 'striations' seen between tines, a comment stated by Ryder (1997) as well. Since utilitarian function is inadequately understood, the placement of combs within the textile toolkit (DeRoche 1995: 76, 100–101; DeRoche 2012) and a consideration of their social implications (e.g., Joy 2011; Chittock 2014) are questionable.

9.4 Long-Handled Comb Morphology

Long-handled combs were created as seamless objects. Saws, knives, files, and sanders, were all likely required to cut, shape, and smooth these combs. Iron Age combs tend to divide into two or three sections for archaeological analysis, the dentate, handle, and terminal. The tines are described as sub-rectangular in cross section and triangular shaped in profile, which are usually spaced evenly across the width of the dentate. Material choice may have impacted the morphology of the combs, and the social importance of bone or antler may have dictated the overall aesthetics. These two considerations are important to include in an object biography approach since they influence the ways we interpret the shape and function of a tool.

[6] Tuohy (1995) mentioned conducting experiments with warp-weighted loom weaving which led her to suspect that the use-wear observed did not match the evidence on the archaeological samples. The parameters of her experiments were not published.

9.4.1 Material Component of Combs

Antler and skeletal bone are comparable in form and structure and can be shaped in a similar fashion. As a raw material, Iron Age people made a clear distinction in how they may have perceived bone and antler (Chittock 2014: 317–318). There are structural differences in antler and bone which depend on whether red or roe deer antler was used, or if a rib or marrow bone was used. Generally, most Iron Age combs were made from red deer antler (Tuohy 1995: 58). Long-handled combs were sometimes crafted from bone, typically made of the rib or long bones of cattle. Hodder and Hedges (1977: 19) indicated that the cancellous tissue, spongy bone located in the cortical cavity of long bones, would be unsuitable as weaving combs, though Reynolds (1972) experimented with reconstructed combs and proved otherwise. Antler, a seasonal commodity, required expertise to track deer when they discarded their antlers, and needed to be well-timed to collect the best pieces. Chittock (2014: 317) indicated that the slightly stronger structure of antler could outweigh the extra associated processing steps. This is reflected in the Danebury combs which are predominantly made from antler. Both practical and social choices may have impacted material selection.

9.4.2 The Effects of Burning on Preservation and Use-Wear

Taphonomic processes, including deposition and anthropogenic alterations, impact archaeological analysis, but burning can significantly impact the color, surface features, and shape of bone artefacts. Burned bone preserves more poorly than otherwise. The process of burning changes the mechanical structure of bones by removing collagen, the organic matter that gives bone its flexible properties (Lyman and Lyman 1994). Burned bone is rendered brittle, though the extent of how brittle the bone becomes is dependent on which species the bone originally belonged to and the age of the individual. Furthermore, "burned bone tends to dissolve more rapidly than fresh bone in acidic solutions" (Lyman and Lyman 1994: 390). As bone is thoroughly heated, often at a higher temperature, they become calcined. Griffitts (2001) noted that changes bone undergoes during burning impacts use-wear studies. Blackening of bone causes the surface to become rounder in appearance and shinier (Griffitts 2001: 186). Calcined bone suffers a reduction in the shininess thereby making surface features more difficult to detect, or disappear completely (Griffitts 2001). Thoroughly calcined bone will dissolve the fastest (Lyman and Lyman 1994: 390). Trace wear can be further eradicated if the calcined bone is interred in acidic soils. The soil matrix offers a key clue regarding the nature of its deposition and is essential for developing the taphonomic history, as well as other anthropogenic processes which form part of an object biography analysis.

9.4.3 *Terminology: Combing or Beating?*

Consistency of terminology creates a semantic repertoire for researchers. The 'comb' in long-handled ('weaving') comb may not have functioned as a comb in prehistory, but it establishes a common terminology. On the other hand, 'weaving' implies a specific function which may not comply with a functional study. Morphologically, combing warps and combing wool can be performed with the same type of tool, though it does not follow that all long-handled combs were used in these ways. However, referring to the morphology also expands the relevancy of the term 'comb' to include non-textile functions for which it may have been suited: scraping fat from or de-hairing hides. Following this, this chapter refers to long-handled combs to emphasize their morphology, rather than perpetuate one specific function. To illustrate the 'combing' action, the prevailing theories of weft beating and fibre processing are discussed in the subsequent section.

9.5 Functional Characteristics of Long-Handled Combs

There are limitations in the functional attributes of tools and the morphological character provides these insights. The morphological shape of comb suggests they were used for a combing action. From the point of view of textile production, there are two such activities, weft beating and fibre processing. The following subsections outline these uses in a theoretical way and shows how those functions might have influenced comb morphology and generated use-wear. These activities do not form an exhaustive list and it is crucial to bear in mind that long-handled combs may have performed this 'combing' action across a variety of craft production activities. However, it is pertinent to begin with their classic association with the textile toolkit for Iron Age Britain and use the evidence generated from these rubric results and discussions as a launching off point.

The three main weaving appliances which may have been technologically available during the British Iron Age include the warp-weighted loom, the band, or tablet, loom, and the vertical two-beam loom (DeRoche 1995: 84–102). Archaeologically, there is overwhelming evidence of the warp-weighted loom in terms of loom weights which are regularly recovered from sites, and a small assortment of thin, bone tablets with drilled holes near the corners, suggesting their use in band weaving (DeRoche 1995: 84). The simplicity of a vertical two-beam loom cannot completely be disregarded as a technologically available loom type during the British Iron Age because the lack of loom weights and its wood construction may indicate a different textile tradition that would have left no archaeological traces (DeRoche 1995: 102). Tuohy (2004: 104) proposed that the presence of long-handled combs suggested that small tapestry looms existed, though this point is now contested (see further below).

On balance, the fibre processing functionality is also considered. Many of the procedures for processing plant and wool fibres do not require elaborate tools. The best extant tool that would be well suited for either type of processing is the long-handled comb. Placing combs within these two operational sequences affords greater flexibility in creating attribute categories for establishing a recording rubric because a range of activities is considered. Understanding how the comb could be used as a weaving or combing implement highlights the key places on the comb for further examination.

9.5.1 *Weft Beating*

This subsection describes ergonomic hand positions with the warp-weighted loom, band loom, and two-beam loom, to consider where observed wear might be located. Multiple ways of holding a comb exist but this section is theoretical in principle. Importantly, a comb is not absolutely required to weave in any of the cases presented, though it may have been helpful in spreading warps to prevent the weft from waisting, as Crowfoot suggested.

The warp-weighted loom produces fabric from the top down whereby the warp is tensioned by loom weights (Grömer 2016: 107–138). Figure 9.2 is an example of how a weaver might hold a comb with this loom type. The curvature of this comb is similar to that commented on by Ling Roth (1918: 134) and interpreted as a negative attribute. Crowfoot suggested that this feature would improve the quality of the cloth by thrusting warps apart and to counteract the tendency to waist (Crowfoot 1945: 158). In this experiment, the curvature incurred no warp abrasion and the waisting tendency was solved by managing the slack in the weft. Though the parameters of the experiment being conducted in the image cannot be attenuated here, experimentation offers a working context by which to investigate functionality and use-wear development more thoroughly.

Fig. 9.2 Replica comb demonstrating the effect of the curvature on warps. Note how the curvature distorts the warp yarns and that multiple warps pass between each tine. (Photo by J. Beamer)

For a weaver-tensioned band loom, the working fabric typically grows from the weaver outwards. Thus, to beat the weft into place, the comb pushes the weft towards the weaver. The best ergonomic hand position is with the comb in reverse grip where the tines point downwards. Due to the narrower width of fabric, beating the fabric into place would be simple and fast.

For a two-beam vertical loom, the working fabric typically grows from the bottom upwards. The comb may also be held in a reverse grip, pushing the weft downwards. Since tapestry weaving does not require the weft to be worked back-and-forth down the full width of the fabric, a long-handled comb may have been an appropriate tool.

In all three cases, the length of the comb tines may have been influenced by the degree of openness of the shed and the fineness of threads may have influenced the spacing between tines. In fact, the weaver in the experiment conducted by Reynolds (1972: 14) mentioned that the shorter tines of a replica long-handled comb were an improvement over the dinner fork he normally used. Depending on the way the fabric grows, the position of the hand and the weaver in relation to the loom will vary, and the resulting use-wear reflects these positions. Additionally, the oils and sweat from a hand could further develop the observed polish in certain areas, like the handle.

The proposed hand and body positions in this section provide useful insight. For example, smooth wear polish may result from rubbing against a non-abrasive surface, such as an oily hand; equally, striations may result from friction against an abrasive surface. Additionally, pressure can enhance the smoothness of surfaces and repetition could also deepen visible wear marks. Noting the location and character of the smoothness or striations might be helpful in distinguishing between weft beating or fibre processing activities.

9.5.2 Fibre Processing

Sheep and wool have been held in special regard during the Iron Age period by archaeologists (e.g., Serjeanston 2007). Infrequently mentioned, flax was also a fibre available during this period. However, how these fibres have been processed in prehistoric Britain has not been explicitly considered. That said, how wool and flax were prepared has been discussed for the Iron Age of Central Europe (Grömer 2016: 63–73). This subsection comments on some situations that wool and flax could have been processed with combs, even while recognizing they are nonessential for fibre processing. Splicing technology existed at British Bronze Age sites (Gleba and Harris 2019). Therefore, we cannot assume that British Iron Age people combed flax. There are no preserved linen fragments in southern Britain so it is difficult to determine when textile producers transitioned from spliced yarns to draft spun yarns. Therefore, if long-handled combs were being used for processing these fibres, the following scenarios propose how they might have functioned.

To remove vegetable matter from washed wool, it may be held in one hand and combed with the other. Placing the wool on the thigh and using an inverted grip used for the comb, the tines can be pressed down into the fibres. A raking motion of the comb being either pulled towards the body or away from it, would succeed in separating the individual fibres and releasing unwanted debris from the wool. A comb can be used to remove the outer stem of flax after retting[7] and drying. Flax can be bent and broken in the hand, then pulled through the tines of a comb in a similar process as described above for wool. The morphology and length of tines may have been influenced by this combing action. Depending on the flexibility of the fibre and the comb material, shorter tines may have been more appropriate than longer tines and result in fewer breakages overall.

The major difference between wear patterns between these two types of usages, weaving and combing, lie in the wear patterns seen between tines. Striations may be the result of the comb being used as a beater, but the activity of combing fibre may more clearly explain this wear pattern than from warp yarn abrasion. Additionally, there may be differences in wear patterns between working with wool or flax. The combing action is possibly more repetitive than weaving and may include applying pressure on the tines as knots are forcibly separated. This may explain tine breakages and evidence of re-cuts (for experiments, see Rösel-Mautendorfer et al. 2012). The tool itself is ambidextrous and the user might switch hands, thereby resulting in symmetrical wear among the inside tines or the nature of polish on the handle—in Tuohy's (1995) study and in the analysis below, wear is rarely symmetrical. However, it is unknow whether these tools were personal or communal, and other parameters of wear are still not well understood.

The ways of holding a long-handled comb expressed in this commentary are suggestive of use and manifold techniques may have been used for holding and using combs. Variable techniques among users certainly impacts the development of use-wear and polish marks alone are insufficient to suggest use (Lyman and Lyman 1994: 344). Many more factors, such as the differences in material qualities of wool and flax, impact the way a comb might have been held and how those material differences altered and modified the surface features of the comb when combing or weaving. Wear resulting from work elements when the comb was not actively being used, in between changing sheds for example, dangling from a cord or holstered into a belt, could also alter surface features. The considerations expressed in this section must be addressed as an integral component for taking comb measurement data. The aim of this study was to establish attribute categories, listed below, for a recording rubric rather than be comprehensive at this preliminary stage. With this rubric, the recorded information for each attribute category is consistent for each comb and each researcher.

---

[7] A processed of controlled rotting.

9.6 Case Study: The Combs of Danebury

Danebury is an Iron Age hillfort located in Hampshire, UK and a total of 71 combs, spanning the whole Iron Age period, were recovered here during the 1969–1988 excavations (Cunliffe 1984; Cunliffe 1991). There were 23 appropriate for an investigation of utilitarian function.[8] After three thorough examinations of the combs, which are curated at Chilcomb House in Winchester (UK), the rubric was checked for accuracy of the method. The differences in observed wear patterns suggested different uses, prompting a need to qualify their functional character as quantifiable data. The *chaîne opératoire* of textile production and an object biography approach were utilized to focus the design of the rubric. Together, these observations could identify technological variation in manufacture and use-wear.

9.6.1 Danebury Hillfort

Located on the chalklands near the River Test and Wallop Brooke, Danebury hillfort resides on a hilltop in a landscape full of associated settlements and other hillforts which date to the same period. Danebury is an enclosed, univallate hillfort with roundhouses skirting the interior perimeter. There is evidence of a wide array of local craft production at the hillfort and it had a permanent occupancy (Davis 2013: 360) that likely functioned independently from its neighbours. The inhabitants relied on mixed farming tactics which included the cultivation of annual cereal crops (Cunliffe 1995: 43) and sheep husbandry (Cunliffe 1984: 475).

9.6.2 Data Selection

The data was selected based on diagnostic characteristics, which included the dentate. Combs are rarely recovered fully intact, so appropriateness was based on whether any functional features were present. The dentate area is the most functionally relevant portion of a comb, so any comb that had this area preserved was selected for analysis. Digital photos and measurements were taken for each comb in the study. All combs were analyzed in the same orientation for consistency and clarity. Both numerical and short text metrics were recorded in an Excel spreadsheet.

[8] Early Pre-Roman Iron Age (800–500 BC): 3, Middle Pre-Roman Iron Age (500–100 BC): 3; Late Pre-Roman Iron Age (100 BC–43 AD): 16; unphased (1).

Table 9.1 The attribute categories devised for the rubric are in the left column, and the description and type of metric for each attribute are in the right column

Attribute category	Description
Length of tine	This is a numerical metric and measures from the base of the tine to the apex.
Interdentate distance	This is a numerical metric and measures the space between tines.
Interdentate space	This is a short text metric and is measured at the dentate. It characterizes the interdentate distance with either U- or V-shaped.
Use-wear	This is a short text metric and uses the identifiers 'smooth' or 'striated'. When 'striations occur, the angle of these marks is noted.
Surface polish	This is a short text metric and describes the state of polish with phrases like 'shiny/glossy' and the location of such evidence.

9.7 Devising the Rubric

A rubric requires attribute categories. These were partially formulated based on previous studies. This chapter has outlined the possible utilitarian functions as expressed by other scholars and five attribute categories were proposed (Table 9.1). These attribute categories provided important information about functionality that the basic parameters, such as mass and overall length, do not provide. Criteria were defined for each attribute category and the metric type use would either be numerical or free short text. To maintain clarity and conciseness, all free text used to record details during the analysis was kept consistent. For example, if 'striations' were noted, the angle of these striations was also noted. The rubric is flexible and can accommodate new attribute categories when the need arises. A summary of the rationale for establishing these additional attribute categories is restated here. The length of tine dictates how far the comb can be pushed through a material before reaching the interdentate space. If used for weaving, this could reveal why the tines are generally short. If used for fibre processing, estimates of time per volume of processed material could be produced. The distance between tines might indicate the thickness of warp threads, if one warp thread passes the interdentate space, though multiple threads could pass between tines as well, as seen in Fig. 9.2. This measurement might also reveal possible techniques for combing fibre. The interdentate space is characterized as being U- or V-shaped. This might relate to the method of manufacture and modification through use and is linked to the interdentate distance.

Wear is a difficult parameter to quantify, but two types of wear patterns exist between tines: smoothed or striated. Completely smooth tines could indicate the comb was never used or was worn smooth by a material like wool; equally, striations may result from manufacture or use. Identifying the source for surface polish is difficult to isolate. Polish has generally been understood as use-wear, but a polished surface can be achieved in a variety of ways. For example, a smoothly polished comb may have a matte surface and achieve a shinier surface from the user holstering the comb into a belt repeatedly over the course of its active life. In this case, the comb has developed a polish beyond a manufacturing trend, but the act of holstering the comb against the body does not constitute utilitarian wear polish. This distinction may be difficult to discern on individual objects but noting its presence on the comb is relevant. Looking at an assemblage of combs has the potential to reveal a pattern of how the tool was manipulated through the cross-comparison of polish. Together, the study of these new attributes allows for a more refined understanding of the utilitarian function of long-handled combs.

9.8 Results of the Rubric

For simplicity, the data has been collated into a table for discussion (Table 9.2). The number of combs deemed appropriate for this analysis, 23, allowed for a thorough investigation with an object biography approach. This methodological approach generates enough data to reveal hidden technological and use-wear patterns that are

Table 9.2 Data generated from the comb rubric

Attribute category	Results of rubric
Length of tine	Partial data from 23 combs. A total of 184 intact tines combined (7-27 mm for complete tines).
Interdentate distance	Partial data from 22 combs: 'narrow' and 'wide' spaces were noted. 8 combs (from 0.3–0.9 mm); 7 combs (1.0 mm–1.4 mm.) 5 combs spanned both ranges. 1 outlier (2.1 mm–3.4 mm).
Interdentate space	Partial data from 23 combs U-shape: 73 instances U*-shape: (probable) 22 instances V-shape: 18 instances V*-shape: (probable) 1 instance Only 2 combs exhibited the V-shape interdentate space exclusively
Use-wear	Partial data from 20 combs 13 had 'smooth' surfaces between the tines 3 combs had 'striated' surfaces only 4 combs had both types of patterns on the same comb The striations of 5 combs were characteristized as 'angled towards handle', one comb noted as 'severe'. Another comb had perpendicular striations
Surface polish	Partial data from 18 combs 6 combs had areas of 'glossy' shine, on the tines or sides of the handle. 1 comb was glossy only on the handle 2 combs were glossy only on the tines 3 combs had glossy areas on both the handle and tines

The attribute categories are in the left column and the data are in the right column. A full version of this table is included in the appendices of author's PhD thesis usually obscured by the basic parameters of mass and overall length. Statistically, these numbers are too small to make significant conclusions about functionality, but the rubric has shown its viability as an analytical tool.

There are limitations of an object biography approach. The fragmentary state of combs meant that data cannot be reliably gleaned from each comb. Some categories, like interdentate space shape. Generated more information per comb than others. Assessing the 23 combs as an assemblage, however, interesting details about manufacturing trends and use-wear could emerge. Therefore, the attribute categories can generate useful data that clarify the range of technological variation and the presence of use-wear, while also foregrounding life histories that convey trajectories of specific combs against the assemblage. These attributes are discussed more thoroughly in the next section.

9.9 Analysis and Discussion

Five new categories of metrics were proposed to assess features of long-handled combs for characterizing their functional attributes. The clarity and conciseness of this approach standardizes the way comb functionality is described so a larger assemblage can be assessed. By converting subjective descriptions into a codified system of numerical and short text, new technological and use-wear patterns were observed. Rather than creating an elaborate typological system based on decoration or style, a simple rubric was devised to capture data that is relevant to utilitarian function. In doing so, the differences between manufacturing trends and use-wear were possible, and the analysis was not restricted to discussions of weaving or fibre processing. The following subsections outline which attribute categories revealed manufacturing trends and use-wear. Through this study, a new attribute category is proposed.

9.9.1 *Manufacture*

Evidence of manufacture trends were observed in the frequency of two attribute categories, interdentate space shapes and interdentate distances. The U-shape (83.3%) was predominant in this assemblage and it is likely that one type of tool was used to create most of the combs recovered from Danebury. Figure 9.3 shows the consistency of tine cutting in SF1329. Five saws were recovered from Danebury (Cunliffe 1984: 351; Cunliffe 1991: 342) and it is feasible that saws are responsible for the consistency of this metric. The replica comb in Fig. 9.1 demonstrates a U-shape appearance and was constructed by one saw. Otherwise, this U-shape would likely be more variable within each specimen and this was not the case. Thinner saws than those which were recovered could be responsible for V-shaped (16.6%) interdentate spaces, although there is insufficient evidence to confirm this

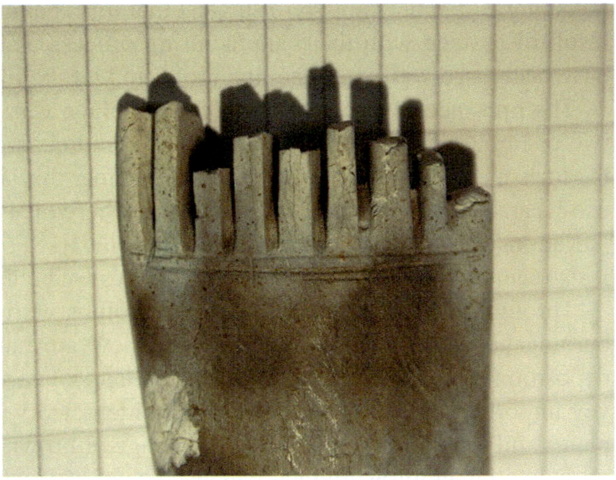

Fig. 9.3 Though comb SF1329 has a stylized interdentate space shape, it is considered U-shaped because of the flattened bottom. Scale is 5 mm. (Photo by J. Beamer)

claim. Knives would also be suitable for creating tines, but most combs have consistent interdentate distances and this aligns well with saws being the likely cutting tool used.

Interdentate distance was the second manufacturing trend. Though these results are more mixed, each specimen had consistently spaced tines. One cutting tool was used to form the tines for each specimen. The selection of the cutting tool was probably based on the width of the comb blank and how the comb was intended to be used. In this case, it probably mattered to its functionality that each tine was spaced evenly across the dentate. This metric also links with interdentate space shape, where the width of the cutting tool influenced whether the finished comb had a U- or V-shaped space.

9.9.2 Use-Wear

Use-wear was the most difficult pattern to discern because there are no known reference collections of long-handled combs by which to compare these visual features. Documenting the presence, location, and intensity of these features, however, are useful when trying to identify the source. Smoothing, for example, can arise as the polishing step of the manufacturing process and while it is being used. Each comb was subjected to a polishing phase, but it is unknown how long it takes for use-wear polish to develop. Additionally, use-wear polish may develop because of how it was worn on the body, not how it was used as a tool. Lastly, use-wear can be altered or obliterated by burning. Despite these caveats, some combs did reveal polished use-wear beyond manufacturing polish.

Comb SF679 was well-polished and the surface was highly reflective. This type of polish was rare in this assemblage and may be an outlier. Two other combs, SF1225 and SF1387, have polish on the tips of the tines. In these three cases, the polishing is distinct enough to suggest that this is evidence of use-wear and not the result of a stage with the manufacturing process. Additionally, SF679 appears to have been used well after one tine broke and left behind a saddle-shaped stub.

The presence of striations in this assemblage was sparse and mainly present in two examples, SF179 and SF540. These striations are cut deeply into the sides of the tines and it could not be determined whether this showed manufacture effects or use-wear. The SF179 comb only had stubs where the tines used to be, so relating the 'striations' to use-wear was impossible. Similarly, SF540 and deep striations though it also seemed like this comb was in the process of being recut when it was discarded.

Through the process of analysing the combs with the rubric, discoloration and burning became possible candidates for new attribute categories. Pre- and post-depositional processes can alter the appearance of combs and both aspects can convey important details about how people interacted with combs. Tannin-containing plants and soil pH can influence how orange or brown the bone might be, and proximity to other cultural artefacts can leach chemicals into surrounding soils. As stated earlier, burning can similarly alter the appearance and intensity of use-wear. These are proposed as new attribute categories because the results of these categories may impact the efficacy of the rubric.

9.9.3 New Attribute Category: Discoloration

The stratigraphic context where combs were deposited may modify the appearance of the combs and impact the quality of the analysis. Two combs, SF260 and SF2702, were stained green from nearby copper alloys leaching into the soils (Dupras and Schultz 2013: 327). Additionally, both combs were burned. These two types of discoloration, staining and burning, expand the rubric and elevate the interpretive potential of combs, especially when burning is an irregular occurrence. Reviewing Tuohy's (1999) comb description again, only 39 of 774 combs were noted as having been burned and 20 of those came from Danebury hillfort. Whether these are anthropogenically-caused or natural will provide additional insight into patterns of behaviour associated with the way combs were treated and could state important details about the surrounding environment. A colorimetic analysis to examine discoloration against a standard metric is also useful for a cross-comparison (Bradfield 2018: 503).

9.10 Interpretation

Morphologically, all combs in this study could have been appropriate for either weaving for fibre processing, and the majority might have been used for either activity. They might have been used for non-textile production activities as well. However, a few combs in the study exhibited characteristics that reveal the possibility of a specific intended use. The consistently spaced tines suggest that they would have been most useful as weft beaters since distortions in the warp are less likely to occur. A specific set of combs, SF1225 and SF1387, had very narrow interdentate spaces and might indicate being used to beat thin warp threads. Likewise, all combs had a minimum degree of polish to make their surfaces smooth. Comb SF679, however, had significant polish over the stub of the broken tine and the wide interdentate spaces is more consistent with fibre processing—wool, in all likelihood. This rubric has greatly enhanced our understand of long-handled comb functionality and improve our skills of deduction. The evidence collected thus far no longer supports Tuohy's interpretation that tapestry looms are evidenced through the presence of combs because they could have been used for a variety of activities. Furthermore, this new characterization of use-wear on combs bolsters the suggestion by DeRoche (1995: 75–76) that combs could be used for either activity.

9.11 Conclusion

The results demonstrate how the rubric can reveal hidden patterns of manufacture and use-wear that a stylistic typology cannot show. Decoration, usually the primary metric featured in these typologies, does not appear to impact functionality (Fig. 9.4). The study of long-handled combs has had a long and varied history and the ideas and critiques presented by previous scholars have fuelled the development of this rubric. Accounting for these ideas and critiques, attribute categories were developed for the rubric that would explore the possibility of two functions, weaving and fibre processing. As a tool, the rubric has demonstrated its efficacy analytically and the discussion of discoloration as a category shows its flexibility in creating new metrics. Creating a reference collection that establishes different types of manufacturing trends, use-wear development, longevity, and the circumstances of discoloration would be exceedingly useful to osseous studies in general.

There was compelling evidence from 23 combs which showed trends of manufacture, and the evidence from use-wear returned unexpected results. By standardizing these attribute categories and assigning them to numerical and short text characteristics, the results can be compared computationally. Select attribute categories relate together, like the interdentate space shape and interdentate distance, adding depth to a comparative analytical approach. Looking at individual attribute categories helps us understand technological variability and looking at them collectively help us understand the potential differences in the ways individual combs may have been used.

As a pilot study, this rubric has increased the interpretive value of long-handled combs and refocused attention on their utilitarian functions. The 'weaving' attribute potentially restricted the way previous scholars considered their functional purpose and it is evident now why 'long-handled combs' was adopted for this chapter. There are several avenues left to pursue that may yield answers to difficult questions, including theories of removing fat or fur from hides. A use-wear reference

Fig. 9.4 Comb SF161 has been decorated, recut, and burned, which has obliterated all traces of use-wear and distorted the decoration. Scale is 5 mm. (Photo by J. Beamer)

collection is a daunting project though experimentation may be the only way to resolve some of the persistent questions. Such a project has great value when considering that bone and antler are very long-lived materials. Recutting combs increases their longevity and may have been imbued with many life histories. In these ways, a functional approach to the use of long-handled combs can reveal facets of their social importance that a stylistic typological study cannot do alone. Furthermore, the circumstances that combs undergo in pre- and post-depositional processes bear on the way a functionalistic study operates. Adding in new attribute categories, like discoloration, will make the rubric established thus far a much stronger tool.

This study coincides with the development of specialist studies in Iron Age research (Haselgrove et al. 2001: 22). Scholarship on British Iron Age textile studies are few and the narrative of textile production at Danebury is far from complete. This comb recording rubric encourages the development of new scientific approaches through experimentation and foregrounded the importance of archived material. The narrative for textile production at Danebury hillfort points towards users who probably used their combs for multiple purposes, but the evidence for specialized combs within the textile production sequence hints that more might be discovered with additional research.

Acknowledgements I would like to thank Helen Chittock and two anonymous reviewers for commenting on an earlier draft of this chapter. All errors remain my own. This research formed part of my doctoral thesis project which has re-evaluated textile tools and their depositional contexts within Iron Age Britain.

References

Anderson J (1871) Notes on the evidence of spinning and weaving in the brochs or pictish towers supplied by the stone whorls and the long-handled 'broch combs' found in them. *Proceedings of the Society of Antiquaries of Scotland* 9:548–561.

Anderson J (1883) *Scotland in pagan times: The Iron Age,* Vol. 1. David Douglas, Edinburgh.

Bailey R (1999) Those weaving combs – yet again. *Archaeological Textiles Newsletter* 28:5–9.

Bradfield J (2018) Some thoughts on bone artefact discoloration at archaeological sites. *Journal of Archaeological Science: Reports* 17:500–509.

Bulleid A, Gray HSG (1911) *The Glastonbury lake village: A full description of the excavations and the relics discovered, 1892–1907*, Vol. 1. Glastonbury Antiquarian Society, Somerset.

Bulleid A, Gray HSG, Munro R, Dawkins WB, Jackson JW, Andrews CW, Reid C (1917) *The Glastonbury Lake Village*, Vol. II. Glastonbury Antiquarian Society, Somerset.

Bulleid A, Gray HSG (1948) *The Meare Lake Village. A full description of the excavations and the relics from the eastern half of the West Village, 1910–1933,* Vol. 1. Taunton Castle, Somerset.

Chittock H (2014) Arts and crafts in Iron Age Britain: Reconsidering the aesthetic effects of weaving combs. *Oxford Journal of Archaeology* 33(3):313–326.

Coles J, Minnitt S (1995) *'Industrious and Fairly Civilized': The Glastonbury Lake village.* Somerset Levels Project and Somerset County Council Museums Service, Somerset.

Coughtrey M (1871) Notes on materials found in a kitchen midden at Hillswick, Shetland, with special reference to long-handled combs. *Proceedings of the Society of Antiquaries of Scotland* 9:118–151.

Crowfoot G (1945) The bone 'gouges' of Maiden Castle and other sites. *Antiquity* 19(75):157–158.

Cunliffe B (1984) *Danebury: An Iron Age hillfort in Hampshire. Vol. 2: The finds.* Council for British Archaeology, London.

Cunliffe B (1991) *Danebury: An Iron Age hillfort in Hampshire. Vol. 4: The site.* Council for British Archaeology, London.

Cunliffe B (1995) *Danebury Iron Age hillfort in Hampshire. Vol. 6: A hillfort community in perspective.* CBA Research Report 102. Council for British Archaeology, London.

Cunnington ME (1923) *The Early Iron Age inhabited site at All Cannings Cross farm Wiltshire: A description of the excavations, and objects found, by Mr. and Mrs. BH Cunnington, 1911–1922.* George Simpson & Company, Devizes.

Davis O (2013) Re-interpreting the Danebury assemblage: houses, households, and community. *Proceedings of the Prehistoric Society* 79:353–375. doi:https://doi.org/10.1017/ppr.2013.16.

DeRoche CD (1995) *Textile production in Britain during the first millennium B.C.* Unpublished dissertation, University of Cambridge.

DeRoche D (2012) England: Bronze and Iron Ages. In: Gleba M, Mannering U (eds) *Textiles and textile production in Europe from Prehistory to AD 400.* Ancient Textiles Series 11. Oxbow Books, Oxford and Oakville, p 444–450.

Dupras T, Schultz J (2013) Taphonomic bone staining and color changes in forensic contexts. In: Pokines J, Symes SA (eds) *Manual of forensic taphonomy.* Taylor and Francis Group, Boca Raton, p 315–340.

Fitzpatrick AP (2020) A face from the British Iron Age: A decorated Iron Age weaving comb from Harwell, Oxfordshire. *Oxford Journal of Archaeology* 39(2):177–188.
Fox AL (1881) XXI. – Excavations at Mount Caburn Camp, near Lewes, conducted in September and October, 1877, and July, 1878. *Archaeologia* 46(2):423–495.
Gleba M, Harris S (2019) The first plant bast fibre technology: Identifying splicing in archaeological textiles. *Archaeological and Anthropological Sciences* 11(5):2329–2346.
Griffitts J (2001) Bone tools from Los Pozos. In: Choyke AM, Bartosiewicz L (eds) *Crafting bone – skeletal technologies through time and space: Proceedings from the 2nd meeting of the (ICAZ) worked bone research group.* BAR 937. Archaeopress, Oxford, p 185–196.
Grömer K (2016) *The art of prehistoric textile making.* Veröffentlichungen der Prähistorischen Abteilung 5. Natural History Museum Vienna, Vienna.
Hallén Y, O'Neill M (1994) The use of bone and antler at Foshigarry and Bac Mhic Connain, two Iron Age sites on North Uist, Western Isles. *Proceedings of the Society of Antiquaries of Scotland* 124:189–231
Haselgrove C, Armit I, Champion TC, Creighton J, Gwilt A, Hill JD, Hunter F, Woodward A (2001) *Understanding the British Iron Age: An agenda for action.* Trust for Wessex Archaeology, Salisbury.
Hedges JW (1973) *Textiles and textile production in Prehistoric Britain.* Unpublished MA thesis, University of Sheffield.
Henshall AS (1950) Textiles and weaving appliances in Prehistoric Britain. *Proceedings of the Prehistoric Society* 16:130–162.
Hodder I and Hedges JW (1977) Weaving combs: Their typology and distribution with some introductory remarks on date and function. In: Collis J (ed.) *The Iron Age in Britain – a review.* Sheffield University Press, Sheffield, p 17–28.
Hodson F (1964) Cultural Grouping within the British pre-Roman Iron Age. *Proceedings of the Prehistoric Society* 30, 99–110.
Jones AM, Diaz-Guardamino M, Crellin RJ (2016) From artefact biographies to 'multiple objects': A new analysis of the decorated plaques of the Irish Sea region. *Norwegian Archaeological Review* 49(2):113–133.
Joy J (2015) 'Things in process': Biographies of British Iron Age pits. In: Hahn HP, Jung M, Kienlin T, Kreuz PA, Hofmann KP, Joy J, Wittekind S, Niehaus M, Boschung D (eds) *Biography of objects.* Wilhelm Fink Verlag, Paderborn, p 125–141.
Joy J (2011) Fancy objects in the British Iron Age: Why decorate? *Proceedings of the Prehistoric Society* 77:205–229.
Keller F (1866) *The lake dwellings of Switzerland and other parts of Europe.* Longmans, Green & Co., London.
Ling Roth H (1918) Studies in primitive looms. *The Journal of the Royal Anthropological Institute of Great Britain and Ireland* 47:113–150.
Lyman RL, Lyman C (1994) *Vertebrate taphonomy.* Cambridge University Press, Cambridge.
MacGregor A (1985) *Bone, antler, ivory, and horn: The technology of skeletal materials since the Roman period.* Croom Helm, London and Sydney.
Marreiros J, Mazzucco N, Gibaja JF, Bicho N (2015) Macro and micro evidence from the past: The state of the art of archeological use-wear studies. In: Marreiros J, Mazzucco N, Gibaja JF, Bicho N (eds) *Use-wear and residue analysis in archaeology.* Springer, Cham, p 5–26.
Reynolds PJ (1972) Experimental archaeology. *Worcestershire Archaeological Newsletter* 9(7):1–18.
Roes A (1963) *Bone and antler objects from the Frisian terp-mounds.* Tjeenk Willink HD and Zoon NV, Haarlem.
Rösel-Mautendorfer H, Grömer K, Kania K (2012) Farbige Bänder aus dem prähistorischen Bergwerk von Hallstatt. Experimente zur Herstellung von Repliken, Schwerpunkt Faseraufbereitung und Spinnen. In: Weller U, Lessig-Weller T, Hanning E, Strugalla-Voltz B (eds) *Experimentelle Archäologie in Europa – Bilanz 2012.* Europäische Vereinigung zur Förderung der Experimentellen Archäologie, Bad Langensalza, p 190–201.

Ryder ML (1997) Fleece types and Iron Age wool textiles. *Archaeological Textiles Newsletter* 25:13–16.
Sellwood L (1984) Objects of bone and antler. In: Cunliffe B (ed.) *Danebury: An Iron Age hillfort in Hampshire. Vol. 2: The finds.* Council for British Archaeology, London, p 371–378.
Serjeanston D (2007) Intensification of animal husbandry in the Late Bronze Age? The contribution of sheep and pigs. In: Haselgrove C, Pope R (eds) *The Earlier Iron Age in Britain and the near Continent.* Oxbow Books, Oxford, p 80–93.
Smith RA (ed.) (1925) *Guide to the Early Iron Age antiquities.* British Museum, London.
Stolpe MH (1874) Sur la découvertes faites dans l'Ile de Björkö. *Congrès internationale d'anthropologie et d'archaéologie préhistoriques, Session 7,* Vol. 2. Stockholm, p 619–640.
Tuohy C (1992) Long-handled 'weaving combs' in the Netherlands. *Proceedings of the Prehistoric Society* 58: 385–387.
Tuohy C (1995) *Prehistoric combs of antler and bone.* Unpublished dissertation, University of Exeter.
Tuohy T (1999) *Prehistoric combs of antler and bone.* BAR 285. Hadrian Books Ltd, Oxford.
Tuohy T (2004) Weaving as a domestic craft at the Iron Age site of Glastonbury Lake Village in Somerset, Britain. *Journal of Wetland Archaeology* 4(1):97–109.
Wheeler REM (1943) *Maiden Castle, Dorset.* Reports of the Research Committee of the Society of Antiquaries of London Number 12. Oxford University Press, Oxford.

Jennifer Beamer has received her PhD from the University of Leicester in 2021, for the thesis project entitled "Re-evaluating textile production and tool deposition in Iron Age Britain". The aim of this project is to consider textile tools within their production sequences as assemblages, and to use aspects of object biography to discern utilitarian traces. Additionally, she uses experimental archaeology to facilitate critical thinking within this research project.

Chapter 10
From Use-Wear to User: Working with Literary Sources on Worn Textile Tools

Magdalena Öhrman

Abstract Focussed on Greek and Roman textile production, this paper argues that attentive reading of textual sources effectively complements our understanding of user-tool relationships in antiquity, when informed by archaeological and experimental investigations alongside the text's literary context. While the class and gender bias of ancient literary sources place them at one remove from many production contexts and literary aesthetics determine their expression and content (often at the expense of information about craft processes, raw materials, etc.), these texts originate in a historical context where textile production could be experienced as part of everyday life. Thus, literary texts provide a valuable bystander's view on Graeco-Roman textile crafts and the information they provide can be fruitfully employed in textile research. This does, however, require a fully interdisciplinary approach, including not only understanding of archaeological and craft-historical evidence but also of literary and stylistic aspects.

Through case studies centred on Greek Hellenistic epigrams featuring workers involved in textile production (*Anthologia Palatina* 6.247; 7.726 and 6.47; 285; 288) the paper tests this hypothesis by examining the extent to which user attitudes displayed in literary texts correlate with user attitudes to tools suggested by archaeological evidence, such as use-wear and tool inscriptions, and experimental archaeological evidence. Through close readings, the paper demonstrates that literary tool users assess and relate to their tools in terms of functionality and efficiency in securing a desired output, as well as economic dependency, even if their depiction is shaped also by other contemporary cultural concerns.

Keywords Greek epigram · Focalisation · Craft process · Agency · Spinning · Use-wear · Textile tools · Votive gifts · Leonidas of Tarent

---

M. Öhrman (✉)
Faculty of Humanities, University of Wales, Trinity Saint David, Lampeter, UK
e-mail: m.ohrman@uwtsd.ac.uk

A. Ulanowska et al. (eds.), *Ancient Textile Production from an Interdisciplinary Perspective*, Interdisciplinary Contributions to Archaeology,
https://doi.org/10.1007/978-3-030-92170-5_10

10.1 Introduction

Ancient literary texts are several steps removed from historical craft practice. Class and gender bias mean that these texts were rarely written primarily by or for people involved in textile production. We are rightly cautious of their ability to convey both factual information on production conditions and of their interest in accurately representing the perspectives or concerns of ancient craftspeople (cf. Wild 2000: 209–211; Gleba 2008: 23–24; Andersson Strand 2018: 20; cf. Bijker 2010: 68). However, when we take the image of textile production in literary texts for what it is – a bystander's view, one shaped equally by historical practice and by literary aesthetics – we gain valuable insights in how textile crafts were contextualised in the Graeco-Roman world.[1]

Careful interpretation of literary texts reveals that textile crafts constituted a frame of reference relatable to all layers in society, as has often been affirmed in relation to the widespread use of textile metaphors in literary texts (esp. Fanfani and Harlizius-Klück 2016; also Harries 1990; Rosati 1999). However, readings of literary texts dealing with craft production, informed by an archaeological perspective that takes into account both experimental and scientific-analytical advances in textile research of recent decades, also indicate that texts preserve a range of information about different production settings and experiences of craftspeople involved. This information is not necessarily overtly stated but conveyed by the stylistic means appropriate to each text. In previous papers, my main concern has been the sensory experience of textile work and its representation in poetry through sound-play and rhythm, which draw on the soundscape and sensory environment of crafts like spinning and weaving in the ancient world (e.g. Öhrman 2017, 2018, 2020).

Here, however, my interest is in a series of interlinked Greek and Roman poetic texts that address the relationships between textile tools and their users. Chronologically, the texts considered range from the Hellenistic period to Late Antiquity: a span of nearly six centuries. My main focus is on two types of references to textile tools in Greek epigram: first, epigrams where the reader is made to see with the craftsperson's eyes and share their attitude to their tools (*Anthologia Palatina* 6.247; 7.726), and, secondly, epigrams where comments on tools are presented as voiced by the tool user themselves (*A.P.* 6.285; 47; 288). Throughout, I examine the extent to which user attitudes displayed in literary texts correlate with user attitudes to tools suggested by archaeological evidence, such as use-wear and tool inscriptions, and experimental archaeological evidence.

In my discussion of the work elements of spinning, I draw on observation of experiments undertaken at the Centre for Textile Research (CTR) in 2018 (within

[1] Translations from Greek and Latin are my own unless otherwise noted. Abbreviations of titles and authors of Classical texts follow the Liddell and Scott *Greek-English Lexicon* and *Thesaurus Linguae Latinae* respectively.

the framework of Eva Andersson Strand's project Practises of Practice) and on numerous discussions with experienced spinners, in particular Ida Demant (Land of Legends at Lejre) and Barbara Köstner (Xanten Archaeological Park). As ever, I am grateful for their patience and willingness to share their expertise.

10.2 Literary Textile Tools: Skill and Functionality

It must first be noted that it is rare for ancient literary texts, whether poetry or prose, to allow textile practitioners any semblance of commentary on how they feel about their tools or their work. Instead, most literary texts mention textile tools and how they are handled primarily to indicate the craftsperson's skill: the opening of Ovid's canonical tale of the weaving contest between Arachne and Minerva in the *Metamorphoses* illustrates this.

> haud mora, constituunt diuersis partibus ambae 4.53
> et gracili geminas intendunt stamine telas;
> tela iugo uincta est, stamen secernit harundo, 4.55
> inseritur medium radiis subtemen acutis,
> quod digiti expediunt, atque inter stamina ductum
> percusso pauiunt insecti pectine dentes.
> utraque festinant cinctaeque ad pectora uestes
> bracchia docta mouent, studio fallente laborem. 4.60 (Ov. *Met*. 4.53–60)

> Without delay, they both assemble the various parts and stretch fine warp on twin looms: the web is tied to the cloth-beam, the rod divides the warp, the weft is inserted in between with sharp pin beaters, which their fingers guide, and the cut teeth beat the weft thread, pulled through [the warp] with the thrust-through pin beater. They both hurry and, with their clothes tied to their waist, move their clever arms, their eagerness belying their effort.

Verbal indications of haste (*haud mora* 'without delay' in Ov. *Met*. 6.53; *festinant* 'they hurry' in 6.59) together with an initially paratactic sentence structure (*constituunt* [...] *et* [...] *intendunt* 'they set up [...] and [...] stretch [the warp]' in 6.53–54) generates a sense of rapid and competent handling of weaving equipment.[2] At the opening and the close of the description, the two weavers are the grammatical subjects of each clause: the weaver's activity is clearly foregrounded. In the middle section, however, the importance of their tools is brought out: here, the tools themselves function as subjects of the verbs *uincta est* and *secernit* ('is bound', 'divides', both in 6.55), *inseritur* ('is inserted', 6.56), and *pauuiunt* ('beat', 6.58). The passive voice of 'is bound' and 'is inserted' implies the weaver's necessary involvement but allows the reader's focus to remain on the work accomplished by their use of tools. Hinting at the productive symbiosis of weaver and tool, the workers' ability to react

[2] The tri-partite description of their handling of loom parts and tools in Ov. *Met*. 6.54–56 has a similar paratactic structure centred on the verbs *uincta est* ('is bound', 6.55), *secernit* ('divides', 6.55), *inseritur* ('is inserted', 6.56).

to sensory information relayed by their tools about the progress of their work (Fregni 2019: 40–41) is reflected by the very structure of the text.

Other texts occasionally use tools as a stand-in for the workers themselves, projecting onto the tools such skills and character traits that the workers demonstrate in relation to their craft (*ambitiosa colus* 'ambitious distaff', Claud. 8.594 or *sollicitas telas* 'busy loom', Claud. *Carm. Min.* 46.15). Workers' bodies are sometimes treated in the same fashion: Minerva and Arachne both have 'skilled arms' (*bracchia docta*, Ov. *Met.* 6.60) and texts on spinning frequently mention the spinner's trained fingers or light-handedness.[3] In such snapshots of user/tool engagements, there is little to indicate any emotional connection between the two beyond the desire for efficient operation of equipment.

This literary emphasis on functionality in user/tool engagements can be related to archaeological evidence that demonstrates the requirement to meet strategic production tasks as guiding a craftsperson's interactions with their tools. This requirement manifests in the archaeological record in a number of ways: textile researchers will be particularly familiar with functionality as a driver in the composition of tool kits, reflected in artefact distribution through groups of tools with differing functional parameters, e.g. weight or size (e.g. Andersson Strand 2003: 50–59; Andersson Strand and Nosch 2015 for key case studies). Adaptations to tools (or traces of tool maintenance) similarly reflect the immediate engagement by the craftsperson with their tools with the purpose of ensuring that production needs are met (Fregni 2019: 40–45; cf. Murphy and Poblome 2013: 204–205). In a recent paper on spindle whorls at Vindolanda and Corbridge, Marta Alberti (2018: 9–10) discusses a spindle whorl with a graffito instruction as to the up- and downside of the whorl. The graffito may have been added by a user to improve functionality and ensure standardised production, reflecting wider patterns of decoration reinforcing craft practice.[4]

The prevalence of functionality as determining the usefulness of a tool (or the skill of its user) even in literary texts that speak primarily to a male, educated elite, is significant. It suggests that, despite their likely lack of direct involvement in production, this group, too, retained a clear awareness of the need for efficient use of resources in textile production both in terms of raw materials and in terms of labour and know-how. Texts that at first glance seem to hold relatively little information on working conditions or attitudes to and assessments of textile work may reveal more if studied together than when individually mined for information.

[3] E.g. Ov. *Met.* 4.36; Stat. *Theb.* 11.401; Tert. *Psych.* 364; Claud. 1.177.

[4] Cf. also Sofianou 2012: 80–82; Marchesini and Migliavacca 2018: 236–238 and Kleibrink 2018: 170–172 on surface decoration on loom weights as guiding the weaver in set-up and pattern replication. Conversely, many pre-firing marks on loom weights are connected to production rather than use, cf. F Beltrán Lloris and M Beltrán Lloris 2012: 136–139 with further bibliography. Gleba 2008: 137–138 highlights the range of different functions of decoration on loom weights.

10.3 Imbued with Meaning: User/Tool Engagements in Context

The relationship of craftspeople with tools and technology at their disposal is, of course, not shaped by functionality alone. There is a wide consensus in archaeological scholarship that user/tool engagements in ancient societies were also influenced by a range of other factors, including wider social and cultural reproduction processes (e.g. Dobres and Hoffmann 1994: 222–235; Ingold 2000; Dobres 2010; Jordan 2015; Lee 2019: 21–26). Indeed, Ovid's weaving contest between Minerva and Arachne illustrates how such broader influences on user/tool engagements are mirrored in our literary sources: the contest is initially motivated by the unwillingness of Arachne, the human, to acknowledge any indebtedness to Minerva, the divine teacher of textile arts. This mythological tale is therefore concerned with the negotiation of social learning (on the concept, Jordan 2015: 20–27) and the tensions surrounding technological development (e.g. Bender Joergensen 2012: 243–248; Wendrich 2012).

While mostly difficult to trace, such overarching influences and their effect on individual craftspeople and their tools do occasionally manifest themselves even in the archaeological record. In workshop settings with multiple workstations of similar type, inscription of names on individual tools suggests strategies for demarcation of each worker's individual space and equipment, but it also hints at potentially longstanding relationships between individual tool users and their tools (Murphy and Poblome 2013: 206–207, cf. Mullen (forthcoming) on spindle whorls with inscribed names in bilingual contexts). In the literary evidence, a small set of texts indicate that long-term use bestowed meaning (beyond the utilitarian and task-specific) on tools to their users, and that this was perceptible to people in their proximity. Two texts connect this to use-wear traces as indicative of long use.

10.3.1 Worn Tools in Claudian's The Rape of Prosperpine

I focus first on a brief mention of tools in Claudian's Late Antique mythological epic on the rape of Proserpine. After Proserpine's capture by Pluto, her mother Ceres searches for her in their empty home, finding only her daughter's abandoned loom:

> [...] desolata pererrat 3.154
> atria, semirutas confuso stamine telas
> atque interceptas agnoscit pectinis artes. 3.156
> [...]
> adtritosque manu radios proiectaque pensa 3.161
> cunctaque uirgineo sparsa oblectamina ludo.
> ceu natam pressat gremio; [...] 3.163 (Claud. *Rapt. Pros.* 3.154–163)

> ...wandering through deserted halls, she recognised the half-ruined warp with its disordered threads and the interrupted work of the pin beater [...] she clasps to her bosom, as if it were her child, the hand-worn pin beaters, the cast aside wool, all the toys scattered in childish play.

To Ceres, the abandoned tools represent her daughter's talent and promise: they evoke an emotional response by being hand-worn (*adtritosque manu radios*, 3.161) and by being scattered as if in play (*uirgineo sparsa oblectamina ludo*, 3.162). The emphasis on wear highlights the extent of Proserpine's long hours of practice at the loom, but *oblectamina* ('toys' or, more literally, 'delights', 3.162) suggests that she perceived this as a source of joy (Öhrman 2018: 97–98). To Ceres, however, the use-wear on the scattered *radii* seems mainly to establish that they belonged to Proserpine – the emphasis is on *her hand* rather than on *worn* – and that she used them in her charmingly childish way, so that Ceres' viewing and handling of them in turn brings her closer to her missing daughter (Gruzelier 1993: 258).

Claudian's reference to Proserpine's handling of her tools has intertextual dimensions unnoted by Ceres. The use-wear traces on Proserpine's pin beaters connect Claudian's passage directly to one other literary text (*A.P.* 6.247) and indirectly to a wider group of Greek epigrams mentioning textile tools. Attitudes to tools displayed in these intertexts are, as we shall see, at odds with Ceres' perception of Proserpine as a vulnerable child, instead casting her as someone who willingly abandons her loom.

10.3.2 Worn Tools and Old Age in A.P. 6.247: Aesinoe the Spinner

The designation of Proserpine's tools as hand-worn has only one immediate parallel,[5] a first or second century CE fictional, dedicatory epigram in Greek but written for a Roman audience and ascribed to Philip (Gow and Page 1968). It lists textile tools dedicated to Athena, which are initially given positive associations:

> κερκίδας ὀρθρολάλοισι χελιδόσιν εἰκελοφώνους,
> Παλλάδος ἱστοπόνου λειομίτους κάμακας,
> καὶ κτένα κοσμοκόμην, καὶ δακτυλότριπτον ἄτρακτον
> σφονδυλοδινήτῳ νήματι νηχόμενον,
> καὶ τάλαρον σχοίνοις ὑφασμένον, ὅν ποτ' ὀδόντι
> ἐπλήρου τολύπη πᾶσα καθαιρομένη,
> σοί, φιλέριθε κόρη Παλλαντιάς, ἡ βαθυγήρως
> Αἰσιόνη, πενίης δῶρον, ἀνεκρέμασεν. (*A.P.* 6.247)

> The pin beater that sings like the early-chattering swallows, the thread-smoothing shaft of Pallas the weaver, the comb that well orders the wool, the hand-worn spindle, swimming with the twirling thread, and the basket of reeds (which every ball of wool she cleaned with

---
[5] By contrast, references to wear traces in dedicated clothing is comparatively frequent, Petsalis-Diomidis 2018: 424–427, esp. 472 n. 33.

> her teeth used to fill) to you, Pallatian maid, who loves the loom, did Aesione, now bowed with age, suspend as the gift of her poverty.

Speaking of similar tools-focused epigrams collected in the sixth book of the *Anthologia Palatina*, Joseph Day suggests that epigrams opening in this fashion make the reader re-enact the initial viewing of dedicated tools and the subsequent reading of the inscription that accompanies them (Day 2019: 25–32). Each epigram thus carries two implicit narratives: that of the reader/viewer approaching the dedicated/described objects, and the backstory of the dedicated objects that they draw from the text.

In Philip's epigram, the tension between the perspective by the observer of the dedicated tool, and that of the original tool user, is developed as the reader progresses through the poem. In the first half of the epigram, each tool is given a brief description, moving from the κερκίς ('pin beater', 6.247.1), to the wool comb (κτένα), to the spindle (ἄτρακτον). The pin beater (κερκίς) sings like morning birds, Pallas' shaft smooths the threads softly (λειομίτους κάμακας, 2),[6] and the comb orders the fibres *well* (κοσμοκόμην, 3, on which Gow and Page 1968: 345). So far, the attributes given to the tools evoke a pleasant, well-working set-up.

However, the text gradually brings the tool user into view by establishing a relationship between the tools and their user: the spindle is δακτυλότριπτον ('hand-worn', 3), positing long-term handling,[7] and the wool thread in the basket has been cleaned by the spinner's teeth (5–6). The reference is to the spinner's regular inspection of just-spun thread and her biting off any unevenness in it (Barber 1992: 70; Spantidaki 2016: 36–37, misunderstood in Dübner 1864–1872: 254; Gow and Page 1968: 345).

It is in the second distich that the spinner becomes the focaliser: with the first use of a present-tense verb form (the participle νηχόμενον, 'swimming', 4), the perspective shifts from that of visitor to a sanctuary viewing dedicated tools. Instead, the reader is drawn into a scene of ongoing work, and into the spinner's own perspective. The description of the spindle as 'swimming' has caused consternation among literary commentators (Gow and Page 1968: 345; Paton and Tueller 2014: 431), but reflects the spinner's point of view, looking down at a drop spindle, hanging loosely and rotating after being twisted.[8] The reader is made to adopt Aesinoe's own perspective on her work.

This emerging focus on Aesinoe's spinning influences the interpretation of the following: unlike Gow and Page (1968: 345), I do not see the next two lines as referring to wool-preparation by combing, but rather continuing the depiction of one sequence of work, i.e. spinning. *A.P.* 6.247.5–6 extends the focalisation through Aesinoe: the spinner's tooth (ὀδόντι, 5) pulls tufts off the thread before the spun wool returns to her wicker basket, filling it afresh (ἐπλήρου [...] καθαιρομένη, 6).

[6] καμαξ is usually taken as a repeated reference to the pin beater, cf. LSJ s.v. καμαξ 5.

[7] For tools used in other crafts as worn, *A.P.* 6.103; 297.

[8] On the importance of focalisation in literary dedicatory epigram generally, Day 2019: 27–32; cf. Petsalis-Diomidis 2018: 424–427.

Complex word order prompts the reader to follow her motions and her gaze, while simultaneously hinting at her motivation. The basket (τάλαρον) is mentioned at the beginning of line 5, as if the spinner considers its content of weighed-out wool before starting her work. Gow and Page (1965: 37) rightly interpret the reference to a καλαθίσκος in *A.P.* 6.160.5–6 as receptacle both for raw wool and finished yarn: the same applies to Aesinoe's τάλαρον here. Thus, the basket she needs to empty of combed wool and fill with tooth-cleaned spun thread (τολύπη, cf. Aristoph. *Lys.* 586), is on her mind from the first. The mention of her biting off any unevenness in the finished thread is only clarified by the participle καθαιρομένη ('cleaned') at the end of the following line (6). The phrase τολύπη πᾶσα ('every ball of wool', 6) is sandwiched between words referring to her filling of her basket (the goal of her work) and the explanatory participle καθαιρομένη (cleaned, 6), which also implies that her yarn is now ready for use. Word order highlights how Aesinoe applies this process to all the thread that she spins.

So far, the spinner's body and tools (and indeed the Philip's syntax and word order) work in concert in the same way as we saw in Ovid's description of Arachne and Minerva setting up their looms (above). By means of the initial tool attributes (6.247.1–3), the tool/user relationship is evaluated in terms of its functionality and how well the task can be completed. The final distich of *A.P.* 6.247 changes our perspective. Here, the dedication to Athena, the 'maiden fond of wool spinning' (φιλέριθε κόρη, 7) contrasts with – and so emphasises – the spinner's old age. The epithet βαθυγήρως ('aged', 7) is placed emphatically at the end of the line, requiring the reader to dwell on Aesinoe's age while the text withholds her name until the next line. Immediately after the name Aesinoe is given, the tools she has dedicated are described as a 'gift of her poverty' (πενίης δῶρον, 8). Thus, it becomes clear that Aesinoe has only been able to eke out a meagre existence by her craft.

In hindsight, both her age and her poverty are heralded already in the third line: the mention of use-wear (δακτυλότριπτον, 'hand-worn') reflects Aesione's long hours of work. Furthermore, her habit of biting off extraneous tufts of wool from the thread with her teeth casts her as an old woman. The intertext of Catullus 64.315, where the crone-like Fates spin in this fashion (Fitzgerald 1995: 163–164; cf. also Ps. Verg. *Eleg. in Maec.* 73–74), would have shaped the perception of Aesinoe's age among Philip's contemporaries. Philip's epigram thus initially traces a hypothetical viewing process of dedicated tools, but gradually draws the reader closer to the experience of the spinner herself. Where the poem is focalised through Aesione herself, the tools are associated not with accomplishment but her all-too limited ability to support herself.

10.3.3 Platthis as Focaliser in Leonidas of Tarentum (A.P. 7.726)

A similar shift in focalisation occurs in a much earlier epigram by Leonidas of Tarentum (early third century BCE), which has exerted considerable influence on Philip (Gow and Page 1968: 345). This funerary epigram shares the focus on the textile worker's poverty as well as issues of focalization with Philip's epigram on Aesinoe (*A.P.* 6.247). It commemorates a weaver named Platthis:

Ἑσπέριον κὴῷον ἀπώσατο πολλάκις ὕπνον
  ἡ γρηὺς πενίην Πλατθὶς ἀμυνομένη
καί τι πρὸς ἠλακάτην καὶ τὸν συνέριθον ἄτρακτον
  ἤεισεν πολιοῦ γήραος ἀγχίθυρος
καί τι παριστίδιος δινευμένη ἄχρις ἐπ᾽ ἠούς
  κεῖνον Ἀθηναίης σὺν Χάρισιν δόλιχον,
ἢ ῥικνῇ ῥικνοῦ περὶ γούνατος ἄρκιον ἱστῷ
  χειρὶ στρογγύλλουσ᾽ ἱμερόεσσα κρόκην.
ὀγδωκονταέτις δ᾽Ἀχερούσιον ηὔγασεν ὕδωρ
  ἡ καλὴ καλῶς Πλατθὶς ὑφηναμένη (*A.P.* 7.726)

Evening and morning-sleep often the old Platthis drove away in her battle against penury, and a melody to go with the distaff and her fellow-worker, the spindle, she sang, neighbour of old age. And a melody at her loom, while she was spinning until daybreak, that long labour of Athena with the Graces, or while, with wrinkled hand on wrinkled knee, lovely she fattened her spindle with sufficient weft thread for her loom. At eighty years of age she crossed the Acherousian stream, beautiful Platthis, who wove beautifully. (Translation adapted from Klooster 2019: 307–309).

This epigram belongs to a series of dedicatory epigrams where Leonidas describes scenes from the working lives of humble craftspeople. These epigrams are typical of the tendency in Hellenistic literature to combine everyday topics and poetic diction, suffused by allusion to epic and tragedy and delighting in novel vocabulary as well as technical terminology (e.g. Gutzwiller 1998: 90; Coughlan 2016: 42). It is not purely an exercise in poetic brilliance: Leonidas seems to explore deliberately the tension between topic and medium, while often displaying a detailed knowledge about the tasks of the working poor (Gutzwiller 1998: 98; Solitario 2015: 47, cf. also Gigante 1971).

The Platthis epigram elevates the commonplace activities of spinning and weaving through its vocabulary. Klooster (2019: 307–309) notes stylistic features of a highly literary register especially in its opening lines: the unusual adjective ἥσπεριον ('evening', 1) and the poetic-sounding πενίην [...] ἄμυνομένη ('as she warded off poverty', 2), and slightly later the circumlocutory phrase 'neighbour of old age' (πολιοῦ γήραος ἀγχίθυρος, 4).

From the general observation that Platthis often worked both early and late, the second distich details her tasks and notes that she sings to accompany her movements. Here, the distaff (ἠλακάτην, 3) is mentioned first, then the spindle (ἄτρακτον, 3). This sequence corresponds to the order in which these tools must be handled as a spinner begins to work: the distaff must be dressed with prepared wool before she starts to spin. Although it recalls the Homeric Athena acting as one of Nausicaa's servant girls (Hom. *Od.* 6.20–40, cf. Solitario 2015: 48 n. 179), the epithet συνέριθος ('fellow worker') also makes sense from the craftswoman's point of view: the rotating spindle works as hard as she does. As in Philip's epigram on Aesinoe, the reader is drawn into the worker's perspective.

In the fifth line, the rotating spindle holds Platthis' attention while she sings to accompany her work. With centrifugal force, the centrally placed δινευμένη ('spinning', formally referring to Platthis' actions but also suggesting the constant rotation of the spindle itself) relegates both the loom (implied by παριστίδιος, 'at the loom') and the coming dawn (ἠοῦς) to the periphery of the reader's perception, now mirroring that of Platthis herself. Yet the image of a woman singing at her work has epic connotations, alluding to Circe's singing by her loom in the tenth book of the *Odyssey* (Hom. *Od.* 10.220–228). This allusion is strengthened by the epic register of the verb δινεύω (e.g. Hom. *Il.* 9.388; *Od.* 4.19, cf. Gow and Page 1965: 376). These allusions begin to overlay the reader's image of Platthis at work. The distancing impact is fully felt in the sixth line, which uses a racing metaphor (Gow and Page 1965: 376) to describe the spinner's long labour as associated with Athena and the Graces. The elevated vocabulary seems well removed from the life of old Platthis (for the discrepancy, Solitario 2015: 49, cf. Zanker 2007: 240–241).

In the next distich (7–8), the text tackles directly the tension between idealisation and realistic depiction. The image of Platthis' aged body (7–8) is juxtaposed with insistence on her beauty (ἱμερόεσσα, 8), and the alliterative repetition of the stem ῥικν- may be interpreted both as implying light-handedness and ease, and as underscoring relentless repetition of monotonous movements (Klooster 2019: 307–308). Editors of the text differ on accentuation, printing either ῥικνὴ (7) as a nominative referring to Platthis herself (to be translated as 'wrinkled' generally; e.g. Dübner 1864–1872; Gow and Page 1965; Page 1975b) or ῥικνῇ, an instrumental dative agreeing with χειρὶ, an emendation to the text suggested by Jacoby and accepted by e.g. Beckby 1965; Klooster 2019: 307–308. To address this question, we must first establish what work element Platthis is performing.

The only verb expanding on Platthis' action is the participle στρογγύλλουσα (8), derived from στρογγύλλω. The verb is rarely attested, however the cognate adjective στρογγυλλός ('round') supports the translation 'make round' (*LSJ* s.v. στρογγύλλω 1). Due to the involvement of the worker's hands and knees, Gow and Page (1965: 376) understand Platthis' work as a finishing practice taking place between spinning and weaving, where the spinner smooths the yarn by rubbing it with their hand against knee or thigh, hypothesised also by Blümer (1912: 120). Spantidaki's evaluation of the iconographic evidence for pre-spinning or spinning scenes does not offer any conclusive evidence for this practice (Spantidaki 2016: 34–37). Based on the sharp focus on spinning in the preceding lines, I would tentatively suggest a different solution. Gow takes the accusative κρόκην ('the thread') as the direct object of στρογγύλλουσα, which Platthis rounds or smooths. If we instead understand Platthis' spindle (ἄτρακτον, 3) as an implied object and κρόκην merely as an accusative of limitation, the verb would refer to Platthis 'fattening up' the spindle (shaft) with regard to the amount of thread it holds, i.e. she would be

Fig. 10.1 Woman spinning while seated. Spun thread rolled onto the shaft of the spindle is displayed with a distinct, rounded profile. The British Museum: 1867 0508.1042

making it rounder by regularly rolling up newly spun thread on it (cf. Fig. 10.1). Reference to this work element follows naturally after the emphasis on the rotating spindle in the fifth line.

Interestingly, experimental evidence shows that spinners do not consistently look down at their hands and knees when they spin with a drop spindle. Spinners may glance down as they first set the spindle in motion, but during drafting, their eyes more often focus on their raised hands or wander around the room. Historically, vase paintings generally display the spinner's eyes as focused on their raised hands, their vision supporting the drafting of new fibres rather than the twisting of the spindle itself (Fig. 10.2). However, spinners are more likely to fix their eyes on their hands holding the spindle shaft when they roll up spun thread, so that during that work element, their field of vision is more likely to encompass the details identified by Leonidas: hands held on or above one's knee. Thus, the close focalisation through Platthis is re-activated in lines 7–8: the reader once again sees with her eyes. This strengthens the case for Jacoby's emendation: when spinning and winding up thread, Platthis is more likely to see and consider the wrinkles on her hand (ῥικνῇ) than make any observation about her general appearance.

Furthermore, by taking Platthis' action in lines 7–8 as rolling up spun thread on the spindle shaft, we also better understand the adjacent mention of her aim to produce weft thread for the loom (for κροκή as weft, Spantidaki 2016: 155). As spinners wind the just-spun thread onto the spindle shaft, they also inspect it for the first time, perhaps adjusting any unevenness by biting it off (as discussed above). This

Fig. 10.2 Woman spinning with a distaff. The fingers of her drafting hand are clearly articulated and her gaze focussed on the new fibres drawn from the distaff. Tondo of fragmentary kylix attributed to the Triptolemos painter. The Metropolitan Museum of Art: 1978.347.5

allows them to ensure that the thread is consistent and suitable for the intended use. In this context, Platthis recalling that she must produce 'weft thread sufficient for her weave' (ἄρκιον ἱστ ῷ...κρόκην, 8) makes excellent sense.

In the final distich, the focalisation shifts again, turning back to a broader perspective on Platthis' life. Yet here, too, the text draws attention to the interweaving of voices and themes: the summary of Platthis' life and death in lines 9 and 10 is pointedly given in highly poetic language (Gow and Page 1965: 376–377) and blurs the time lines of the poem: Klooster (2019: 308–309) suggests that the reader is prompted to reminisce about Platthis' youth. At the same time, the sound play in ἡ καλὴ καλῶς ('who [sc. wove] beautifully', 10) recalls the clanking noises from a loom and allows the weave Platthis has been preparing to materialise at long last (for the soundscape of ancient weaving, see Öhrman 2020). Sound-mimicking draws our attention to the practical, hands-on background of Platthis' craft displayed earlier in the epigram. Only with the phrase 'beautiful Platthis, who wove beautifully' (10) as a final reference to her work do the literary and the realistic perspectives mesh: the weave of Platthis' life is only completed as the epigrammatist completes his poem. To my mind, rather than claiming that they are one and the same, this conclusion acknowledges that there is a gap between the lived and the literary experience of poverty and textile-making (on this tension across Leonidas' tools epigrams, cf. also Solitario 2015: 47).

The association between long hours, old age, and textile work recalls Philip's epigram on Aesinoe (*A.P.* 6.247). The epigram on Platthis does not contain any overt comments on tools voiced by Platthis herself, but the focalisation of the text through her indirectly ascribes feelings about her tools to her (e.g. above on συνέριθον, 'fellow worker'). To Platthis, textile work denotes long hours (esp. 1, 6), takes a toll on her body (7), and is directed toward a specific goal (6–8) that has allowed her a simple but long life (9–10). The spindle might afford her an income sufficient to survive, but her life is also described in terms of hardship.

10.4 The Tool User's Voice: Craftswomen in Epigrams Featuring Direct Speech

The portraits of Aesinoe and Platthis belong in a group of epigrams where textile craftspeople define their tools as a means to make a humble living. Beyond the two already discussed, seven dedicatory epigrams from the *Greek Anthology* mention textile tools in this context (*A.P.* 6.39; 47; 48; 160; 174; 285 and 288), ascribing to tool users attitudes that vary from the resentful and restless to the trustful (*A.P.* 6.39; 174; 288). In this section, I focus on three epigrams that explicitly lend the craftspeople a voice.

10.4.1 Women Rejecting Their Craft: Nikarete (A.P. 6.285) and Bitto (A.P. 6.47)

Two textile-related epigrams adapt the pattern for epigrams on craftspeople dedicating their tools as old age force them to give up their craft (e.g. *A.P.* 6.247; Gutzwiller 1998: 93) by juxtaposing the drudgery of textile work with the quick and easy money supposedly gained by prostitution (Zanker 2007: 239–240). These epigrams are vivid snapshots of craftswomen claiming the centre-stage to vent their frustration and reject their previous life, contrasting sharply with the diligence and regret displayed by other epigrammatic craftswomen like Aesinoe and Platthis. Yet the emphasis on play with literary *topoi* belonging to the emerging tradition of crafts-focussed epigrams means that, although they may feature the craftswomen's own voices, those voices are also engaged in a specifically *literary* dialogue.

Nikarchos, assumed to be writing in the established Roman Empire of the second century CE (Schatzmann 2012: 21), lets a woman called Nikarete make a bonfire of her textile tools:

ἡ πρὶν Ἀθηναίης ὑπὸ κερκίσι καὶ τὰ καθ΄ἱστῶν
  νήματα Νικαρέτη πολλὰ μιτωσαμένη
Κύπριδι τὸν κάλαθον τά τε πηνία καὶ τὰ σὺν αὐτοῖς
  ἄρμεν΄ ἐπὶ προδόμου πάντα πυρῆς ἔθετο,
ἔρρετε, φωνήσασα, κακῶν λιμηρὰ γυναικῶν
  ἔργα, νέον τήκειν ἄνθος ἐπιστάμενα.
εἵλετο δὲ στεφάνους καὶ πηκτίδα καὶ μετὰ κώμων
  ἡ παῖς τερπνὸν ἔχειν ἐν θαλίαις βίοτον
εἶπε δε, παντός σοι δεκάτην ἀπὸ λήμματος οἴσω,
  Κύπρι· σὺ δ΄ἐργασίην καὶ λαβὲ καὶ μετάδος. (*A.P.* 6.287)

> Nikarete, who before lived by the pin beaters of Athena and set up many a weave on the loom, put the wool basket, the spools and all her tools on the fire in front of the house, for Kypris. 'Farewell', she called, 'you hunger-bringing works of wretched women, that wither the flower of youth'. The girl then took up garlands and the lyre, and with song went to have a pleasant life in festivities. She said, 'I will give you a tenth of my profit, Kypris: take your share of the work, and give a share also.'

Nikarete describe her tools as 'hunger-bringing tools of wretched women' (κακῶν λιμηρὰ γυναικῶν, *A.P.* 6.285.5–6) that 'wither the flower of youth' (6), implying that that textile work is endlessly time-consuming, yet never yields sufficient income for a comfortable life. Nikarete is described as a girl (παῖς, 8, cf. Zanker 2007: 239), yet her agency suggests that she is well into adulthood: first, rather than display her as a weaver or spinner in general, Nikarchos establishes that she is experienced in the complex task of heddling (νήματα…πολλὰ μιτωσαμένη, 2).[9] Her equipment is suitably extensive, ranging from wool-basket and spools to 'the things that go with them' (3), as well as the *kerkis* alluded to in the first line of the epigram. Most importantly, she claims space for herself not merely at home but in the public streetscape (ἐπὶ προδόμου 'in front of the house', 4). Finally, she confidently assumes that her new career will make her a profit and haggles with Kypris about the tithe she will owe her (9–10). The reader might conclude that Nikarete is no longer quite so young as she would like to think.

This is supported by a similar, earlier epigram by Antipater of Sidon (late second century BCE, e.g. Coughlan 2016: 38; Ambühl 2019), assumed as a model for Nikarchos (e.g. Zanker 2007: 239–240). In Antipater's epigram, the dedicant Bitto is a widow of forty years, who also rejects her previous profession as a weaver in favour of prostitution:

> κερκίδα τὴν φιλάοιδον Ἀθηναίῃ θέτο Βιττώ
> ἄνθεμα λιμηρῆς ἄρμενον ἐργασίης,
> εἶπε δὲ χαῖρε, θεά, καὶ τήνδ'ἔχε· χήρη ἐγὼ γὰρ
> τέσσαρας εἰς ἐτέων ἐρχομένη δεκάδας
> ἀρνεῦμαι τὰ σὰ δῶρα, τὰ δ' ἔμπαλι Κύπριδος ἔργων
> ἅπτομαι· ὥρης γὰρ κρεῖσσον ὁρῶ τὸ θέλειν. (*A.P.* 6.47)

> I, Bitto dedicate to Athena the melodious pin beater, an offering and tool of her hunger-bringing work. She said: 'Hail, goddess, and have this; for I, a widow in my fortieth year, forswear your gifts and instead take up the works of Kypris; I see that desire is stronger than the bloom of youth.'

While these epigrams may reflect professional choices available to poor women (e.g. McKechnie 2005: 70–71; Fischer 2013: 220; Sebilotte-Cuchet and Dillon 2013: 224–225; Strong 2016: 109–110), Nikarchos and Antipater create, in the first instance, a literary juxtaposition. These texts not only adapt the epigrammatic *topos* of the dedicant craftsperson, but also tap into a *topos* of disgusted fascination with

[9] Gow and Page 1965: 427 translates the phrase simply as 'weave', cf. LSJ s.v. μιτόομαι. My interpretation relies on the specific meaning of μίτος as 'heddle' in combination with the downward motion implied by καθ΄ἱστῶν, recalling how the weaver fixes the starting border to the cloth beam during set-up.

sexually active older women, widespread both in Hellenistic and Roman literature. This *topos* is particularly relevant in epigram (Richlin 1984) and the statement that desire is stronger than age in the final line of the Bitto epigram explicitly links to it, adding an ironic undertone to this and other textile-worker-turned-prostitute epigrams. Nikarete and Bitto (featured in *A.P.* 6.47; 48) defiantly reject their textile tools *after* a reasonably long period of working with them. They are no longer young. They are also shown as pursuing their career change as much for their own sexual satisfaction (marked as questionable) as for financial gain. To a literary audience already familiar with the horror image of the crone lover, they must seem bound to fail.

Other poetic references to textile work as a last resort confirm this impression. The anonymous epigram *A.P.* 6.283 speaks of an old woman, formerly courted by wealthy lovers, now earning a meagre living by the loom. In the Augustan period, Tibullus' elegiac narrator urges his erring mistress to be faithful, lest she be forced to rely on wool-work in her old age (Tib. 1.6.75–76; 81–84). He gives as a warning example the fate of a one-time prostitute, whose only recourse in old age is her wool-working skills (1.6.77–80). Like Nikarchos' Nikarete, Tibullus' old weaver works on a rented loom (Flower Smith 1964: 322; Maltby 2002: 278) and has heddling, not weaving, as her specific task. Tibullus also notes that while she was once able to catch the eye of potential lovers, her eyesight is now failing so much that she cannot see that the wool she is preparing is less than shiny white (Öhrman 2017: 284–286; 2020). The implication is that despite her efforts, the old woman's work is of an insufficient standard to ensure her continued livelihood (cf. Lee-Stecum 1998: 202). She is depicted as an object of pity, mired in poverty despite constant work. Her tools are strongly associated with a constant struggle for a livelihood and highlight her tenuous financial position. Thus, the intertextual analysis of the motif of women rejecting textile work to pursue a more profitable and enjoyable career in prostitution suggests that women like Nikarete and Bitto have little hope of success either in life or in literature.

10.4.2 Virtuous Voices: Family Daughters on Their Tools

The juxtaposition of textile work and prostitution taps into another common, cultural field of associations for female textile work, that of moral virtue: a *topos* inverted in the epigrams just discussed. Many tools-focussed dedicatory epigrams imply that textile workers are poor but morally good, in the way that we have already seen in Leonidas' portrait of Platthis. The dedicant, Telesilla, described as φιλοεργός ('industrious') in *A.P.* 6.160.7 is another example. Dedicants in these epigrams are defined by their familial connections, particularly to their parents.[10] Even here,

---

[10]Daughters in *A.P.* 6.39.1–2; 160.7; 288.1; 289.1–2. Cf. also *A.P.* 6.286; 287. The pattern is inverted in *A.P.* 6.284 and adapted in *A.P.* 6.174 (Coughlan 2016: 44–45).

however, tools are a means to income. This is illustrated by another epigram by Leonidas, featuring a dedication of textile tools by four sisters:

> αἱ Λυκομήδευς παῖδες Ἀθηνὼ καὶ Μελίτεια
> καὶ Φιντὼ Γλῆνίς θ΄, αἱ φιλοεργόταται,
> ἔργων ἐκ δεκάτας ποτιθύμια τόν τε πρόσεργον
> ἄτρακτον καὶ τὰν ἄτρια κριναμέναν
> κερκίδα, τὰν ἱστῶν μολπάτιδα, καὶ τὰ τροχαῖα
> πανία †κερταστὰς τούσδε ποτιρρόγεας†,
> καὶ σπάθας εὐβριθεῖς †πολυάργυρα τὼς δὲ† πενιχραί
> ἐξ ὀλίγων ὀλίγην μοῖραν ἀπαρχόμεθα.
> τῶν χέρας αἰέν, Ἀθάνα, ἐπιπλήσαις μὲν ὀπίσσω,
> θείης δ΄εὐσιπύους ἐξ ὀλιγησιπύων. (*A.P.* 6.288)

> We, the industrious daughters of Lycomedes, Atheno, and Meliteia and Phinto and Glenis, destitute, offer as a tithe from our work these welcome gifts, the laborious spindle and the pin beater that separates the warp threads, songstress of the loom, and the running spools…and heavy weaving swords…a little portion from our meagre holdings. Ever after, Athena, fill our hands, and render us rich in bread out of our current poverty. (Translation adapted from Coughlan 2016: 44–45).

Identified by a patronymic, the moral fibre of the sisters is immediately established with the superlative 'most industrious' (φιλοεργόταται, 2). It is underscored to the reader first by the number of tools they dedicate (spindle, pin beater, spools, and a weaving sword, 4–7) and finally by their statement that they give a small gift from what little they have (8). Yet these young sisters, too, speak of their tools in terms of functionality and income generation. The adjective πρόσεργον (3) describing the spindle is particularly interesting. Most commentators tentatively translate this *hapax legomenon* as 'laborious' (e.g. Gow and Page 1965: 350; Coughlan 2016: 43). Solitario (2015: 50–51) interprets it as a synonym for συνεργός and thus an indication of craftsperson and tool working in tandem toward a common goal: as in texts discussed above, the user/tool engagements are understood in terms of functionality. Consideration of the related noun, τὸ πρόσεργον, adds another dimension. It refers to 'extra work' (*LSJ* s.v. πρόσεργον II), although this, too, relies primarily on one passage (*I.G.* VII 307.8; 63). However, the verb προσεργάζομαι occasionally has the sense 'earn in addition' (*LSJ* s.v. προσεργάζομαι, Xen. *HG* 3.1.28). The meaning of cognate words encourages the reader of Leonidas' epigram to consider the dedicated spindle as a means for the sisters to earn money, however limited the income.

This section on comments directly voiced by spinners and weavers on their tools shows that regardless of whether epigram authors are sympathetic to the hardships of female textile workers or make their dreams of a better life the subject of literary, intertextual irony, their tools are given the same associations. Textile workers like Aesinoe and Nikarete may be accepting of their lot or frustrated with it, but they acknowledge the importance of their tools in remarkably similar ways. In Greek epigram, textile tools are nearly always related to women's ability to support themselves financially. In Hodder's terminology of entanglement (e.g. Hodder 2012), the relationship between user and tools in these Greek epigrams fits on a sliding scale from enabling dependence to dependency and entrapment.

10.5 Parallels in the Material Record: Texts on Tools

There are parallels for these literary user attitudes to tools in the material record. A remarkable loom weight find examined by Beltrán Lloris and Beltrán Lloris (2012) illustrates this final point. The first-century loom weight from Caesaraugusta bears an inscription linking diligent textile work to both marital and material good fortune. The inscription had been made before firing, so that it is authored *for* rather than *by* weavers. Two sides of the weight are inscribed (A, B below). Beltrán Lloris and Beltrán Lloris print the following text:

> telas multas ṭexat, uirum bon<um> inueniat. A
> ama lateres! fac(i)mus fausta et fel[icia]. B

> May she weave many webs, (so that) she may find a good husband. A
> Love the loom weights! We make fortunate and happy things. B

Although conceived before firing, the text itself makes clear that the intended audience is female: a woman using the weights in her loom. The inscription on the A face makes an obvious association between textile work and status as a wife, but both *bonus* ('good', of the prospective husband) and the formula *fausta felicia* ('fortunate and happy [things]') evokes broadly defined prosperity, if not luxurious wealth.[11] As in the literary texts discussed above, the promise of prosperity and good fortune is linked specifically to the tools, to the extent that the loom weights are explicitly given a voice. The use of first person plural ('we make') is deliberately ambiguous: it might refer to a set of loom weights, or to the productive unit of loom weights and craftswoman working in consort to secure both happiness and financial stability. The fact that the inscription occurs on a single weight (rather than on a set) draws the weaver closer: she is necessary to complete the 'we'. This means that the weaver is made complicit in the attitudes expressed: the shifting syntax of the inscription grants agency to both user and tool.

10.6 Conclusion

Where ancient texts comment on how users of textile tools relate to their equipment, they overwhelmingly depict these user/tool engagements as driven by functionality and need. This tallies with interpretations of archaeological evidence that privileges functionality, for example in the understanding of assemblages. Yet we can also glimpse, in a handful of texts, the emotional attachments that develop among craftspeople from such reliance on their tools to meet their basic needs: the hand-worn spindle may be both income-bringer and fellow-worker, and may

---

[11] *OLD* s.v. *bonus* 6 and 10b (*res bonae* 'wealth'); *ThLL* s.v. *faustus* 1 and 2, e.g. Arnob. *Nat.* 4.5; Plaut. *Pers.* 632.

prompt both gratitude and frustration. The comments of literary tool users also provide a forceful reminder that engagements with tools by craftspeople are embedded in larger societal contexts, determined by familial or community relationships, gender and economic class.

The correlation between archaeological material, experimental and literary evidence suggests that – regardless of who may have performed it – real rather than idealised textile work had a stronger presence in the lives of ancient readers (i.e. primarily the educated elite) than we have sometimes assumed. Only if this is the case is it meaningful for authors to make the experiences of spinners and weavers (their attitudes to their tools, or their visual and sensory perspective on them while working) the subject of highly artful literary play.

As readers keen to engage with such passages from a craft-historical perspective, we need to embrace the fact that literary sources, perhaps particularly poetry, must be interpreted in their own aesthetic and historical context to be understood fully. It would be an error to assume that prostitution was an easy and enjoyable career for middle-aged women on the basis of the ironically undermined portrait of Nikarete, but it is only analysis of literary parallels that demonstrates this conclusively.

This presents significant issues: scholarship on ancient literature is no less dense and difficult to assess for the non-expert than use-wear or dye analysis is to the philologist, and it is as susceptible to scholarly trends as archaeological theory. Experimental and experiential approaches have been slow to make their way into Classical philology, which means that the availability of translations and commentaries that take craft-historical information into account poses a particular problem, especially in regard to non-canonical or little-studied texts. The potential to draw not only on documentary sources but on complex literary texts for corroboration and contextualisation of research based on archaeological material or experimental work should nonetheless encourage us to rise to these challenges.

Acknowledgements Without the input of experienced craftspeople, no academic could attempt to discuss the representation of craft processes. As many times before, I mention with gratitude Ida Demant (Land of Legends) and Barbara Köstner (Xanten Archaeological Park) as well as Jennifer Beamer (Leicester). Lampeter colleagues have offered valuable feedback on versions of this paper: thanks especially to Louise Steel and Ruth Parkes. I am also indebted to Benjamin Cartlidge (Liverpool) and to the staff at the Fondation Hardt pour l'étude d'antiquité classique in Geneva for helping me access scholarship when the Covid-19 lockdown prevented travel. Finally, I am grateful to my colleague and partner Errietta Bissa: at once the harshest critic and the most patient reader. Needless to say, remaining errors are my own.

References

Ancient sources

Claudian De raptu Proserpinae:

Claudii Claudiani carmina. JB Hall (ed.) (1985). Bibliotheca scriptorum graecorum et romanorum teubneriana. BG Teubner Verlagsgesellschaft, Leipzig.

Ovidius Metamorphoses:

P. Ovidi Nasonis Metamorphoses. RJ Tarrant (ed.) (2004). Oxford Classical Texts. Oxford University Press, Oxford.

Epigrams from the Anthologia palatina (Leonidas, Nikarchos, and Philip):

Epigrammata graeca. Page DL (ed.) (1975). Oxford Classical Texts. Oxford University Press, Oxford.

Modern works

Alberti M (2018) The construction, use, and discard of female identities: Interpreting spindle whorls at Vindolanda and Corbridge. *Theoretical Roman Archaeology Journal* 1. https://doi.org/10.16995/traj.241.

Ambühl A (2019) Moving to Rome: Antipater of Sidon, Archias of Antiocheia, Philodemus, and others. In: Henriksén C (ed.) *A companion to ancient epigram.* John Wiley and Sons, Malden, p 407–421.

Andersson Strand E (2003) Textile production in Scandinavia during the Viking Age. In: Bender Jørgensen L, Banck-Burgess J, Rast-Eicher A (eds) *Textilien aus Archäologie und Geschichte. Festschrift für Klaus Tidow.* Wachholz Verlag, Kiel, p 46–62.

Andersson Strand E (2018) Early loom types in ancient societies. In: Siennicka M, Rahmstorf L, Ulanowska A (eds) *First textiles. The beginnings of textile manufacture in Europe and the Mediterranean.* Ancient Textiles Series 32. Oxbow Books, Oxford and Philadelphia, p 17–29.

Andersson Strand E, Nosch M-L (eds) (2015) *Tools, textiles and contexts. Investigating textile production in the Aegean and Eastern Mediterranean Bronze Age.* Ancient Textiles Series 21. Oxbow Books, Oxford and Philadelphia.

Barber EJW (1992, third edition) *Prehistoric textiles. The development of cloth in the Neolithic and Bronze Ages with special reference to the Aegean.* Princeton Paperback, Princeton.

Beckby H (ed.) (1965) *Anthologia graeca. Buch VII-VIII.* Tusculum Bücherei. Ernst Heimeran Verlag, Munich.

Beltrán Lloris F, Beltrán Lloris M (2012) Ama lateres! Sobre una pesa de telar cesaraugustana relativa al lanificium. *Sylloge epigraphica Barcinonensis* 10:127–148.

Bender Jørgensen L (2012) Writing craftsmanship? Vocabularies and notation systems in the transmission of craft knowledge. In: Wendrich W (ed.) *Archaeology and apprenticeship. Body knowledge, identity, and communities of practice*. University of Arizona Press, Tucson, p 240–254.

Bijker WE (2010) How is technology made?—That is the question! *Cambridge Journal of Economics* 34:63–76.

Blümer H (1912) *Technologie und Terminologie der Gewerbe und Künste bei Griechen und Römern*. Zweite, gänzlich umgearbeitete Auflage. Vol I. 2 edn. B. G. Teubner, Leipzig.

Coughlan T (2016) Dialect and imitation in late Hellenistic epigram. In: Rengakos A, Sistakou E (eds) *Dialect, diction, and style in Greek literary and inscribed epigram*. De Gruyter, Berlin and Boston, p 37–70.

Day J (2019) Reading inscriptions in literary epigram. In: Kanellou M, Petrovic I, Carey C (eds) *Greek epigram from the Hellenistic to the Early Byzantine Era*. Oxford University Press, Oxford, p 19–34.

Dobres M-A (2010) Archaeologies of technology. *Cambridge Journal of Economics* 34:103–114.

Dobres M-A, Hoffmann CR (1994) Social agency and the dynamics of prehistoric technology. *Journal of Archaeological Method and Theory* 1:211–258.

Dübner F (ed.) (1864–1872) *Epigrammatum anthologiae palatinae*. Firmin-Didot, Paris.

Fanfani G, Harlizius-Klück E (2016) (B)orders in ancient weaving and archaic poetry. In: Fanfani G, Harlow M, Nosch M-L (eds) *Spinning fates and the song of the loom. The use of textiles, clothing and cloth production as metaphor, symbol and narrative device in Greek and Latin literature*. Ancient Textiles Series 24. Oxbow Books, Oxford and Philadelphia, p 61–100.

Fischer M (2013) Ancient Greek prostitutes and the textile industry in attic vase painting ca. 550–450 BCE. *The Classical World* 106:219–259.

Fitzgerald W (1995) *Catullan provocations. Lyric poetry and the drama of position*. University of California Press, Berkeley.

Flower Smith K (ed.) (1964) *The elegies of Albius Tibullus*. Wissenschaftliche Buchgesellschaft, Darmstadt.

Fregni EG (2019) Looking over the shoulder of the Bronze Age metalsmith: recognising the crafter in archaeological artefacts. In: Burke C, Spencer-Wood SM (eds) *Crafting in the world: Materiality in the making*. Springer Nature, Charm, p 37–50.

Gigante M (1971) *L'edera di Leonida*. Collana di filologia classica, vol 1. Morano, Naples.

Gleba M (2008) *Textile production in pre-Roman Italy*. Ancient Textiles Series 4. Oxbow Books, Oxford.

Gow ASF, Page DL (eds) (1965) *The Greek anthology. Hellenistic epigrams. Volume II. Commentary and indexes*. Cambridge University Press, Cambridge.

Gow ASF, Page DL (eds) (1968) *The Greek anthology. The garland of Philip and some contemporary epigrams. Volume II. Commentary and indexes*. Cambridge University Press, Cambridge.

Gruzelier C (ed.) (1993) *Claudian de raptu Proserpinae. Edited with introduction, translation and commentary*. Oxford Classical Monographs. Claredon Press, Oxford.

Gutzwiller K (1998) *Poetic garlands. Hellenistic epigram in context*. University of California Press, Berkeley.

Harries B (1990) The spinner and the poet: Arachne in Ovid's *Metamorphoses*. *Proceedings of the Cambridge Philological Society* 216:64–82.

Hodder I (2012) *Entangled: An archaeology of the relationships between humans and things*. John Wiley and Sons, Malden.

Ingold T (2000) *The perception of the environment. Essays on livelihood, dwelling and skill*. Routledge, London and New York.

Jordan P (2015) *Technology as human social tradition. Cultural transmission among hunter-gatherers*. University of California Press, Berkeley.

Kleibrink M (2018) Textile utensils from Francavilla Marittima (Lagaria), Italy. In: Busana MS, Gleba M, Meo F, Tricomi AR (eds) *Textiles and dyes in the Mediterranean economy and society. Proceedings of the VIth international symposium on textiles and dyes in the Ancient Mediterranean world (Padova – Este – Altino, Italy 17–20 October 2016).* PURPUREAE VESTES VI. Libros Portico, Valencia, p 167–176.

Klooster J (2019) Leonidas of Tarentum. In: Henriksén C (ed.) *A Companion to ancient epigram.* John Wiley and Sons, Malden, p 303–317.

Lee-Stecum P (1998) *Powerplay in Tibullus. Reading elegies book one.* Cambridge University Press, Cambridge.

Lee R (2019) Made to remake the world: the bronze age tool and the 'idea of craft'. In: Burke C, Spencer-Wood SM (eds) *Crafting in the world: Materiality in the making.* Springer, Cham, p 19–36.

Maltby R (ed.) (2002) *Tibullus: elegies. Text, introduction and commentary.* Francis Cairns, Leeds.

Marchesini S, Migliavacca M (2018) The inscribed loom weights from Monte Loffa, Monte Lessini (VR, Italy): can we crack the code? In: Busana MS, Gleba M, Meo F, Tricomi AR (eds) *Textiles and dyes in the Mediterranean economy and society. Proceedings of the VIth international symposium on textiles and dyes in the Ancient Mediterranean world (Padova – Este – Altino, Italy 17–20 October 2016).* PURPUREAE VESTES VI. Libros Portico, Valencia, p 231–240.

McKechnie P (2005) Beau nonde and demi-monde in Alexandria, 323–116 BC. *L'antiquité classique* 74:69–82.

Mullen A (forthcoming) A sociolinguistic and archaeological epigraphy: Translingualism, uncertainty and Roman inscribed spindle whorls. In: Cousins E (ed.) *(Un)set in stone: Embodied epigraphy.*

Murphy EA, Poblome J (2013) Technical and social considerations of tools from Roman period ceramic workshops at Sagalassos (Southwest Turkey). Not just tools of the trade? *Journal of Mediterranean Archaeology* 25:197–217.

Öhrman M (2017) Listening for licia. A reconsideration of Latin licia as heddle leashes. In: Gaspa S, Michel C, Nosch M-L (eds) *Textile terminologies from the Orient to the Mediterranean and Europe, 1000 BC to 1000 AD.* Zea Books, Lincoln, p 278–287.

Öhrman M (2018) Textile work in shared domestic spaces in the Roman house: the evidence from Latin poetry. *Fasiculi Archaeologiae Historicae* 31:93–101.

Öhrman M (2020) The soundscape of textile work in the Roman world: Old sources and new methods. In: Cooper K (ed.) *New approaches to ancient material culture in the Greek and Roman world.* Monumenta graeca et romana 27. Leiden, p 183–200.

Page DL (ed.) (1975) *Epigrammata graeca.* Scriptorum classicorum bibliotheca oxniensis. Oxford University Press, Oxford.

Paton WR, Tueller MA (eds) (2014) *The Greek anthology. Revised by Michael A. Tueller.* The Loeb Classical Library. Harvard University Press, Cambridge.

Petsalis-Diomidis A (2018) Undressing for Artemis: sensory approaches to clothes dedications in Hellenistic epigram and in the cult of Artemis Brauronia. In: Kampakoglou A, Novokhatko A (eds) *Gaze, vision, and visuality in ancient Greek literature.* De Gruyter, Berlin and Boston, p 464–492.

Richlin A (1984) Invective Against Women in Satire. *Arethusa* 17:67–80.

Rosati G (1999) Form in motion: weaving the text in the *Metamorphoses.* In: Hardie P, Barchiesi A, Hinds S (eds) *Ovidian transformations. Essays on the Metamorphoses and its reception.* Cambridge Philological Society, Cambridge, p 240–253.

Schatzmann A (ed.) (2012) *Nikarchos II: Epigrammata. Einleitung, Texte, Kommentar.* Hypomnemata. Unversuchungen zur Antike und zu ihrem Nachleben. Vandenhoeck and Ruprecht, Göttingen.

Sebilotte-Cuchet V, Dillon J (2013) Women as "wool-workers" and "sex-workers" in Athens (fourth century BC). *Clio Women, Gender, History* 38:219–227.

Sofianou C (2012) The loomweights of the Hellenistic settlement at Trypitos, Siteia. In: Tzachli I, Zimi E (eds) *Textiles and dress in Greece and the Roman East: A technological and social approach.* Ta pragmata, Athens, p 77–86.

Solitario M (2015) *Leonidas of Tarentum between cynical polemic and poetic refinement*. SemRom. Seminari Romani di cultura graeca. Edizioni Quasar, Rome.
Spantidaki S (2016) *Textile production in classical Athens*. Ancient Textiles Series 27. Oxbow Books, Oxford and Philadelphia.
Strong A (2016) *Prostitutes and matrons in the Roman world*. Cambridge University Press, Cambridge.
Wendrich W (2012) Archaeology and apprenticeship: body knowledge, identity, and communities of practice. In: Wendrich W (ed.) *Archaeology and apprenticeship: Body knowledge, identity, and communities of practice*. University of Arizona Press, Tucson, p 1–19.
Wild JP (2000) Textile production and trade in Roman literature and written sources. In: Cardon D, Feguère M (eds) *Archéologie des textiles*. Édition Monique Mergoil, Montagnac, p 209–213
Zanker G (2007) Characterization in Hellenistic epigram. In: Bing P, Steffen Bruss J (eds) *Brill's companion to Hellenistic epigram. Down to Philip*. Brill, Leiden, p 233–250.

Magdalena Öhrman is Senior Lecturer in Classics at the University of Wales Trinity Saint David, UK. Trained as a classical philologist in Sweden and Germany, her current research interests are in the area of Graeco-Roman textiles and textile production, textile terminology and technology, and in sensory experience in past societies. In 2016–2018, she held a Marie Skłodowska-Curie fellowship at the Centre for Textile Research (University of Copenhagen) for the project "Textile Reflections: Multi-Sensory Representation of Textile Work in Latin Poetry and Prose".

Part IV
Traditions and Contexts: Fibres, Fabrics, Techniques, Uses and Meanings

Chapter 11
Lime Bast Winning: Know-How, Labour Input and Quantity Needed for the Production of Two Selected Neolithic Finds

Sabine Karg

Abstract Chronostratigraphic analysis of plant remains from archaeological sites in the Circum-Alpine region dated between 4300 and 2600 BC prove that plant fibres played an important role in the daily life of prehistoric people. Numerous objects made of plant fibres are preserved, such as strings, ropes, containers, nets and clothes. Fibre identification shows that many objects are made of bast from trees. The amount of raw material needed for the production of these objects must have been enormous, and the production process was very labour intensive. In addition, the exploitation of tree bast must have had an impact on the forest composition around the settlements. Experiments with lime bast extraction in a forest on the Danish island of Zealand allow us to calculate the yield of bast and the extrapolation of the human labour that is necessary to harvest and process lime bast. Although only few tools are required for manufacturing objects from plant fibres, we could demonstrate that Neolithic brushes made of thorns from sloe or hawthorn were most probably used for splitting tree bast into fine and regular strips that could then be twisted or even spun into threads. Moreover, we calculated the amount of bast that is needed for the production of two selected Neolithic objects.

Keywords Neolithic · Europe · Lime bast · Textile finds · Experimental archaeology · Brushes · Shoes · Nets

S. Karg (✉)
Free University of Berlin, Berlin, Germany
e-mail: Sabine.Karg@fu-berlin.de

A. Ulanowska et al. (eds.), *Ancient Textile Production from an Interdisciplinary Perspective*, Interdisciplinary Contributions to Archaeology,
https://doi.org/10.1007/978-3-030-92170-5_11

11.1 Introduction

Neolithic wetland settlements in the European Circum-Alpine region provide incredibly good preservation conditions for organic material. Additionally, the sites can be dated absolutely with the aid of dendrochronological methods. Chronostratigraphic analysis of textile remains from these sites that date from 4300 to 2600 BC indicate that bast fibres played an important role in the daily life of prehistoric people. Threads, ropes, nets and even clothing such as hats and shoes were made of bast fibres (Rast-Eicher and Dietrich 2015). Tree bast fibre strips were spun using spindles, which is proved by the finds of several Neolithic spindle whorls with still attached threads (Médard 2010; Grabundžija et al. 2021).

These Neolithic people had at their disposal great knowledge about the properties of all kinds of plant material. They knew, for example, that bast from lime trees offer much better properties than bast from other trees. Their lives must have been organised according to the time of the year, including when it was best to harvest and process such natural goods such as tree bast (Karg and Weber 2019).

11.2 Definition of Tree Bast and the Properties of Lime Bast

Tree bast fibres are formed between the outer part of a tree, the so-called bark, that is dead and the inner part of the tree that carries water, which dissolves minerals from the roots and provides mechanical strength to the plant (Fig. 11.1). The bast is mainly composed of thick layers of cellulose fibrils, each of which are arranged differently and have a variable thickness. The fibrils stick together with a kind of glue consisting of different ratios of pectin and lignin (Schweingruber and Börner 2018). Therefore, the bast still sticks to the bark after the harvest. The bark must be soaked in water for several weeks in order to dissolve this glue. Not until then can the bast strips be removed from the bark, and the bast is ready for further processing.

The quality and length of bast fibres as well as their diameter are variable depending upon the plant species and also on the growing conditions. Lime bast provides a

Fig. 11.1 A freshly felled lime tree: (1) bark, (2): bast layers sticking together, (3) wood. (Photo by S. Karg)

unique combination of volume, tensile strength, elongation and limited water absorbance (Myking et al. 2005). In addition, lime tree bast is soft and can be easily plied. These particular properties, and especially its superior strength, are assumed to be the reason for its millennia-long use. A lime tree produces 10–12 layers of bast, with the most recent and smoothest layers close to the inner part of the woody stem, and the coarsest layers close to the bark.

11.3 The Timing of Lime Bast Winning

Bast from lime trees must be harvested in the spring, when the water transport in the trunks has just started. At that time, the bark with the attached bast layers can be easily removed from the stems. Young lime trees apparently deliver better and softer bast than old trees. However, thick branches from older trees can also be used. Old photographs from rural Sweden demonstrate that lime trees were coppiced for this reason (Granlund 1943). Every 5–10 years a c. 2 cm thick bast layer can be harvested from coppiced trees according to Schjølberg (1988, cited in Myking et al. 2005: 68).

11.3.1 The Lime Bast Experiment at the Viking Ship Museum on Zealand, Denmark

In 2006, a reconstruction of the Viking Age cargo vessel Skuldelev 3 was successfully finished and the ship called "Sif Ege" set sail in the Roskilde Fjord. This project was performed at the Viking Ship Museum in Roskilde with the help of volunteers and the late shipbuilder Ole Magnus, who recorded the experiment. The following data concerning the production of the ropes made of lime bast for the ship was gathered from this report (Magnus 2006):

In June 2005, the following amount of lime bast was harvested:

- From 58 lime trees with an age of 42 years, 600 m bark with attached bast was removed;
- The bark was taken only from stems and thick branches (minimum thickness of an arm);
- After the retting, washing and drying process 126 kg of useable bast was produced;
- 2.5 hours of work was required for the production of 1 kg lime bast: removing the bast from the bark, retting and sorting for further production.

11.3.2 Processing of the Bark and the Bast

The bark was then placed into a water hole with stagnant water for 2 months until August. The practical experience of the volunteers who had previously produced ropes showed that the highest amount of the harvest could be achieved when the bark with thin bast layers was placed in the bottom of the water hole, where the water is coldest, and the bark with the thick bast layers right at the top, where the water warms up more quickly. After the retting process, the bark was taken out of the water and the bast was removed from the bark, then thoroughly washed and dried. During winter, the bast layers were separated and split by hand into 2–3 cm narrow strips. The bast strips were then used for making ropes.

11.4 Experiments with a Possible Neolithic Tool for Splitting the Bast

Small brushes made of the thorns of sloe (*Prunus spinosa* L.) or hawthorn (*Crataegus* spec.) are frequently excavated in Neolithic wetland sites (Karg 2020). Figure 11.2a shows a 6 cm-long bundle of thorns from sloe bushes that was excavated at the site of Hornstaad-Hörnle IA at Lake Constance in southern Germany dated to c. 3900 BC. Until recently, the exact function of the thorn bundle was not known. Therefore, in the summer of 2019 we performed an experiment with a reconstructed brush (Fig. 11.2b). It turned out that the brush is perfectly suited to split tree bast into fine and regular strips (Fig. 11.2c). We suggest that this tool was used for the production of tree bast strips.

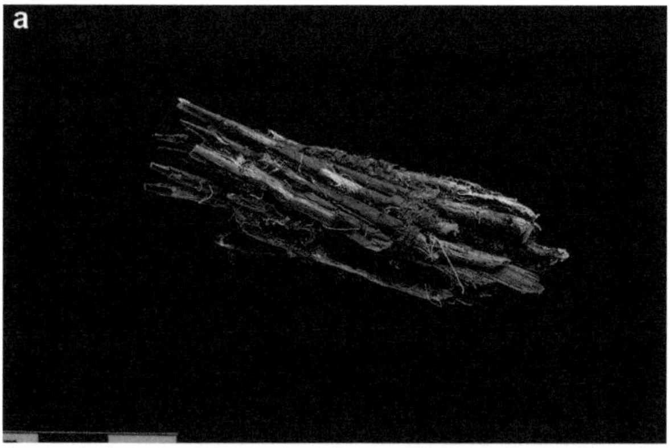

Fig. 11.2 (**a**) Original brush from the archaeological site of Hornstaad-Hörnle IA dated to c. 3900 BC. (Photo by Y. Mühleis, Landesamt für Denkmalpflege im Regierungspräsidium Stuttgart). (**b**) Reconstructed brush by Manfred Pfeifer made of sloe thorns. (Photo by S. Karg) (**c**) Using the reconstructed brush in the Geschichtserlebnisraum Roter Hahn e.V. Lübeck to make narrow lime bast strips. (Photo by S. Karg)

Fig. 11.2 (continued)

11.5 Quantification of Tree Bast for the Production of Two Selected Neolithic Objects

In Neolithic wetland sites of the Circum-Alpine region many objects made of lime bast were recovered during archaeological excavations, for example several shoes and nets. Two such shoes were discovered at the site of Allensbach-Strandbad dated to c. 3000 BC (Czarnowski and Karg 1986); one is shown in Fig. 11.3. This shoe matches the modern size 38, and the weight of the reconstruction made by Anne Reichert is 120 g. The extrapolation shows that 1 kg lime bast is needed to produce four pairs of shoes, which corresponds to half of the bast from a single tree of average age and diameter. For eight pairs of shoes the bast of a whole tree had to be processed.

Likewise lime bast was used during the Neolithic for making fishing nets. Such finds were discovered for example at three lake dwelling sites dated to the fourth millennium BC: Maur-Schifflände (Greifensee), Meilen-Feldmeilen/Vorderfeld (Lake Zürich) and at the important site of Wetzikon-Robenhausen (Pfäffikersee).

Fig. 11.3 Original shoe from the archaeological site of Allensbach-Strandbad dated to c. 3000 BC. (Photo by M. Schreiner, Archäologisches Landesmuseum Konstanz)

All three sites are located in the Canton Zürich in Switzerland (Rast-Eicher and Dietrich 2015). In collaboration with Manfred Pfeifer of the Geschichtserlebnisraum Roter Hahn in Lübeck, Germany, such a net was reconstructed by using lime bast (Fig. 11.4). The net measures 70 × 30 cm, and its weight is 43 g. The original size of such fishing nets was definitely bigger. If we assume that the size of a Neolithic net was e.g. three times larger than the reconstruction, the bast of one single lime tree would have been needed for the production of eight fishing nets.

11.6 The Availability and Decline of Lime Trees in the Neolithic

The manifold number of Neolithic objects made of lime bast suggests that many trees were needed, and that the exploitation of bast caused an irreversible transformation in forest composition, at least in closer vicinity to the settlements. This mutual relationship between forest and human impact has already been quoted by Kurt Altorfer (2010: 86). He writes that "… lime trees do not naturally grow in dense forest formations, and lime bast had not been available in huge quantities. It

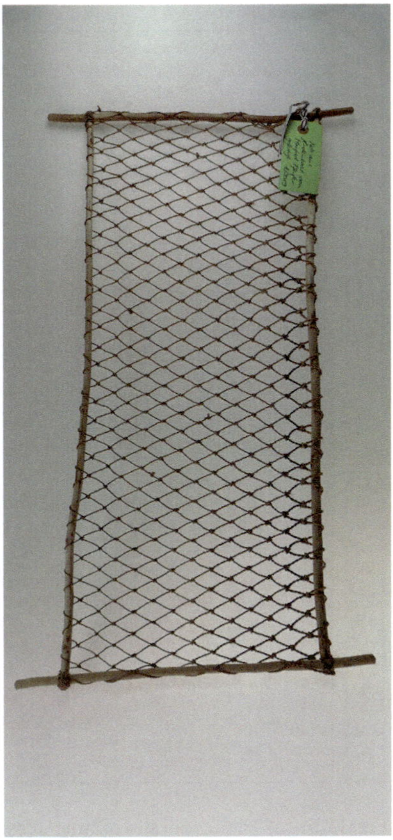

Fig. 11.4 Reconstructed net made of lime bast by Manfred Pfeifer, Lübeck. (Photo by J. Müller-Edzards)

is therefore questionable if already during Neolithic times a systematic forest management was executed". The withdrawal of lime trees in the vicinity of Neolithic villages could probably also have led to a bottleneck in the supply of this important raw material.

Lime trees spread throughout Central Europe during the Atlantic period, i.e. from the seventh millennium BC onwards (Lang 1994). As seen in Fig. 11.5a, b, the extent of lime tree forests declined between the fourth and second millennium BC (sixth and fourth millennium BP = before present) in Europe (for details see Giesecke et al. 2016, 2017). A closer look at high-resolution pollen diagrams in the vicinity of Neolithic settlements would probably clarify whether or not this decline

a

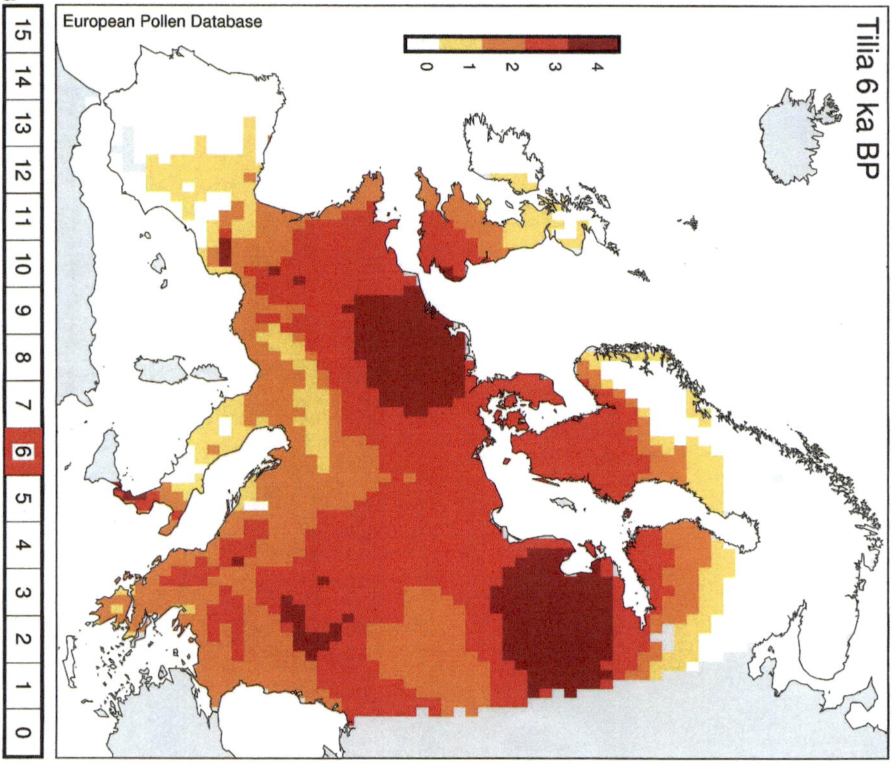

Fig. 11.5 (**a**) The cover ratio of lime trees during the sixth millennium BP (before present) according to the European Pollen Database, after Giesecke et al. 2017. (**b**) The cover ratio of lime trees during the fourth millennium BP (before present) according to the European Pollen Database, after Giesecke et al. 2017

can be explained by human impact or by the naturally caused displacement of the formerly dominating trees, oak and lime, by beech trees. In any case, we can state that Neolithic wetland sites were generally abandoned after only one decade (Ebersbach 2010). A probable cause for this pattern of abandonment is the depletion of important raw materials, such as lime bast.

b

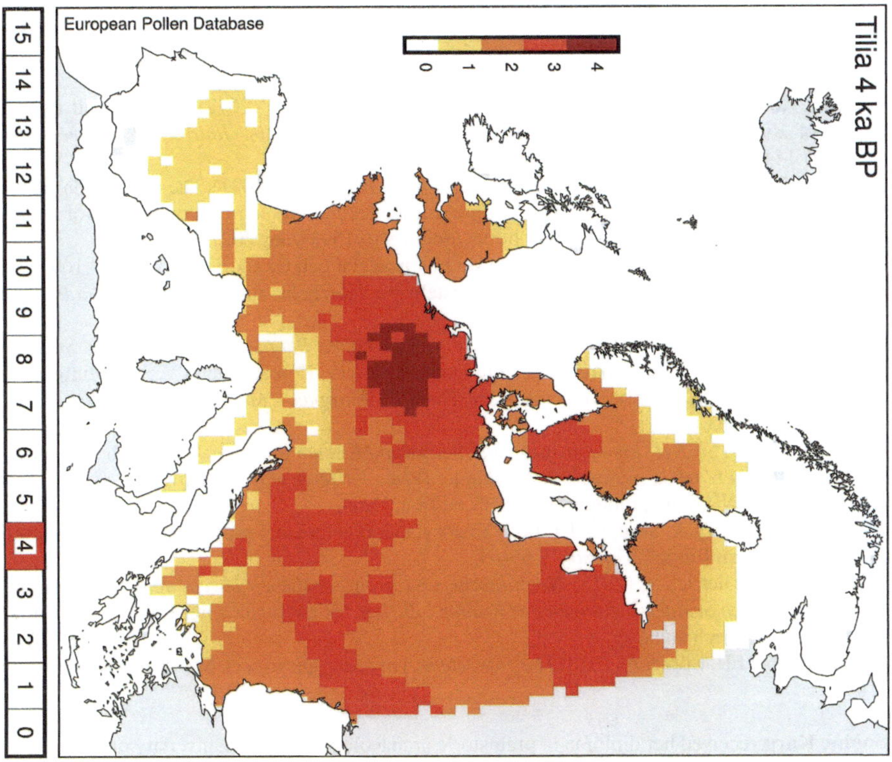

Fig. 11.5 (continued)

Acknowledgements I am grateful to Morten Ravn from the Viking Ship Museum in Roskilde, Denmark for making the unpublished report of Ole Magnus available. Manfred Pfeifer from the Geschichtspark Roter Hahn in Lübeck is a fantastic collaboration partner when acting on the practical implementation of ideas. Emily Schalk improved the English text and Jan Müller-Edzards helped with Fig. 11.1.

References

Altorfer K (ed.) (2010) *Die prähistorischen Feuchtbodensiedlungen am Südrand des Pfäffikersees. Eine archäologische Bestandesaufnahme der Stationen Wetzikon-Robenhausen und Wetzikon-Himmerich*. Monographien der Kantonsarchäologie Zürich 41, Zürich and Egg.

Czarnowski E, Karg S (1986) Neue archäologische Untersuchungen am Strandbad Allensbach. *Allensbacher Almanach* 37:6–8.

Ebersbach R (2010) Vom Entstehen und Vergehen – Überlegungen zur Dynamik von Feuchtbodenhäusern und -siedlungen. In: Matuschik I, Strahm C, Eberschweiler B, Fingerlin G, Hafner A, Kinsky M, Mainberger M, Schöbel G (eds) *Vernetzungen. Aspekte siedlung-*

sarchäologischer Forschung. Festschrift für Helmut Schlichtherle zum 60. Geburtstag. Lavori Verlag, Freiburg i. Br., p 41–50.

Giesecke T, Brewer S, Finsinger W, Leydet M, Bradshaw RHW (2016) Interpolated maps of pollen percentages in Europe over the last 15,000 years, links to PDF files. *PANGAEA*. https://doi.org/10.1594/PANGAEA.865096.

Giesecke T, Brewer S, Finsinger W, Leydet M, Bradshaw RHW (2017) Patterns and dynamics of European vegetation change over the last 15,000 years. *Journal of Biogeography* 44(7):1441–1456. https://doi.org/10.1111/jbi.12974.

Grabundžija A, Schlichtherle H, Leuzinger U, Schier W, Karg S (2021) The interaction of distant technologies: Bridging Central Europe using a techno-typological comparison of spindle whorls. *Antiquity* 95(381):627–647. https://doi.org/10.15184/aqy.2021.6.

Granlund J (1943) Lindbast och träbast. En studie i material och teknik. *Folk-liv* 7:166–169.

Karg S (2020) Textilhandwerk in der Jungsteinzeit. Hecheln aus Dornen. *Archäologie in Deutschland* 1:7.

Karg S, Weber E (2019) *Heilsam, kleidsam, wundersam. Pflanzen im Alltag der Steinzeit.* Sonderheft der Zeitschrift Archäologie in Deutschland 11. Konrad Theiss Verlag, Stuttgart.

Lang G (1994) *Quartäre Vegetationsgeschichte Europas. Methoden und Ergebnisse.* G. Fischer, Jena.

Magnus O (2006) *Sif Ege. Lindebastprojekt. Vikingship Museum Roskilde.* Unpublished report.

Médard F (2010) *L'art du tissage au néolithique. IVème – IIIème millénaire avant J.-C. en Suisse.* Collection CNR monographies 30. CNRS edition, Paris.

Myking T, Hertzberg A, Skrøppa T (2005) History, manufacture and properties of lime bast cordage in northern Europe. *Forestry* 5:65–71.

Rast-Eicher A, Dietrich A (2015) *Neolithische und bronzezeitliche Gewebe und Geflechte. Die Funde aus den Seeufersiedlungen des Kantons Zürich.* Monographien der Kantonsarchäologie Zürich 46, Zürich and Egg.

Schweingruber FH, Börner A (2018) *The plant stem. A microscopic aspect.* Springer Nature, Charm.

Sabine Karg received her diploma in prehistoric archaeology from the University of Tübingen in 1987 and her PhD in botany from the University of Basel in 1994. Since then she has continuously worked at different European research institutions and museums and since 1987 she teaches courses in archaeobotany and plant resources in antiquity. Her main interests are the long-term development in the interplay of humans and their natural environment, subsistence strategies in the past, the transition from hunter-gatherer societies to farming communities, ancient know-how of plant fibre technologies, trade with food products in the past, and food traditions.

Chapter 12
Food for Thought or Threads for Weaving: Can We Identify the Uses for Ancient Flax Seeds Discovered in the Southern Levant?

Deborah Cassuto , Andrea Orendi , and Itzhaq Shai

Abstract *Linum usitatissimum* L. was, and still is, cultivated for either flax fibres or for oil-rich linseeds. Depending on their intended purpose, flax plants differ in height, branching and quantity of capsules. Likewise, the linseeds themselves vary in size: linseeds for oil production are larger and heavier than those for flax plants purposed for fibres. Archaeobotanical studies of ancient linseeds have succeeded in identifying these motives, by comparing the lengths and width of waterlogged seeds. However, carbonization of archaeobotanical macro remains, found in southern Levantine contexts, often results in shrinkage and deformation of seeds and limits the meaningfulness of morphometric analyses. At Tel Burna (c. 30 km southwest of Jerusalem), linseeds found in Iron Age strata, along with groups of loom weights representing weaving on warp-weighted looms, may point towards flax cultivation for the production of linen. Excavations concentrating on the Iron Age IIA context (950-900 BCE) outside the fortification wall, exposed an area of destruction containing hundreds of charred linseeds in proximity to complete storage vessels. These carbonized linseed finds from Tel Burna may provide the basis for establishing a method to distinguish whether these archaeobotanical macro remains are residues of flax cultivation intended for oil or for textile production.

D. Cassuto (✉)
Martin (Szusz) Department of Land of Israel Studies and Archaeology, and the RIAB Minerva Center, Bar Ilan University, Ramat Gan, Israel

Albright Institute of Archaeological Research, Jerusalem, Israel
e-mail: debicass@aiar.org

A. Orendi
ArchaeoConnect GmbH, Department of Archaeological Sciences, Archaeobotany Laboratory, Tübingen, Germany
e-mail: andrea.orendi@archaeoconnect.de

I. Shai
Institute of Archaeology, Ariel University, Ariel, Israel
e-mail: itzhaqsh@ariel.ac.il

A. Ulanowska et al. (eds.), *Ancient Textile Production from an Interdisciplinary Perspective*, Interdisciplinary Contributions to Archaeology,
https://doi.org/10.1007/978-3-030-92170-5_12

Keywords Archaeobotany · Iron Age · Flax · Linseed · Tel Burna · Southern Levant

12.1 Introduction[1]

The absence of tangible textiles in archaeological contexts of the southern Levant poses a conundrum for scholars interested in reconstructing the impact of textile production and consumption on daily life in antiquity.[2] For this reason, comprehensive studies of ancient textiles have benefitted from incorporating data from complementary research fields and resources from multiple disciplines.[3] One such field is that of archaeobotany. Flax seeds, also known as linseed (*Linum usititatissimum* L.), have been found in archaeological contexts throughout the ancient Near East, and have frequently been referred to as evidence of flax cultivation for textile production. The discovery of cultivated linseed (*Linum usititatissimum* L.) throughout the ancient Near East has demonstrated that flax cultivation had a significant role in the early development and spread of Neolithic agriculture (Zohary et al. 2013). Flax, however, is not only cultivated for the long fibres stored in its stems, but also for its oil-rich seeds, as Zohary, Hopf and Weiss (2013: 101) point out at the beginning of their discussion on flax: "Flax, (…) was a principle oil and fibre source in the Old World and probably the earliest cultivated plant for weaving clothes." Changes in the biometric measurements of flax seeds since their early cultivation (Karg 2011; Karg et al. 2018) have been interpreted as reflecting general shifts in the primary consumption patterns of flax plants, whether cultivated for fibres or for oil-rich seeds (Karg et al. 2018).

The archaeobotanical study of flax seeds has enabled researchers to reconstruct ancient crop management styles related to the cultivation of flax in the southern Levant (Riehl 2008, 2009; Riehl and Shai 2015; Orendi 2020b). The decisions behind crop management are socially determined, taking into consideration the economic benefits of a particular type of produce, environmental factors, such as soil and precipitation, and cultural factors which vary depending on politics/economy

[1] This study was funded by the Israel Science Foundation Grants No. 522/16 and 257/19 (I.S.)

[2] We would like to thank the organisers of the session at the 25th EAA annual meeting in Bern (2019), and the editors of this publication, for the opportunity to share our work here. We are also grateful to the readers for their valuable comments and to Agata Ulanowska for her assistance in preparing this paper for publication.

[3] One of the first to bring in ethnographic studies to interpret archaeological textile tools has been Crowfoot (1931), followed by Barber's (1991) tome which integrates archaeological, ethnographic, iconographic, and written sources to piece together the history of textile production, Vogelsang-Eastwood (2000), and notably the growing corpus of research and publications produced by the Centre for Textile Research (CTR), https://ctr.hum.ku.dk/articlesbooks/.

and ecological aspects of a given geographic region, and/or period in time (Riehl 2008, 2009). The study of linseed found in Iron Age strata at Tel Burna, in the southern Levant,[4] provides an opportunity to contribute to our understanding of the role of flax cultivation in the local agricultural economy and to assess whether they provide evidence for textile production, or alternatively, for linseed oil production.

12.2 *Linum ustitatissimum* L. – A Most Useful Plant

The Latin species name for flax, *Linum ustitatissimum* L., identifies the serviceability and versatile consumption of this plant. The plant's seeds are pressed for oil, and the fibres found in the stems are processed into linen threads for producing textiles. Although archaeobotanical studies appear to indicate that early flax was most likely cultivated as an oil crop, and written sources describe applications of linseed oil for hair dressing, for illumination, preservatives, lubricants, and medicine (Kislev et al. 2011),[5] the overwhelming majority of textual and iconographic evidence referring to flax in the ancient Near East deals with it as the source for linen threads. Textual evidence for the production of linen cloth has been found throughout the ancient Near East in administrative, royal and cultic documents where it is frequently referred to as a luxury item, e.g., Mesopotamia (Bongenaar 1997; Zawadzki 2006, 2010, 2013), Ugarit (Vita 2010), Hattusa (Vigo 2010), and Mycenae (Rougemont 2007; Del Freo et al. 2010). A depiction of flax preparation for weaving features in the Sumerian love poem on the Courtship of Inanna and Dumuzi (Kramer and Wolkstein 1983). In Egypt, where linen was a prime commodity and the conditions for preservation of linen textiles is high, linens feature in commodity labels (linen lists) from Protodynastic tombs, appear in wall paintings and wooden funerary models depicting cultivation and processing from as early as the Old Kingdom (Vogelsang-Eastwood 2000; Jones 2010). In the southern Levant administrative dockets were found at Hazor requisitioning fine linen garments for Mari (Horowitz and Oshima 2002, 2006, 2010; Horowitz and Wasserman 2000, 2004). In the Masoretic text references are made to flax, *pišta/pištim*, referring to the linen threads, finished garments (e.g. Lev. 13:47–48; Deut. 22:11) and plants in the fields (e.g. Ex. 9:31), lamp wicks (e.g. Isa. 43:17) and the drying of stalks on rooftops (Jos. 2:6).[6]

[4] The traditional dating of the Iron Age II in the southern Levant spans the time between ~1000–586 BCE (see for example Lee et al. 2013; and contra Finkelstein and Piasetzky 2011 in which the beginning of this period should be lower by several decades).

[5] Renfrew (1985) notes the prominence of references in Akkadian texts to oil from sesame seeds over flax seeds.

[6] 1 Chronicles 4:21 refers to a family/guild of linen *būtz*-workers located at Beit-ashbea in the eastern Shephelah. *Būtz*, another term used in the Old Testament for flax, apparently referred to such a superior textile that in later translations, the name *byssum* or *byssus* was translated as fine

It is important to keep in mind that this predominance for linen textiles over linseed oil may more likely be indicative of the attention given to textiles, particularly high-quality linen textiles, and the roles they played in ancient societies, than of an agricultural preference for growing flax for fibres rather than for seeds.

The species *Linum usitatissimum* L. specifies the cultivated flax which belongs to the family of Linaceae. The genus *Linum* contains some 200 species which grow mainly in subtropical and temperate climate zones (Dambroth and Seehuber 1988: 15). Flax is a dioecious plant which self-pollinates like most of the founder crops (Zohary et al. 2013). However, insect borne cross-pollination is also possible (Diederichsen and Richards 2003). Notably, evidence for local cultivation of *Linum usitatissimum* L. has not been found in any of the pollen records studied to-date in the southern Levant, from the Sea of Galilee (Kinneret) (Baruch 1986), the Dead Sea (Litt et al. 2012) and the northern Golan Heights (Neumann et al. 2007). Whereas it is possible that flax was not cultivated in these areas, textual evidence from the Roman period accentuates the high-quality linens which were produced in the area of the Sea of Galilee (e.g. *Diokletians Preisedict* tr. Lauffer 1971: 169–170; Safrai 1985). It is, however, also likely that the heavy pollen of *Linum* could not reach the wetland sediments via wind, and what is more, that the range of insects which could have transported flax pollen is limited (Leuzinger and Rast-Eicher 2011).

A herbaceous annual plant, flax is characterized by a thin and upright stalk growing to a height of between 50 and 120 cm. As it is a variable plant, contingent to climate, soils and cultivation methods, different varieties of flax have evolved (Dambroth and Seehuber 1988: 16; Herbig and Maier 2011).

The species is subdivided into four convarieties, two of which represent the plants cultivated for either fibres or oil: the convariety *mediterraneum* for its larger oil-rich seeds and the taller growing convariety for *elongatum* for fibres (Dambroth and Seehuber 1988: 16). There are also intermediate varieties with moderate growth and seed size which can be used for both purposes. Notably, in these cases, when cultivated for the best quality fibres, the flax should be harvested before the seeds have completely matured (Barber 1991).

The convariety *mediterraneaum*, is the shorter of the two (<70 cm) and has multiple branches increasing the number of seed capsules per plant (up to 100 capsules containing 10 seeds each). The seeds, which weigh between 5 and 15 mg contain 30 to 44% oil found in the cells of the epidermis and the cotyledons (van Zeist 1985; Dambroth and Seehuber 1988: 23; Diederichsen and Richards 2003; Herbig and Maier 2011; Kislev et al. 2011).

The pressed linseed oil can be used for lighting, lubricating or for medicinal purposes. It is also known for its quick-drying qualities and has been used in varnish mixtures. The oil produced from cold pressing spoils quickly and must be used immediately or stored in well-sealed vessels. Likewise, the seeds themselves require proper storage if not immediately used, as they spoil easily in warm weather and

---

linen, silk and was even adopted as the term for the prestigious sea silk, filaments of the fan shell (Maeder 2017).

contact with water produces a toxic compound hydrogen cyanide (Renfrew 1973: 124; Kislev et al. 2011).[7]

The fibres are procured from the fibrous cells found in the stem of the *elongatum* convariety which grows taller (>70 cm) and contains smaller seeds (weighing 3 to 5 mg). They are sown densely to minimize branching and hence have fewer than 30 capsules (Diederichsen and Richards 2003; Herbig and Maier 2011; Kislev et al. 2011).[8]

To access the long strong fibres from the plant stems, the harvested flax stems require a lengthy process of drying, retting (rotting in either designated baths or in dewy fields) to break down the gluey pectin which encases the fibres, and then for the stems to be dried again. The dried stems can easily be transported and stored, later to be hackled, combed and spun to produce fine threads. The properties of fine linen thread, fine, smooth, cool, and light weight, make flax ideal for garments worn in warm weather (Harris 2009, 105–6, Table 18.2).

12.3 Cultivation – Flax vs. Oil

Flax is sensitive to extreme temperatures and requires substantial amounts of water, from either rainfall or irrigation. In the southern Levant, it is a winter-annual, sown in autumn and harvested in spring. However, it is also known to deplete soil nutrients, necessitating crop rotations (Helbaek 1959; Charles 1985; Borowski 1987; Zohary et al. 2013).

As modern flax requires different quantities of water for different purposes, scholars maintain that the requirements for ancient flax were similar. Fibre flax requires c. 700 mm of annual water supply and should be sown densely to inhibit branching and encourage long stems. It grows best in well-drained loams and requires continuous weeding in the early stages of growth (Renfrew 1973).

For the highest yield of quality fibres, the plants must be pulled when the stem begins to brown and just before the seeds ripen. While plants cultivated for their seeds require less water (minimum 300 mm per year), they are generally cultivated in thinner stands enable the plants to grow side branches resulting in more capsules and hence more linseed. In this case, the capsules would need to be ripe for oil processing and harvesting would entail only cutting off the capsules from the stems. However, both harvesting methods can be combined if the intent is to utilize both the oil and the fibres (Riehl 2009; Kislev et al. 2011).

[7] In addition, today seeds are consumed for their protein-high nutritional value, while the by-product of hot-pressing, namely the press-cakes, are high in protein and used today for fodder.

[8] Renfrew (1973: 123) notes that the differences in seed size in ancient contexts may actually be the result of artificial irrigation in drier regions.

12.4 Chronological and Regional Distribution of Archaeobotanical Flax and Linen Finds

Once domesticated, flax cultivation underwent additional developments, particularly during the 3000 years of the Bronze and Iron Ages in the southern Levant (Orendi 2020b). Excavations throughout the region have recovered significant quantities of flax remains (Riehl 2009; Orendi 2020b), mainly represented by charred linseed, with few capsules and stem fragments. A drastic decrease in flax finds takes place from the Early Bronze Age II/III to the Early Bronze Age IV (Riehl 2009).[9] This scarcity of flax finds lasts until the end of the Late Bronze Age, corresponding with a shift in the locations of sites with flax finds from the semiarid and arid regions to areas with higher modern mean precipitation. At the onset of the Iron Age, we observe an increase in linseed recovered (Orendi 2020b). Notable amongst these are the thousands of charred seeds found at Tel Beth Shean (Kislev et al. 2009) and Tell Deir Alla (van Zeist and Heeres 1973), and now most recently in an Iron Age IIA context at Tel Burna.

Interestingly, there is a similar trend in the archaeological evidence for linen textiles dating to the Bronze and Iron Ages. Nonetheless, the fluctuations in the linseed and linen textile finds appear to reflect the same decreasing and increasing trends in flax cultivation during the Bronze and Iron Ages (Orendi 2020a).

Historically, the cultivation of flax, and in turn the production of linen textiles, was affected by variations in climatic and socio-cultural developments such as the climatic deterioration in the Early Bronze Age IV (Riehl 2008), and the Egyptian suzerainty over Canaan in the Late Bronze Age (Orendi 2020a).

Notwithstanding the multiple applications for flax, the definitive aims for its cultivation in the ancient southern Levant remain unclear. Therefore, a broader investigation of the charred flax seeds found at Tel Burna could shed some light on this matter.

12.5 The Site of Tel Burna

The site of Tel Burna is located in the region known as the Shephelah (Fig. 12.1), a strip of land between the southern coastal plain and the Judean mountain range (Uziel and Shai 2010; Shai et al. 2012, 2014). Throughout the Bronze and Iron Ages, this region was densely populated. Major Bronze and Iron Age sites such as Lachish, Ekron, Tell eṣ-Ṣâfī/Gath, Khirbet Qaiyafa, Azekah and Beth Shemesh are no more than 10–30 kilometers from the site.

The site transcends the northern bank of Nahal Guvrin, one of several east-west routes from the sea to the mountains, and is surrounded by fields of rich brown and

[9] The Early Bronze Age II/III dates from 3000 to 2400 BCE, subdivided into the Early Bronze Age II from 3000 to 2850 and the Early Bronze Age III from 2850 to 2400.

Fig. 12.1 Map of the Shephelah

Fig. 12.2 Tel Burna overview with fields and Nahal Guvrin. (Photography Griffin)

pale rendzina soils, suitable for extensive agricultural activities, as it is still farmed today (Fig. 12.2). Geomorphological studies on three wadis south of Tel Burna (Lachish, Adorayim, Shiqma) point to alluvial events during the Chalcolithic/Early Bronze period forming rich alluvial soils along the wadi beds (Rosen 1986), which would most likely be the same for Nahal Guvrin.

The lower Shephelah is located within the sub-humid belt of the southern Levant with lower precipitation than to the north, which makes this strip of land particularly susceptible to slight shifts in precipitation. Modern annual precipitation rate in the Shephelah ranges between 400 and 600 mm. Calculated palaeo-rainfall at the Soreq-Cave, in the eastern Shephela, lies between 430 and 510 mm for the Late Bronze/Iron Age and corresponds to the mean annual precipitation rates of modern times (Bar-Matthews and Ayalon 2004: Fig. 12). The measurements of $\delta^{18}O$ isotopes taken from the Soreq Cave show no significant amplitudes of increased or decreased rainfall during the Late Bronze and Iron Ages (Bar-Matthews and Ayalon 2004: Fig. 10a). Given that the analysis of ^{13}C-isotope on barley grains from Iron Age II levels at Tel Burna indicates that there was ample water supply for the barley plants (Riehl and Shai 2015: Fig. 3), it can be presumed that in drier seasons the barley fields were irrigated tapping water from the adjacent river (Riehl and Shai 2015: 526). In a similar manner, despite the annual precipitation being only adequate for the rain-fed agriculture of flax for oil, irrigation methods could have been employed to water flax for fibre.

Archaeological work at Tel Burna, which began in 2010, has focused on five areas (Fig. 12.3), covering different parts of the summit, the lower plateaus on both the eastern and the western sides of the tell and two areas along the slope. Area B1 is located on a terrace on the western side of the summit. The finds in this area are well dated to the Late Bronze Age (Shai et al. 2015; Shai and McKinny 2020) and will not be presented here.

Area A1 located on the NE corner of the summit was opened to expose more of the fortification wall which was visible on the surface. Excavations have shown that the fortifications were built in the Iron Age IIA and functioned continuously until at

Fig. 12.3 Tel Burna excavation areas. (Photography Griffin)

least the end of the Iron Age IIB. In the Iron Age IIC the inner wall went out of use (Shai et al. 2012). In light of this study, it should be mentioned that a small installation containing c. 40 loom weights, well dated to the ninth century BCE, was discovered within the wall (Shai et al. 2012: 147). Area A2 is located approximately mid-center of the area enclosed by the summit. In this area a large building with several phases from the Iron Age IIA through the Iron Age IIB has been exposed. In a large courtyard, a concentration of c. 30 loom weights, along with a plastered drainage installation, suggests that different stages of textile production took place here (Shai 2017: 50). The building was sealed by poorly preserved Iron Age IIC architecture and a series of silos (Shai 2017). The Iron Age gate was discovered on the southern face of the summit in Area G. Area B2 is located on the north-west corner of the summit, and the goals here were to date the fortification and to have a stratigraphic section that connected the Late Bronze Age remains in Area B1 with the summit. The stratigraphic sequence includes poorly preserved Persian Period architecture, an Iron Age IIC domestic building, the Iron Age IIA and IIB fortification wall, as well as an early Iron Age destruction level west of domestic remains (outside and earlier than the fortification system) (McKinny et al. 2019: 9–10).

In Area B2, west of the fortification wall, excavations revealed the remains of an early Iron Age IIA building which had been destroyed in a heavy fire. This building activity pre-dates the fortification system, as the wall in Area B2 seals this destruction. The finds include early Iron Age IIA storage, cooking and serving vessels (see for example: McKinny et al. 2019: Fig. 110 for the storage jar assemblage). In addition, the botanical remains found in this context are extremely rich and among the various species was a concentration of linseed.

12.5.1 The Linseed Data from Tel Burna

Ongoing systematic archaeobotanical sampling has been conducted at the Tel Burna excavations since 2011. The sampling has covered most of the contexts exposed during the various seasons of excavation. This has resulted in more than 200 archaeobotanical samples providing macro remains from the Late Bronze and Iron Age (Shai et al. 2014; Orendi et al. 2017). The analysis of the seeds and fruits from the Iron Age strata attests to the agricultural economy of the Judahite town. The archaeobotanical material is characterized by a wide variety of cultivated plants comprising several demanding crops such as flax (*Linum usitatissimum* L.). The cultivation of flax plants is attested by both the discovery of thousands of charred flax seeds, as well as the remnants of dodder seeds (*Cuscuta* sp.). The flax dodder (*Cuscuta epilinum* Weihe) is a typical parasite plant found together with cultivated flax. In Hjelmquist's (1950) proposal that the history of flax weeds be included in any discussion concerning the origins of flax, he identifies *Cuscuta epilinum* Weihe as one of the earliest species of flax weeds in the Ancient Near East. This species, he notes, appears in isolated areas "remote from the coast" (Hjelmquist 1950: 282) in the southern Levant, which would suggest inland areas such as the Shephelah.

This suggestion is supported in the *Flora Palaestina* (Feinbrun-Dothan 1977/78: 47) where the cultivated flax fields of the Shephelah are noted as the local habitat for *Cuscuta epilinum* Weihe. As for the dodder seeds from Tel Burna, due to their carbonization, it was not possible to identify their exact species. Nevertheless, it is clear that the dodder finds represent flax weed plants growing in the flax fields, as the linseed and the dodder seeds were found within the same contexts and samples (Orendi et al. 2017: 177).

The high presence of charred linseed throughout the Iron Age strata attests to the high value of this crop at Tel Burna (Table 12.1). In the first years of excavation, flax seeds were found primarily in the storage silos dating to the Iron Age IIC dispersed across the summit. The flax remains were either stored inside the silos and/or by chance entered the silos during occupation of the site and/or following its abandonment. Notably, the number of seeds per context is low. Nonetheless, the linseed finds from Tel Burna contribute to understanding flax cultivation in the southern Levant during the Bronze and Iron Ages, as linseed finds in this area are quite rare (Orendi 2020b).

In the course of the 2017 and 2018 seasons of excavation, thousands of charred flax seeds were exposed in two localities (Area B2 and Area A1) in the destruction level of the Iron Age IIA directly located outside of the enclosure wall. Smashed vessels adjacent to the seeds may have provided storage facilities for them. The analysis of the latest samples is ongoing, hence total numbers of the macro remains from these samples are still lacking (Table 12.1).

Table 12.1 The carbonized linseeds from Tel Burna listed by their find contexts

Context	N of samples	N of linseeds	Total N of macro remains from context	Period
Silo 21205	4	5	147	Iron Age IIC
Silo 21217	1	1	33	Iron Age IIC
Silo 32103	11	97	1777	Iron Age IIC
Silo 31404	1	1	13	Iron Age IIC
Silo 31207	1	1	24	Iron Age IIC
Pit 52911	1	1	49	Iron Age IIB
Vicinity of Tabun 62211	1	1	12	Iron Age IIB
Ash layer 62215	1	5	12	Iron Age IIB
Destruction layer B2		Several thousand		Iron Age IIA
Destruction layer A1		Several thousand		Iron Age IIA

N Number of macro remains

The objective for the cultivation of flax, whether to produce linseed for oil or threads for linen textiles, at Tel Burna must be discussed, relying on the data drawn from the charred seeds themselves. As mentioned above the sizes of the seeds differ between the modern varieties of flax plants grown for oil and for textile production, and it is assumed that the size of the seed also points to the intended use of flax plants in ancient times. Biometric analyses of waterlogged flax seeds have resulted in comprehensive differentiation of seeds deriving from oil and fibre flax plants from the Neolithic lake shore sites of southwest Germany (Herbig and Maier 2011). According to the length and breadth measurements of the waterlogged linseed it was possible to assign the smaller seeds (breadth on average: 1.7 mm; length on average: 3 mm) to textiles plants and the larger seeds (breadth on average: 2.1 mm; length on average: 3.5 mm) to oil plants (Herbig and Maier 2011: Fig. 3, 4).

Metric analysis on charred linseed poses difficulties as the carbonization process deforms the seeds. The degree of deformation by bloating and shrinking of the seeds depends on the intensity of heat, the duration of heat, and the oil content of the seeds (Figs. 12.4 and12.5; Helbaek 1959: 107).

Fig. 12.4 Image of a bloated linseed (Tel Burna, Area B2, Iron Age IIA) caused by carbonization compared to a modern uncarbonized linseed. (Reference collection HEP-INA Arch.bot. Kat nr. 6662, photo by A. Orendi)

Fig. 12.5 Image of a carbonized (Tel Burna, Area B2, Iron Age IIA) and a modern linseed. (Reference collection HEP-INA Arch.bot. Kat nr. 6662, photo by A. Orendi)

Helbaek (1959: 108) defines a shrinking effect of 15% caused by carbonization. Therefore, he advises handling comparable measurements of charred linseed from different sites with care. Nevertheless, for now the comparison of the charred linseed material from Tel Burna with other (charred)[10] linseed finds from Near Eastern sites is the most promising way to get conclusions about the usage of the cultivated flax plants at Iron Age Tel Burna.

Helbaek's (1959: 108–110, Tab. 8) morphometric analyses of various linseed finds from Syria, Iraq and Egypt lead to the assumption that the seed size in semi-arid and arid environments reflect whether the flax plants were cultivated by irrigation or were rain-fed. The seeds come from different sites (e.g., Brak, Ur, Fayum, Hama) and periods of time, and were discovered in various degrees of preservation (charred, dried, and imprints of linseed). But Helbaek (1959: 109) concluded that the seeds with increased length and width measurements (length: 4.21–5.03 mm; width: 2.32–2.56) originate from localities where irrigation was the primary cultivation method of arable fields, as in the Nile Valley or in the plains of the Euphrat-Tigris Valley. As the flax seeds from Tell Brak (length: 4.04 mm; width: 2.20 mm) and Arpachiyah (length: 3.84 mm; width: 1.83 mm) were smaller, Helbaek (1959:119) considered that the agriculture at both of these sites, in the foothills of

[10] There are also sites from the Near East with linseed finds which are not charred.

Syria and Iraq, would have been rain-fed (Helbaek 1959: 119). However, recent studies on the archaeological and archaeobotanical remains from Tell Brak, have raised the possibility that irrigation was, in fact, employed in the area of the Upper Khabur basin (Deckers and Riehl 2007; Hald and Charles 2008). Nonetheless, Helbaek refrained from associating linseed size with cultivation objectives, despite the availability of modern records for cultivation of flax in Iraq which document linseed oil (Guest 1930) and in Egypt, which documented textiles (Helbaek 1959:107) as primary objectives for flax cultivation.

The average values of the charred flax seeds from Tel Burna (Table 12.2; Fig. 12.6) lie between the values for the rain-fed and for the irrigated flax fields in Iraq and Egypt. When adding the shrinkage factor of 15% the values for Tel Burna increase to an average length of 4.77 mm. This value falls into the measurements of the irrigated flax plants, but shows no indication as to whether the plants were cultivated for oil or textile production. If the flax fields at Tel Burna were rain-fed, the mean length of the seeds might point to cultivation for oil, as the mean length is greater than the values for rain-fed seeds from the Iraq sites which nowadays is dedicated to producing flax seeds for oil (Helbaek 1959).

When compared with linseed finds from the various European sites, the finds from Tel Burna show increased dimensions. At the Neolithic lake-shore sites in southwestern Germany, implementing morphometric analyses, it was possible to distinguish between fibre and linseed flax (Herbig and Maier 2011). However, the dimensions of the waterlogged seeds of both varieties are, on average, almost 1 mm smaller than the charred flax seeds from Tel Burna (Herbig and Maier 2011: Fig. 3; Tab. 2). Further morphometric studies, this time on charred flax seeds, from Roman Medina, Italy (Bosi et al. 2011) and from Iron Age Uppåkra 2:25, Sweden (Larsson 2013), indicate the objectives of flax cultivation at these sites. In light of the small dimensions of the seeds (mean length: 2.63 mm; mean width: 1.40 mm), corroborated by written sources, it was concluded that the seeds from Roman Modena, which were discovered in a drainage canal, derive from flax plants cultivated for textile production (Bosi et al. 2011). The mean values for the charred seeds from Uppåkra 2:25 were 3.86 mm in length and 2.02 mm in breadth. The seeds were

Table 12.2 Measurements of 50 carbonized linseeds from the Iron Age IIA (Area B2) destruction layer of Tel Burna

	Measurement	Calculated value with shrinkage factor of 15%	Measurement	Calculated value with shrinkage factor of 15%
Average length/width in mm	4.15	4.77	1.94	2.23
Minimum length/width in mm	3.39	3.90	1.59	1.83
Maximum length/width in mm	4.70	5.41	2.31	2.66

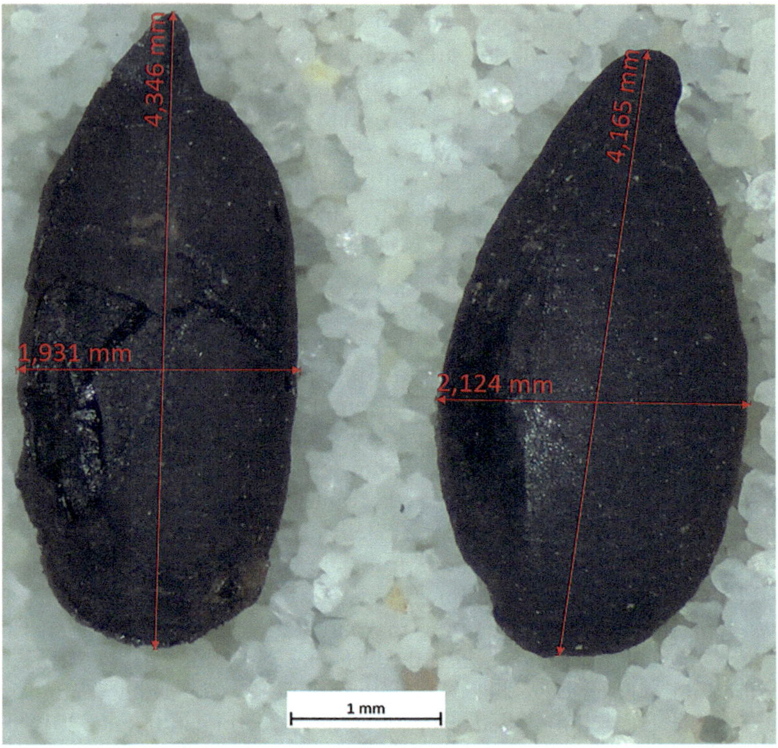

Fig. 12.6 Two carbonized linseeds from Tel Burna (Area B2, Iron Age IIA) with annotated measurements. (Photo by A. Orendi)

lumped together into seed conglomerates which had been exposed in different levels of a pit filling. One of the conglomerates was made up of charred seeds of Camelina (*Camelina sativa* L.), also known as Gold-of-Pleasure, which is known as an oil producing plant. As the process for hot-pressing the seeds for oil requires the seeds to be pre-roasted (Ertuğ 2000: 181), Larsson (2013: 516) suggests that these seed conglomerations were the failed result of pre-roasting the seeds, i.e. the hot oil released from the oil-bearing seeds had glued the seeds together. Finally, the seed conglomerations were discarded into the pit. Similar lumps of consolidated linseed have been found at Tel Beth Shean (Kislev et al. 2011: 581) and, presented here, at Tel Burna. Nevertheless, these lumps do not attest definitively to the production of linseed oil.

The conglomerate of linseed found in Area B2 at Tel Burna came from a destruction layer which dates to the early Iron Age IIA. Charring experiments (Gustafsson 2000; Märkle and Rösch 2008) have shown that flax seeds have a greater probability of surviving the carbonization process under reducing conditions depending on the duration of heat exposure, even with high temperatures of up to 600°C, while in oxidizing conditions the flax seeds completely combust within 1 h of temperatures up to 400°C. Evidence for extensive conflagration in the context where they were found indicates that the carbonization and conglomeration of the seeds was the outcome of a random occurrence, contrary to the intentional heating of the seeds for pressing. This was also likely the case for the storage of charred lentils (*Lens culinaris* MEDIK) found adjacent to the linseed accumulation.[11]

By contrast, the flax seeds found at Tel Beth Shean were discovered in basins, presumably for storage. Morphometric analysis on the linseed material from Tel Beth Shean resulted in larger seeds than at Tel Burna. The mean length of the charred seeds is 4.72 mm and the mean width is 2.41 mm (Kislev et al. 2009: Tab. 17.4). The charred remains were stored in basins made of bricks (Locus 28817 and 88844) located in Houses SK and SG (Panitz Cohen and Mazar 2009), both of which are attributed to the Egyptian presence at Tel Beth Shean and in the southern Levant during the Late Bronze Age. The basin from House SK was divided into three compartments. Inside one of the compartments masses of charred flax seeds were found (Locus 28817; Panitz Cohen and Mazar 2009: 147, Photo 4.68). The Storage Context 28817 consisted of more than 500,000 seeds which were significantly clean of any weeds. In addition, more than 4000 charred flax seeds (Locus 28814) were found scattered around Compartment 28817. Traces of burning within the whole room in which the basin for House SK was located indicate that the destruction of this room and the charring of the linseed material were caused by conflagration (Panitz Cohen and Mazar 2009: 148), comparable to the linseed context in Area B2 at Tel Burna.

These seeds may have been stored for several reasons, for example to be sown in the following year (Kislev et al. 2011: 582). Another reason for storing linseed may be pressing. As pressed linseed oil becomes rancid quickly, well-stored linseed would have been preserved for pressing oil as needed (Kislev et al. 2011: 583); as is suggested by the conglomerated charred linseed and fig seeds found adjacent to a hearth at Rās en-Numēra for food preparations (White et al. 2014: 372).

The stored linseed at Beth Shean may, in fact, be indicative of linen production, as is suggested by the rare discovery of several fragments of locally produced linen wrapped around a hoard of hack silver (Shamir 2009: Photo 11.4, 608), and by the textile tools (Panitz-Cohen and Yahalom-Mack 2009: 737–41), including a significant corpus of spinning bowl fragments (Martin 2011: 45, Fig. 18). Barber (1991:72–73) notes that spinning bowls were used exclusively for processing linen threads. The earliest evidence in the ancient Near East for these vessels comes from Egypt.[12] The appearance of these vessels in the Late Bronze Age strata in the southern Levant, has been associated with a strong Egyptian presence in the region (particularly at Beth Shean), and indicates the introduction of a new technique for processing flax fibres (Dothan 1963: 104). It is interesting to note that textual

---

[11] If this had been a failed attempt at hot-pressing the seeds for oil, we question why the debris would have been left so in an activity or storage area and not removed to pits such as suggested by Larsson (2013: 516).

[12] Spinning bowl fragments have recently been identified by Svilar (2017) in Late Neolithic contexts in the central Balkans.

evidence from the Hellenistic and Roman periods indicates that the city of Scythopolis (Beth Shean) was well known for its high-quality linens, so much so that they were protected under the Price Edict of Diocletian *Edictum de pretiis rerum venalium* (301 AD) (see tr. Lauffer 1971: 169–170; Cassuto 2016: 279, cf. Shamir et al. in this volume).

Besides Tel Burna and Tel Beth Shean, there are some other sites which provide charred flax seeds connected to the destruction of features and installations in combination with evidence of conflagration. At Tell Deir ‘Alla thousands of charred flax seeds were stored in two rectangular features constructed of unburnt mud brick material (van Zeist and Heeres 1973: Tab. 1; van der Kooij and Ibrahim 1989: 35). These features were located in two rooms (EE334 und EE335; Phase IX, Iron Age IIB) which consisted almost entirely of shallow pits (van der Kooij and Ibrahim 1989: 87, Fig. 108). According to van der Kooij and Ibrahim (1989: 82, Fig. 108), the building, including the rooms above, was spontaneously destroyed, possibly by an earthquake, which was accompanied by a fire. Next to the linseed were found charred flax stems (van Zeist and Heeres 1973: 27), directly indicating the cultivation of this crop in the arable fields for the stems, i.e., for linen fibres. The mean dimensions of the seeds from Tell Deir ‘Alla (mean length: 4.20–4.29 mm; mean width: 2.18–2.23 mm) are slightly larger than those from Tel Burna. Van Zeist and Heeres (1973: 27) conclude that the flax plants were grown in irrigated fields, first, in light of the semi-arid environment of the central Jordan Valley, and secondly, following Helbaek’s observations of the seeds’ size and irrigated fields.

Similar scenarios are mentioned for Early Bronze Age Jericho. Few flax seeds were found in buildings (Trench III. phase lxii-lxiii; E III-IV, phase H-G), which were apparently destroyed by an earthquake and accompanying fires (Kenyon 1981: 204, 328; Hopf 1983). With an average length and breadth of 3.70 × 1.62 mm the seeds are much smaller than the ones from Tel Burna and Tel Beth Shean, although one must keep in mind the small number of seeds (N = 7) measured. The environmental conditions at Jericho would only allow the cultivation of flax by irrigation. The small size of the seeds stands in contrast to Helbaek’s suggestion that irrigated flax plants produce larger seeds. Hopf (1983) therefore argues that the flax was used for both purposes: textile and oil production.

The cultivation of flax at Arad needed artificial water supply as well. Yet, no installations nor any water sources in near vicinity of the site were detected which point to irrigation at Early Bronze Age Arad (EBA II; Stratum II). Therefore, it is not impossible that the linseed from Arad was imported from areas with adequate environmental conditions. The measurements for the Arad linseed (3.80 × 2.00 mm) are close to those of the seeds from Jericho, some of which were lumped together (Sample no. 1157/3057) with herbaceous remains, suggesting cooking or baking of linseed for oral consumption (Hopf 1978).

12.5.2 Food for Thought or Threads for Weaving?

At Tel Burna we have two different sets of linseed finds. One comes from the late Iron Age IIC silos which contained few random charred flax seeds. The other comes from the early Iron Age IIA destruction layer containing two storage finds of charred linseed (Area B2 and Area A1). The cultivation of flax appears to have been an important factor in the agricultural landscape of Iron Age Tel Burna. Morphometric analysis of the flax seeds provides the basis for comparisons with other archaeological sites and contributes to the discussion on the objectives for the cultivation of linseed at the site. The mean length for the charred linseed remains from Tel Burna, 4.15 mm, falls between the smaller flax seeds from the European sites and the larger seeds from the irrigated fields of the Near Eastern sites (Table 12.3). Taking into consideration shrinkage due to carbonization, adding the calculated shrinkage factor for carbonized seeds of 15% to the seeds from Tel Burna, the non-carbonized preservation of the seeds is 4.77 mm. Morphometric datasets of recent flax varieties show that the mean length of fibre flax seed is 4.28 mm (range of seed length 3.51 to 5.14 mm), for intermediate flax varieties it is 4.41 mm (range of seed length 3.15 to 5.41 mm) and for large seeded flax (for oil production,), it is 5.35 mm (range of seed length 5.04 to 5.81) (Karg et al. 2018: Tab. 3.3). This measurement falls below the modern large seeded linseed varieties with a mean length of 5.35 mm (seed length varies between 5.04 and 5.81 mm; Karg et al. 2018: Tab. 3.3). Hence the carbonized linseed from Tel Burna fits best with the modern flax and intermediate flax varieties, even when adjusted for shrinkage.

When comparing ancient flax seeds, the discussion turns to morphometric differences as evidence of irrigation of flax fields at the drier Near Eastern sites (Helbaek 1959: Tab. 8; van Zeist and Heeres 1973: 27). Since the cultivation of fibre flax requires about 700 mm water supply during growing period, and Tel Burna is located within the 400–500 mm isohyet, it would have been necessary to add artificial watering, irrigation, to the flax fields in order to cultivate them for fibre (Riehl 2008: S48). At the Early Bronze Age of Jericho and Arad, where the annual rainfall is significantly low the crops would have undoubtedly been irrigated, the flax seeds fall out of the range of the large seed measurements, and it is assumed that the flax was cultivated for multiple purposes.[13] The measurements from Tel Burna best fit with the values from Tel Beth Shean and Tell Deir ‘Alla. However, the contextual evidence for the flax seeds found at Beth Shean leans heavily towards flax production for linen, as presented above.

It has been suggested that charred seeds are evidence for hot-pressing, a method for extracting the oil from the seeds (Ertuğ 2000). However, as is the case at Tel Burna, the evidence for conflagration in the area where the large numbers of seeds were found would most certainly negate this assumption. Nonetheless, it cannot be excluded that flax plants were cultivated at the site and that their seeds would have

[13] The literature on these two sites does not make a clear statement on the consumption of these plants – for the seeds from Jericho it is noted that they may be for either flax or oil (Hopf 1983), and for Arad perhaps oil (Hopf 1978).

Table 12.3 The mean length of linseeds. For Tell Deir 'Alla, Tel Burna, and Nimrud the calculated values are adjusted using shrinkage factors of 13%/25% resp. 15%

Location	Site	Period	Preservation	Flax variety	Water supply	Mean length in mm
Europe	Medina/Italy	Roman	Charred	Fibre		2.63
Europe	SW-Germany	Neolithic	Waterlogged	Fibre		3.00
Europe	SW-Germany	Neolithic	Waterlogged	Oil		3.50
Near East	Jericho	Early Bronze Age	Charred	Fibre and oil	Irrigation	3.70
Near East	Arad	Early Bronze Age	Charred	Maybe oil	Rain-fed/ imported?	3.80
Near East	Arpachiyah	Chalcolithic	Imprint		Rain-fed	3.84
Europe	Uppåkra/Sweden	Roman Iron Age	Charred	Oil		3.86
Near East	Brak	Chalcolithic	Imprint		Rain-fed/ irrigation	4.03
Near East	Tel Burna	Iron Age	Charred		Rain-fed/ irrigation	4.15
Near East	Ur	Chalcolithic	Imprint		Irrigation	4.21
Near East	Tell Deir 'Alla	Iron Age	Charred			4.29
Near East	Nimrud	Iron Age	Charred		Irrigation	4.41
Near East	Tel Beth Shean	Early Iron Age	Charred			4.72
Near East	Tel Burna, calculated value 15%	Iron Age	Charred		Rain-fed/ irrigation	4.77
Near East	Tell Deir 'Alla, calculated value 13%	Iron Age	Charred			4.93
Near East	Tell Deir 'Alla, calculated value 25%	Iron Age	Charred			5.72
Near East	Nimrud, calculated value 15%	Iron Age	Charred		Irrigation	6.20

The flax variety and the water supply are note only when the literature clearly mentions the usage and cultivation technique for the flax plants. **Arad** (Hopf 1978); **Arpachiyah, Brak, Ur** (Helbaek 1959); **Jericho** (Hopf 1983); **Medina/Italy** (Bosi et al. 2011); **Nimrud** (Helbaek 1966); **SW-Germany** (Herbig and Maier 2011); **Tel Beth Shean** (Kislev et al. 2009); **Tell Deir 'Alla** (van Zeist and Heeres 1973); **Uppåkra/Sweden** (Larsson 2013)

been preserved in storage, either for their oil content, or for their future cultivation, growing future crops for fibre or for oil (Kislev et al. 2011). It is anticipated that the completed examination of the latest caches of linseed collected from Tel Burna, along with data from impending charring experiments on modern local linseed varieties, compiled into a broader regional study of charred linseed finds, shall augment the available data on flax cultivation and consumption in the Iron Age Shephelah.

12.6 Conclusions

The study of ancient textiles profits greatly from a multi-disciplinary approach. The extremely poor preservation of organic textiles and production tools creates a large gap in our ability to understand the roles textiles and textile production played in ancient societies. Ancient flax seeds discovered in archaeological contexts may indicate the cultivation of flax for fibres. Variously, flax plants have been grown for their oil-rich seeds. Biometric studies of ancient flax seeds in European contexts have associated changes in the sizes of the waterlogged seeds with a shift from flax plants cultivated for oil to flax plants cultivated for fibres. The preserved linseed in archaeological contexts in the southern Levant is different, as here the seeds have undergone carbonization, causing shrinkage. The ongoing study of quantities of flax seeds found at Tel Burna provides the data for biometric comparisons with linseed finds from various contexts, both ancient and modern, used to assess whether the plants were grown for fibres or for seeds. Despite the evidence for the cultivation and consumption of both flax for fibre and for oil, morphometric analysis of Tel Burna's flax seeds has not allowed definitive conclusions as to the intended uses for the flax plants at the site. As the results of our study are inconclusive at this stage, it is hoped that a multivariate analysis, incorporating period and context as well as morphometric measurements, will promote a better understanding of the cultivation and consumption of flax in the Iron Age southern Levant.

References

Ancient Sources

Diokletians Preisedikt. Lauffer S (ed., tr.) (1971). Texte und Kommentare 5. De Gruyter, Berlin.

Masoretic Text

Fisch H (2000) *The Holy Scriptures*. Korén Publishers, Jerusalem.

Akkadian Text

Kramer SN, Wolkstein D (1983) *Inanna, queen of heaven and earth: Her stories and hymns from Sumer.* Harper and Row Publishers, New York.

Modern Works

Barber EJW (1991) *Prehistoric textiles: The development of cloth in the Neolithic and Bronze Ages with special reference to the Aegean.* Princeton University Press, Princeton.

Bar-Matthews M, Ayalon A (2004) Speleothems as palaeoclimate indicators, a case study from Soreq cave located in the Eastern Mediterranean region, Israel. In: Battarbee RW, Gasse F Stickley C (eds) *Past climate variability through Europe and Africa.* Kluwer Academic Press, Dordrecht, p 363–391.

Baruch U (1986) The late Holocene vegetational history of Lake Kinneret (sea of Galilee), Israel. *Paléorient* 12(2):37–48.

Bongenaar ACVM (1997) *The Neo-Babylonian Ebabbar temple at Sippar: Its administration and its prosopography.* Nederlands Historisch-Archaeologisch Instituut, Leiden.

Bosi G, Rinaldi R, Bandini Mazzanti M (2011) Flax and weld: Archaeobotanical records from Mutina (Emilia Romagna, Northern Italy), dated to the Imperial Age, first half 1st century A.D. *Vegetation History and Archaeobotany* 20:543–548.

Borowski O (1987) *Agriculture in Iron Age Israel.* Eisenbrauns, Winona Lake, Indiana.

Cassuto D (2016) Textile production implements. In: Syon D (ed.) *Gamla III – the Shmarya Gutmann excavations 1976–1989. Finds and studies part 2.* Israel Antiquities Authority Reports. Israel Antiquities Authority, Jerusalem, p 261–282.

Charles MP (1985) An introduction to the legumes and oil plants of Mesopotamia. *Bulletin on Sumerian Agriculture* 2:39–61.

Crowfoot GM (1931) *Methods of hand spinning in Egypt and the Sudan.* Bankfield Museum Notes 12. F. King and Sons Ltd., Halifax.

Dambroth M, Seehuber R (1988) *Flachs. Züchtung, Anbau und Verarbeitung.* Ulmer, Stuttgart.

Deckers K, Riehl S (2007) Fluvial environmental contexts for archaeological sites in the Upper Khabur basin (north-eastern Syria). *Quaternary Research* 67:337–348.

Del Freo M, Nosch M-L, Rougemont R (2010) The terminology of textiles in the Linear B tablets, including some considerations on Linear A logograms and abbreviations. In: Michel C, Nosch M-L (eds) *Textile terminologies in the Ancient Near East and Mediterranean from the third to the first millennia BC.* Ancient Textile Series 8. Oxbow Books, Oxford and Oakville, p 338–373.

Diederichsen A, Richards, K (2003) Cultivated flax and the genus *Linum* L. Taxonomy and germplasm conservation. In: Muir AD, Westcott ND (eds) *Flax. The Genus Linum.* Routledge, London, p 22–54.

Dothan T (1963) Spinning bowls. *Israel Exploration Journal* 13:97–112.

Ertuğ F (2000) Linseed oil and oil mills in Central Turkey. Flax/Linum and Eruca, important oil plants of Anatolia. *Anatolian Studies* 50:171–185.

Feinbrun-Dothan N (1977/78) *Flora Palaestina 3. Ericaceae to compositae text.* Israel Academy of Sciences, Jerusalem.

Finkelstein I, Piasetzky E (2011) The Iron Age chronology debate: Is the gap narrowing? *Near Eastern Archaeology* 74:50–54.

Guest E (1930) *The cultivation and marketing of linseed.* Agricultural bulletin 21. Department of Agriculture. Government Press, Iraq.

Gustafsson S (2000) Carbonized cereal grains and weed seeds in prehistoric houses – an experimental perspective. *Journal of Archaeological Science* 27:65–70.

Hald MM, Charles M (2008) Storage of crops during the fourth and third millennia B. C. at the settlement mound of Tell Brak, northeast Syria. *Vegetation History and Archaeobotany* 17(Supplement 1):S35–S41.

Harris S (2009) Smooth and cool, or warm and soft: Investigating the properties of cloth in prehistory. In Andersson Strand E, Gleba M, Mannering U, Munkholt C (eds) *North European symposium for archaeological textiles X*. NESAT X. Oxbow Books, Oxford and Oakville, p. 104–112

Helbaek H (1959) Notes on the evolution and history of *Linum*. *KUML*:103–129.

Helbaek H (1966) Appendix I. The plant finds from Nimrud. In: Mallowan MEL (ed.) *Nimrud and its finds*. Volume III. Dodd, Mead and Company, New York, p 613–620.

Herbig C, Maier U (2011) Flax for oil or fibre? Morphometric analysis of flax seeds and new aspects of flax cultivation in Late Neolithic wetland settlements in southwest Germany. *Vegetation History and Archaeobotany* 20:527–533.

Hjelmquist H (1950) The flax weeds and the origin of cultivated flax. *Botaniska Notiser* 2:257–298.

Hopf M (1978) Plant remains, strata V–I. In: Amiran R, Ilan O (eds) *Early Arad: The Chalcolithic settlement and Early Bronze Age city*. Israel Exploration Society, Jerusalem, p 309–320.

Hopf M (1983) Jericho plant remains. In: Kenyon KM, Holland TA (eds) *Excavations at Jericho IV. The pottery phases of the tell and other finds*. British School of Archaeology in Jerusalem, London, p 576–621.

Horowitz W, Oshima T (2002) Two more cuneiform finds from Hazor. *Israel Exploration Journal* 52:179–186.

Horowitz W, Oshima T (2006) *Cuneiform in Canaan: Cuneiform sources from the Land of Israel in ancient times*. Israel Exploration Society, Jerusalem.

Horowitz W, Oshima T (2010) Hazor 16: Another administrative docket from Hazor. *Israel Exploration Journal* 60(2):129–132.

Horowitz W, Wasserman N (2000) An Old Babylonian letter from Hazor with mention of Mari and Ekallātum. *Israel Exploration Journal* 50:169–174.

Horowitz W, Wasserman N (2004) From Hazor to Mari and Ekallātum: A recently discovered Old-Babylonian letter from Hazor. *Amurru* 3:335–344.

Jones J (2010) The 'linen list' in Early Dynastic and Old Kingdom Egypt: Text and textile reconciled. In: Michel C, Nosch M-L (eds) *Textile terminologies in the Ancient Near East and Mediterranean from the third to the first millennia BC*. Ancient Textile Series 8. Oxbow Books, Oxford and Oakville, p 81–109.

Karg S (ed.) (2011) FLAX – new research on the cultural history of the useful plant *Linum usitatissimum* L. *Vegetation History and Archaeobotany* 20.6.

Karg S, Diederichsen A, Jeppson S (2018) Discussing flax domestication in Europe using biometric measurements on recent and archaeological flax seeds – a pilot study. In: Siennicka M, Rahmstorf L, Ulanowska A (eds) *First textiles: The beginnings of textile manufacture in Europe and the Mediterranean*. Ancient Textiles Series 32. Oxbow Books, Oxford and Philadelphia, p 31–38.

Kenyon KM (1981) *Excavations at Jericho III. The architecture and stratigraphy of the tell, text*. British School of Archaeology in Jerusalem, London.

Kislev ME, Simchoni O, Melamed Y, Maroz L (2009) Food and industrial crops. In: Panitz Cohen N, Mazar A (eds) *Excavations at Tel Beth-Shean 1989–1996. Volume III. The 13th–11th BCE strata in Areas S and N*. The Israel Exploration Society, Jerusalem, p 764–771.

Kislev ME, Simchoni O, Melamed Y, Maroz L (2011) Flax seed production: Evidence from the early Iron Age site of Tel Beth Shean, Israel and from written sources. *Vegetation History and Archaeobotany* 20:579–584.

Larsson M (2013) Cultivation and processing of *Linum usitatissimum* and *Camelina sativa* in southern Scandinavia during the Roman Iron Age. *Vegetation History and Archaeobotany* 22:509–520.

Lee S, Ramsey CB, Mazar A (2013) Iron Age chronology in Israel: results from modelling with a trapezoid Bayesian framework. *Radiocarbon* 55: 731–740.

Leuzinger U, Rast-Eicher A (2011) Flax processing in the Neolithic and Bronze Age pile-dwelling settlements of eastern Switzerland. *Vegetation History and Archaeobotany* 20:535–542.

Litt T, Ohlwein T, Neumann FH, Hense A, Stein M (2012) Holocene climate variability in the Levant from the Dead Sea pollen record. *Quaternary Science Reviews* 49:95–105.

Maeder F (2017) Byssus and sea silk: A linguistic problem with consequences. In: Landenius Enegren H, Meo F (eds) *Treasures from the sea: Sea silk and shellfish purple dye in Antiquity*. Ancient Textiles Series 30. Oxbow Books, Oxford, p 4–19

Martin M (2011) *Egyptian-type pottery in the Late Bronze Age southern Levant*. Contributions to the chronology of the Eastern Mediterranean 29. Denkschriften der Gesamtakademie 69. Verlag der Österreichischen Akademie der Wissenschaften, Wien.

Märkle T, Rösch M (2008) Experiments on the effects of carbonization on some cultivated plant seeds. *Vegetation History and Archaeobotany* 17(Supplement 1):S257–263.

McKinny C, Tavger A, Cassuto D, Sharp C, Suriano M, Ortiz S, Shai I (2019) Tel Burna after a decade of work: The Late Bronze and Iron Ages. *Near Eastern Archaeology* 83:4–15.

Neumann F, Schölzel C, Litt T, Hense A, Stein M (2007) Holocene vegetation and climate history of the northern Golan heights (Near East). *Vegetation History and Archaeobotany* 16:329–346.

Orendi A (2020a) *Entwicklung und Bedeutung agrarischer Ressourcen in der Südlevante der Bronze- und Eisenzeit*. Dissertation, Eberhard Karls Universität, Tübingen. https://doi.org/10.15496/publikation-40980.

Orendi A (2020b) Flax cultivation in the Southern Levant and its development during the Bronze and Iron Age. *Quaternary International* 545:63–72.

Orendi A, Smejda L, McKinny C, Cassuto D, Sharp C, Shai I (2017) The agricultural landscape of Tel Burna: Ecology and economy of a Bronze Age settlement in the southern Levant. *Journal of Landscape Ecology* 10(3):165–188.

Panitz Cohen N, Mazar A (2009) *Excavations at Tel Beth-Shean 1989–1996 Volume III. The 13th–11th BCE strata in Areas S and N*. The Israel Exploration Society, Jerusalem.

Panitz-Cohen N, Yahalom-Mack N (2009) Textile-related objects and a basket imprint. In: Pantiz-Cohen N, Mazar A (eds) *Excavations at Tel Beth-Shean 1989–1996 Volume III. The 13th–11th Century BCE strata in Areas N and S*. Israel Exploration Society, Jerusalem, p 737–741.

Renfrew JM (1973) *Palaeoethnobotany: The prehistoric food plants of the Near East and Europe*. Columbia University Press, New York.

Renfrew JM (1985) Finds of sesame and linseed in ancient Iraq. *Bulletin on Sumerian Agriculture* 2:3–66.

Riehl S (2008) Climate and agriculture in the ancient Near East: A synthesis of the archaeobotanical and stable carbon isotope evidence. *Vegetation History and Archaeobotany* 17:S43–S51.

Riehl S (2009) Archaeobotanical evidence for the interrelationship of agricultural decision-making and climate change in the Ancient Near East. *Quaternary International* 197:93–114.

Riehl S, Shai I (2015) Supra-regional trade networks and the economic potential of Iron Age II sites in the Southern Levant. *Journal of Archaeological Science: Reports* 3:525–533.

Rougemont F (2007) Flax and linen textiles in the Mycenaean palatial economy. In Gillis C, Nosch M-L (eds) *Ancient textiles: Production, craft and society*. Ancient Textile Series 1. Oxbow, Oxford, p 46–49.

Rosen A (1986) *Quaternary alluvial stratigraphy of the Shephela and its paleoclimate implications*. Geological Survey Israel, Jerusalem

Safrai Z (1985) *The Galilee in the time of the Mishna and Talmud* (2nd edition). Shoreshim College, Jerusalem (in Hebrew).

Shai I (2017) Tel Burna – a Judahite fortified town in the Shephelah. In: Lipschits O, Maeir AM (eds) *The Shephelah during the Iron Age. Recent archaeological studies*. Eisenbrauns, Winona Lake, Indiana, p 45–60.

Shai I, Cassuto D, Dagan A, Uziel J (2012) The fortifications at Tel Burna: Date, function and meaning. *Israel Exploration Journal* 62:141–157.

Shai I, Dagan A, Riehl, Orendi A, Uziel J, Suriano M (2014) A private stamped seal handle from Tell Bornat/Tel Burna, Israel. *Zeitschrift des Deutschen Palästina-Vereins* 130:121–157.

Shai I, Uziel J, McKinny C (2015) Late Bronze Age cultic activity in ancient Canaan: A view from Tel Burna. *Bulletin of the American Schools of Oriental Research* 374:115–133.
Shai I, McKinny C (2020) Canaanite votive offerings and their significance within their contexts at Tel Burna. *Israel Exploration Journal* 70:1–17.
Shamir O (2009) Appendix: A linen textile fragment. In: Panitz-Cohen N, Mazar A (eds) *Excavations at Tel Beth-Shean 1989–1996. Volume III. The 13th–11th century BCE strata in Areas N and S.* Israel Exploration Society, Jerusalem, p 608–611.
Svilar MM (2017) Searching for Late Neolithic spinning bowls in the Central Balkans. *Starinar* 67:21–32.
Uziel J, Shai I (2010) The settlement history of Tel Burna: Results of the surface survey. *Tel Aviv* 37:227–245.
Van der Kooij G, Ibrahim MH (1989) *Picking up the threads: A continuing review of excavations at Deir Alla.* University of Leiden, Leiden.
van Zeist W (1985) Pulses and oil crop plants. *Bulletin on Sumerian Agriculture* 2:33–37.
van Zeist W, Heeres AH (1973) Paleobotanical studies of Deir 'Alla, Jordan. *Paléorient* 1:21–37.
Vigo M (2010) Linen in Hittite inventory texts. In: Michel C, Nosch M-L (eds) *Textile terminologies in the Ancient Near East and Mediterranean from the third to the first millennia BC.* Ancient Textile Series 8. Oxbow Books, Oxford and Oakville, p 290–322.
Vita J-P (2010) Textile terminology in the Ugaritic texts. In: Michel C, Nosch M-L (eds) *Textile terminologies in the Ancient Near East and Mediterranean from the third to the first millennia BC.* Ancient Textile Series 8. Oxbow Books, Oxford and Oakville, p 323–337.
Vogelsang-Eastwood G (2000) Textiles. In: Nicholson PT, Shaw I (eds) *Ancient Egyptian materials and technology.* Cambridge University Press, Cambridge, p 268–298.
White CE, Chesson MS, Schaub, RT (2014) A recipe for disaster: Emerging urbanism and unsustainable plant economies at Early Bronze Age Ras an-Numayra, Jordan. *Antiquity* 88:363–377.
Zawadzki S (2006) *Garments of the gods: Studies on the textile industry and the pantheon of Sippar according to the texts from the Ebabbar archive.* Orbis Biblicus et Orientalis 218. Academic Press Fribourg Vandenhoeck and Ruprecht, Fribourg.
Zawadzki S (2010) Garments in non-cultic context (Neo-Babylonian Period). In: Michel C, Nosch M-L (eds) *Textile terminologies in the Ancient Near East and Mediterranean from the third to the first millennia BC.* Ancient Textile Series 8. Oxbow Books, Oxford and Oakville, p 409–429.
Zawadzki S (2013) *Garments of the gods. Vol. 2: Texts.* Orbis Biblicus et Orientalis 260. Academic Press Fribourg Vandenhoeck and Ruprecht, Fribourg.
Zohary D, Hopf M, Weiss E (2013) *Domestication of plants in the Old World.* Oxford University Press, Oxford.

Deborah Cassuto is a postdoctoral fellow at the Martin (Szusz) Department of Land of Israel Studies and Archaeology Bar Ilan Univeristy, for the Tell eṣ-Ṣâfī/Gath Excavations project, an associate fellow at the Albright Institute of Archaeological Research and a member of the RIAB Minerva Center, Israel. The focus of her research has been textile production in archaeological contexts in the Iron Age southern Levant. She has been a staff member of the Tel Burna Archaeological Project since 2009.

Andrea Orendi holds a PhD in archaeobotany. The focus of her research was the analysis of agricultural resource use and resource development in the Bronze and Iron Age southern Levant. Since 2011, she has been responsible for the archaeobotanical research at Tel Burna. She presently works as archaeobotanical specialist in the Archaeobotanical Laboratory of the excavation company ArchaeoConnect GmbH in Tübingen, Germany.

Itzhaq Shai is Professor of Archaeology at Ariel University, Israel. He has been the director of the Tel Burna Archaeological Project since 2009. His research focuses on the Bronze and Iron Ages in the southern Levant.

Chapter 13
The Relationship between Textile Remains in a Hoard of Alexander II Zabinas Coins and Loom Weights Discovered at Hellenistic Tell Iẓṭabba (Beth She'an, Nysa-Scythopolis), Israel

Orit Shamir, Achim Lichtenberger, and Oren Tal

Abstract This paper discusses a recently discovered coin hoard in the German-Israeli Tell Iẓṭabba Excavation Project. The copper-alloy coins date from the time of Alexander II Zabinas (129/128–124/123 BCE). The linen cloth in which the hoard was wrapped is discussed in light of the site history and archaeology – a Seleucid-founded settlement that was destroyed by John Hyrcanus in 108/107 BCE. We examined the relationship between the fabric and the truncated-pyramid loom weights, which were found at the site in considerable numbers as the area of Beth She'an was famous for the cultivation of flax and the production of linen for centuries. Is it possible that this linen fabric was woven on a loom using the loom weights found at the site? Or could it have been manufactured elsewhere?

Keywords Linen · Textile · Loom weights · Coin hoard · Hellenistic period · Beth She'an · Israel

---

O. Shamir (✉)
Department of Museums and Exhibits of the Israel Antiquities Authority, Jerusalem, Israel
e-mail: orit@israntique.org.il

A. Lichtenberger
Institut für Klassische Archäologie und Christliche Archäologie/Archäologisches Museum of Münster University, Münster, Germany
e-mail: lichtenb@uni-muenster.de

O. Tal
Department of Archaeology and Ancient Near Eastern Cultures of Tel Aviv University, Tel Aviv, Israel
e-mail: orental@tauex.tau.ac.il

A. Ulanowska et al. (eds.), *Ancient Textile Production from an Interdisciplinary Perspective*, Interdisciplinary Contributions to Archaeology,
https://doi.org/10.1007/978-3-030-92170-5_13

13.1 Introduction

Beth She'an is a city in the north of Israel, which played an important role in history due to its geographical location at the junction of the Jordan Rift Valley and the Jezreel Valley. It is also mentioned in the Bible (e.g., Joshua 17:11, 16; Judges 1:27; 1 Kings 4:12; 2 Samuel 31:12). The area around Beth She'an played a prominent role in the cultivation of *Linum usitatissimum* L. throughout time (Boertien 2013: 37; Shamir 2015: 13; Orendi 2020).

The earliest loom weights discovered in the area of the Beth She'an Valley were found at Beth she'an (Tell el-Ḥuṣn), dating to the Middle Bronze Age (Yahalom-Mack 2007) and later at Tel Reḥov, dating to the twelfth century BCE (Mazar 2019: 123) and at Tel 'Amal, dating to the tenth century BCE (Shamir 2013). They are gypsum piriform loom weights of some 84.4–518.7 g in weight, averaging 242.9 ± 80.3 g. Later, eighth century BCE doughnut loom weights weighing some 43.28–630.11 g and averaging 273.1 ± 82.2 g were documented in Building 28636 on the biblical mound of Beth She'an (Tell el-Ḥuṣn) (Shamir 2006). The area of Beth She'an is relatively humid for the preservation of textiles, but where they have been preserved at the site, they have normally adhered to metal objects. The earliest textiles (minute fragments) from biblical Beth She'an (Tell el-Ḥuṣn) are made of linen and date to the twelfth century BCE (Shamir 1992, 2009). Flax seeds (linseeds) from the Early, Middle and Late Bronze Age are recorded in the excavations, as well as over a million charred flax seeds (*Linum usitatissimum* L.) dating to the Iron Age IA (twelfth century BCE, Simchoni et al. 2007; Kislev et al. 2011: 581; Simchoni and Kislev 2012; Orendi 2020, cf. Cassuto et al. in this volume). According to Kislev et al. the linseed was used for consumption, for the production of linseed oil, or as seed for subsequent sowing (2011: 582). At a nearby site, Tel Reḥov, c. 1300 loom weights dating from the twelfth to the eighth century BCE were found (Mazar 2019, 2020). There is also evidence from medieval times: a late tenth century CE linen Fatimid coin purse from the excavations on the southern outskirts of Beth She'an (Kool et al. 2011).

Linen became a major economic factor in many parts of the Land of Israel from the second half of the second century CE, and the area of Beth She'an seems to be prominent among these. Sources from the second century CE describe flax in terms of its agricultural importance as well as its industrial significance. It is also described as one of the main export crops of the region. The Greek author Pausanias (5.5.2) implies that the flax of the Hebrews was of a particularly high quality. The price edict of Diocletian (Graser 1940; Avi-Yonah 1962; Wild 1970: 128–129; Lauffer 1971: 328, 351) lists the flax of Beth She'an as the most expensive kind; like all the other listed products of Scythopolis it is categorized as first class. Other sources from Late Antiquity such as the *Expositio totius mundi et gentium* (30–31) and the *Codex Theodosianus* (10.20.8), as well as Rabbinic literature (yQid II.e-62c; yAZ 1.2-39c) testify to the notability of textiles from Scythopolis or at least the importance of textile production. We have to assume that the natural conditions of plentiful water resources and fertile soils, as well as a large specialized urban population contributed to the favourable conditions for textile production that applied not only to Late Antiquity but also to other earlier and possibly later periods.

13.2 The Site

Tell Iẓṭabba is the main settlement of Seleucid Nysa-Scythopolis (Beth Sh'ean). The ancient settlements of Beth She'an are enclosed by a chain of hills north of the River Ḥarod – the watercourse of which crosses the city (Fig. 13.1). To the south of the river is Tell el-Ḥuṣn, to the east is Tell Ḥammam, where one of the main cemeteries of the Roman and Byzantine city is located, and to the west lies Tell Iẓṭabba, which consists of three hills sloping steeply on the southern sides toward River Ḥarod and more moderately on the sides looking toward the plain north of the city. While the two western hills of Tell Iẓṭabba are dominated by Byzantine remains, namely the Kyria Maria monastery, the Church of the Martyr, the Church of Andreas, and the Byzantine city walls, which cross the mound to its full extent, Hellenistic remains have been found on the eastern hill, as shown by the excavations carried out

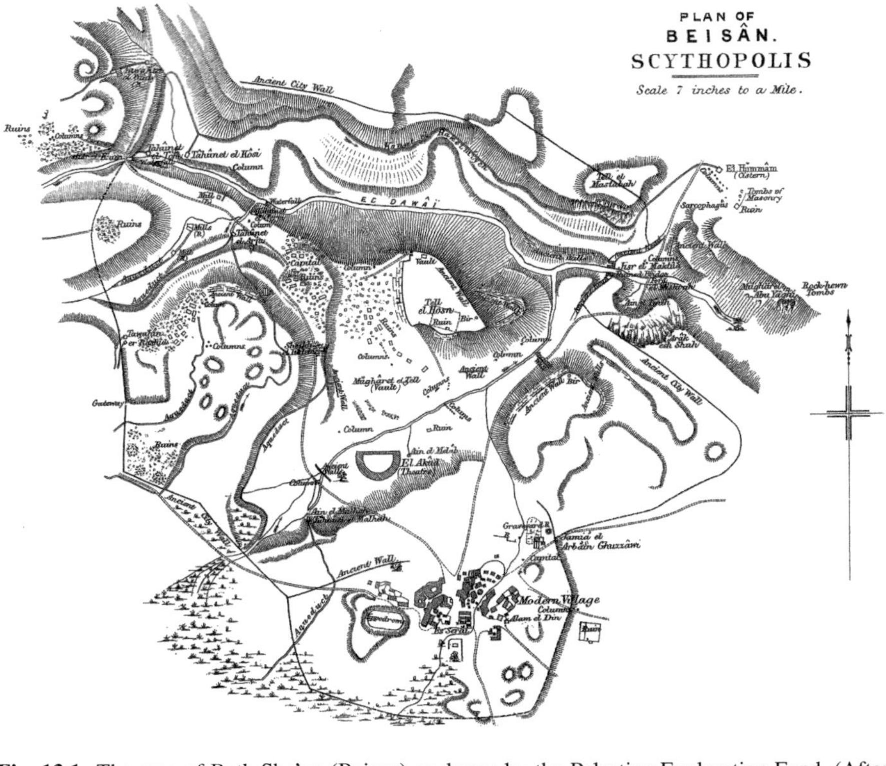

Fig. 13.1 The area of Beth She'an (Beisan) as drawn by the Palestine Exploration Fund. (After Conder and Kitchener 1882, in front of p. 105)

there since the 1960s (Zori 1962; Landau and Tzaferis 1979; Mazor and Bar-Nathan 1998; Mazor 2008; Mazor et al. 2018). Since 2019, the site has been investigated by a joint German-Israeli team that aims at a systematic understanding of the settlement history and economics of the Seleucid site (Ebeling et al. 2020, 2021; Lichtenberger et al. 2020; Lichtenberger and Tal [with an appendix by Shamir] 2020).

The Seleucid settlement is of major interest for the study of the Hellenistic period in the region (Fig. 13.2). Nysa-Scythopolis belonged to a group of Graeco-Roman cities that were supported by the Hellenistic dynasties – first the Ptolemies and later the Seleucids – and these cities formed a group named Decapolis (Tal 2011; Lichtenberger 2020). One of the major formation phases of these cities is in the time of Antiochos IV, who founded several cities in the region in the 160s BCE (Lichtenberger 2008). These urban centres were surrounded by less urbanized entities, such as, e.g., the Jewish state of the Hasmoneans or the Nabataeans, and came into conflict with them. The Seleucid settlement of Nysa-Scythopolis was destroyed by the Hasmonean John Hyrcanus I in 108/07 BCE, providing us with a fixed *terminus ante quem* for the settlement (Josephus *Ant.* 13.280; cf. Finkielsztejn 1999). Since it was only founded about half a century earlier, all the material culture of the city stems from a short period of time.

Nysa-Scythopolis was one of the sites founded by Antiochos IV and it is generally assumed that Greek settlers peopled the newly founded city. This can also be

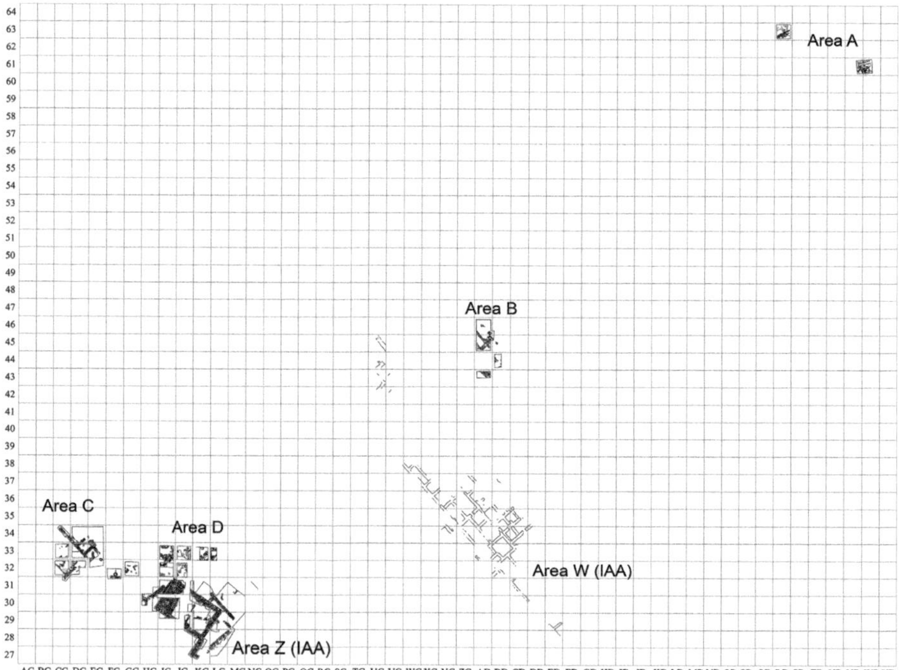

Fig. 13.2 Site plan of Tell Iẓṭabba (East) of the German-Israeli Tell Iẓṭabba Excavation Project with an indication of Areas A, B, C, D and W and Z of the Israel Antiquities Authority. (Plan by S. Pirsky)

inferred from the material culture of the Hellenistic settlement that was excavated by the Israel Antiquities Authority in the 1990s and now in the ongoing German-Israeli project. The pottery and especially the amphorae suggest Greek consumption habits of a community that was well connected with the Mediterranean region (Mazor et al. 2018). Glass, stucco and plaster remains as well as metal objects indicate that the community oriented towards the Hellenistic lifestyle. Until now, no textile remains have been excavated from the Seleucid settlement and, therefore, the new evidence that we present in this paper sheds light not only on the aspect of consumption habits but also on the economy of the Seleucid site.

Archaeobotanic remains from the Seleucid settlement were only sampled and analysed in our excavation project. Of the total 370 identified seeds from the 2020 excavations, only four *Linum usitatissimum* seeds were identified from Loci 338 (Basket 3312: two seeds), 410 (Basket 4038: one seed) and 434 (Basket 4204: one seed, the analysis was undertaken by Andrea Orendi, Tübingen). Only the seeds from Locus 338 are definitely from the Seleucid settlement and stem from the silo of a house.

13.3 The Hoard

During the 2020 excavation campaign of the German-Israeli Tell Iẓṭabba Excavation Project, a lump of several copper-alloy coins was unearthed in Area D (Fig. 13.3). It was found in Locus 414 (Basket 4070) which is located in Square IC31. The locus is a fill close to surface level in an area of domestic structures of the Seleucid settlement and contained Hellenistic pottery (see Ebeling et al. 2021). After cleaning, the hoard turned out to be a bundle wrapped in a textile (a white-coloured linen), containing ten coins altogether (cf. Lichtenberger and Tal [with an appendix by Shamir] 2020).

All the coins were minted by the same Seleucid king, namely Alexander II Zabinas (129–123 BCE), and eight of the ten were of the same type (albeit with varying dates). In our numismatic analysis, we have suggested that the coins were assembled, wrapped in the textile and buried possibly towards the end of Alexander II Zabinas's reign. This burial date also gives a *terminus ante quem* for the textile.

13.3.1 The Textile Remains of the Hoard

The coin hoard found at Tell Iẓṭabba (East) was wrapped in two layers of white linen textile in an excellent state of preservation. In some parts of the bundle, only one layer was preserved. This is due to corrosion of the metal.

The textile (Figs. 13.4 and 13.5) is made of flax, undyed white colour, which was obtained in a long process of bleaching. Due to the copper alloy corrosion, some of the threads were light green. The threads are s-spun (anti-clockwise), medium spun.

Fig. 13.3 Left: lumped hoard before cleaning, right: during cleaning, bottom: the hoarded coins after cleaning. (Photos by S. Flit and M. Lavie)

It was woven in a medium-density plain-weave technique. The number of threads per cm is slightly higher in one system and, since excavated textiles found in the Land of Israel are characterized by slightly larger number of warps then wefts, we have recognized this system as the warp. We counted the threads per cm in three areas and found some differences due to the textile being stretched during wrapping. The number of threads per cm is 15 × 11, 14 × 10, 11 × 10 (the first number is the warp and the second the weft). The threads were uneven in thickness, 0.4–0.6 mm in diameter (measured by N. Sukenik). There are no signs of wear and tear, sewing

Fig. 13.4 The linen textile during cleaning and dismantling process. (Photo by M. Lavie)

Fig. 13.5 The white linen textile in an enlargement. (Photo by M. Lavie)

or patching, or weaving faults. No edges were preserved, probably because it was cut from a bigger piece of cloth.

13.3.2 Analysis

These textile remnants are of an exceptional value to researchers, as linen finds from the north of the country are rarely encountered (Shimoni and Shamir 1994).

Material and Origin Flax was the sole material used for manufacturing textiles in the southern Levant until the Middle Bronze Age (Shamir 2014, 2015). The earliest woollen textiles in this region were found at Jericho. They date to the Middle Bronze Age II, which is when the warp-weighted loom was first introduced into this area (Crowfoot 1960: 521; Shamir personal observation, University College London; Shamir 1996). The domestication of flax was a prerequisite for the production of textiles. Experimental fibre extraction from wild flax in Israel demonstrated that the surfaces of this plant are too heavily textured to produce spinning thread suitable for textiles (Abbo et al. 2014). The quality of the flax fibre improved over time, as indicated by the fine quality of textiles uncovered in Chalcolithic contexts. The earliest linen threads in the southern Levant may have been identified on a 10,000-year-old comb from the Murabba'at caves, near the Dead Sea (Schick 1995; Shamir 2020: 27). Flax was grown in the Pre-pottery Neolithic period and continued through the Roman period (Weiss and Zohary 2011: 249; Zohary et al. 2012: 1–2), primarily for the production of fibres for textiles as well as for the seeds which were used for dietary, medicinal and other purposes. Linen textiles were discovered in the Land of Israel from the Pre-pottery Neolithic period onwards (Shamir 2020). Flax cultivation is known at Beth She'an and the Jordan Valley at least from the Late Bronze Age (Shamir 1992, 1996: 142; 2006), as attested by two flax seed caches with 621,800 seeds from the thirteenth–eleventh centuries BCE (Kislev et al. 2009: 769). In the excavations at Tell Bet She'an, a million charred flax seeds were found dating from the early Iron Age (Kislev et al. 2011). Tiny fragments of linen were found in Bet She'an dating from the twelfth century BC (Shamir 2009).

These textiles from Bet She'an were probably made on the warp-weighted loom, which was in use in the Land of Israel from the Middle Bronze Age till the first century CE (Shamir 2015). Truncated-pyramid loom weights, weighing c. 200–300 g, were found in large numbers at Tell Iẓṭabba (East) (see below). It is likely then that the linen in which the coin hoard was wrapped was produced at the site or at least in the region.[1] We will examine the question of origin below when we try to find relationship between textile remains and the loom weights discovered at the site. Our example was not imported from Egypt, since the number of threads per cm observed in Egyptian textiles is usually higher, and Egyptian textiles generally are warp-faced and made of thinner threads. Fine linen textiles were common at Ptolemaic Egypt (Papadopoulou 2016: 194).

Spinning The linen is s-spun. The traditional spinning direction in Israel, from the Neolithic period to medieval times, was s-spun. In general, z-spun textiles were imported to the Land of Israel.

Weaving Plain weave was very common in the Land of Israel from the Chalcolithic period onwards. Usually, the number of threads per cm in linen textiles is slightly higher in the warp than in the weft. The number of threads per cm (15 × 11, 14 × 10, 11 × 10) is similar to finds from the Qumran Caves, which yielded only linen textiles. In Cave 1, the highest number of threads recorded was 18 × 18 per cm and 20 × 16 per cm, while the lowest was 10 × 8 per cm. In Cave 11 the highest number of thin threads per cm – 26 warp threads and 20 weft threads per cm – and the lowest count was 8 × 8 threads per cm (Shamir and Sukenik 2011; Sukenik et al. 2019). The thread density does not measure up to that of the Chalcolithic textiles from the Judean Desert, which have as many as c. 46 warp threads and 30 weft threads per cm, and are thinner and more delicate (Shamir 2015).

[1] Still, since textiles (esp. linen) were presumably precious, we cannot exclude the possibility that our piece predates the Hellenistic period.

Use The textile could be a cut from a garment because it is bleached. In ancient times garments were too costly to be thrown away when worn. When the garment could no longer be mended, it was cut into pieces and used as patches or for other purposes. It is similar to the Qumran bleached textiles. Bleached-linen textiles are not very common among Hellenistic-Roman textiles. Bleaching was used for garments and scroll wrappers (Shamir and Sukenik 2011; Sukenik et al. 2019). During the bleaching process, the natural colour of the linen changed from greyish-brown to white. The textiles were soaked in cleaning and whitening chemicals, including sodium carbonate, wood ash, or various plants, such as *Anabasis setifera* or *Mesembryanthemum nodiflorum*, with a high sodium content. The fabric was then exposed to sunlight for several weeks to complete the bleaching process. The procedure took several days and involved significant effort. Bleaching was quite common in Egypt, where there was specialization in the production of linen, most of which was used for clothing (Forbes 1956: 27–30; Hall 1986: 9). Tomb paintings and models from Egypt illustrate the repetitive process of putting the wet linen fabric into ash, rubbing it with detergent, and so on, before finally laying it out to bleach in the hot sun. The bleaching process is also depicted on wall paintings at Pompeii, as is the fulling process (Flohr 2013: 113–115).

Although it is difficult to know the extent of bleaching in the textile industry in the Land of Israel, we can assume that textiles used for simple purposes such as sacks or shrouds were not bleached but left in their natural colour. Bleached textiles have been found only at a few sites of the Hellenistic and Roman periods in Israel, including ‘En Raḥel on the Spice Route (Shamir 1999: 96), El-Masaia Cave in Naḥal Deragot (Shamir and Schick 2019: 202), and Qumran (Shamir and Sukenik 2011; Sukenik et al. 2019). More than 65% of the textiles in the Qumran Cave 11 assemblage are white, which was probably the result of bleaching.

13.4 Loom Weights from Tell Iẓṭabba

In the course of the two excavation seasons of the German-Israeli Tell Iẓṭabba Excavation Project three Hellenistic-period truncated-pyramid loom weights were found (Fig. 13.6). All came from Area C (Fig. 13.7), which is located some 50 m west of the previously excavated Israel Antiquities Authority’s Area Z, where additional domestic structures were unearthed. A destruction level in one of the three rooms contained many pottery vessels, including Rhodian and Knidian amphorae (Ebeling et al. 2020, 2021).

Two loom weights came from Locus 314 (Baskets 3095, 3100) and another from Locus 309 (Basket 3037). Both represent destruction levels in one of the rooms (R2) of the domestic structures excavated in Area C, from which many pottery vessels were retrieved. These weights can be added to many others found during the 1990s Israel Antiquities Authority’s excavations at the site, which were then destroyed by a fire at the storage compound of the excavations. Some 300 loom weights came

Fig. 13.6 The loom weights from the German-Israeli Tell Iẓṭabba Excavation Project. (Photo by S. Flit)

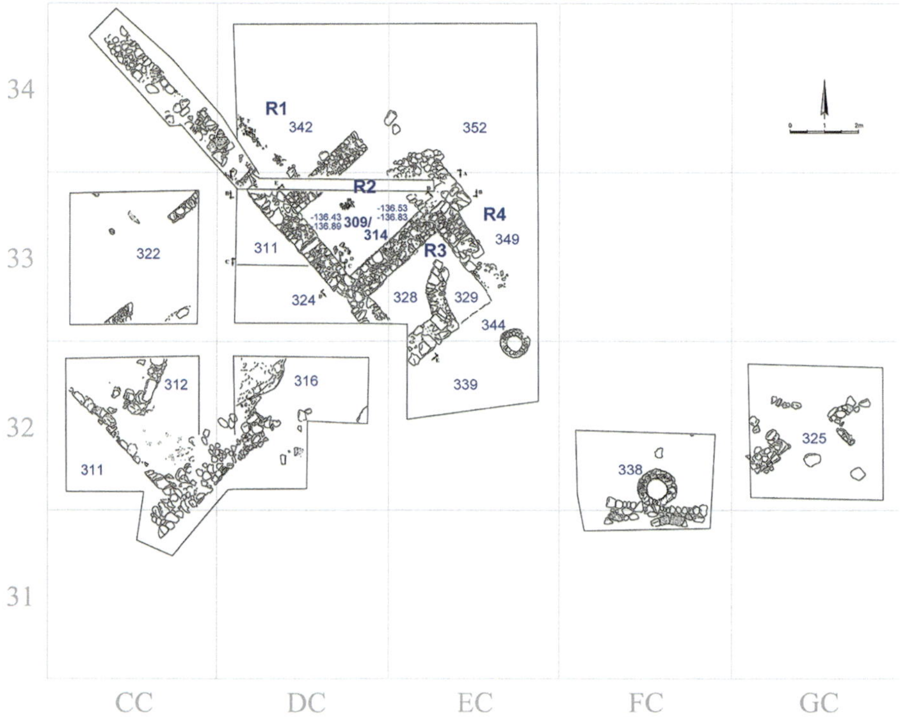

Fig. 13.7 Plan of the German-Israeli Tell Iẓṭabba Excavation's Area C, with indication of loom weights provenance. (Plan by S. Pirsky)

Table 13.1 Tell Iẓṭabba loom weights

No.	Area	L	B	Shape	H (cm)	D (cm)	W (g)	Perforation D (cm)	Preservation	Comments
1	C	314	3100	Pyramidal	8	6 × 6	310	0.4	Complete	Fingerprint on upper part (dent impression)
2	C	314	3095	Pyramidal	8.5	5.7 × 5	200	0.5	Nearly complete	Upper part not preserved (c. 20 g missing)
3	C	309	3037	Pyramidal	7.5	5.5 × 5	250	0.5	Nearly complete	Part of the lower section not preserved (c. 30 g missing)

from Area W; these had been found in four concentrations (100, 80, 80, 40) and were associated with loom remains recovered from ground floor rooms destructed in 108/107 BCE (G. Mazor and W. Atrash personal communication). Their base dimensions differ, 5.4–5.9 ´ 5.3–5.7 cm, as do their height (7.5–7.8 cm) and weight (214.3–234.1 g). In addition, 13 ring-shaped stone spindle whorls were discovered, weighing 4–13 g (G. Mazor and W. Atrash personal communication).

Our three loom weights were weighed and measured, and are presented below with maximum heights and diameters, as well as minimum diameters of perforations. The measurements were carried out with a digital caliper. The four panels of the truncated pyramid are not always equal. The horizontal perforation is in the upper third of the loom weight; the hole is not necessarily centred. The truncated-pyramid loom weights from Tell Iẓṭabba vary in size, weight, and shape (with either a rectangular or square base) – indicating that they were not mass-produced in moulds as in some Hellenistic or Roman Mediterranean assemblages. They were made of unfired clay that was partly fired when the site was burned down during the Hasmonean conquest. They range in weight from 220 g to 310 g (Table 13.1). The perforation is 0.4–0.5 cm in diameter. The loom weight from Basket 3100 has a fingerprint on the upper part.

13.5 Spindle Whorls

Thirteen basalt ring spindle whorls were found during previous excavations (G. Mazor and W. Atrash personal communication). Their weight range is 4–13 g. The thread produced using these spindle whorls could be similar or finer than the thread used to weave the excavated textile. It is impossible to know whether they were used to spin wool or flax (see Shamir 1996; Cassuto 2016: 271–276).

13.6 Discussion

References In Israel, pyramidal loom weights are also commonly found at Hellenistic and Early Roman sites, such as Tel Anafa (Herbert 1979: 79–80, Fig. 17; Larson and Erdman 2018); Ashdod, Stratum 3a (Dothan and Freedman 1967: Fig. 17); Samaria-Sebaste (Reisner et al. 1924: 343, Pl. 68; Crowfoot 1957: 399–400, Pls 92a, 23–26); Caesarea Maritima (Holland 2009: 76, nos 279–280); Shikmona, Stratum H, Room HD (Elgavish 1968: 35–36, Pl. XXV: 255); Jerusalem, City of David (Shamir 1996), Shilo (Brandl 1993: 235, Figs. 9.6, 5, 9.8, 5), Ramat Ha-Nadiv (Kol-Yaakov 2000: 490, nos. 77–78, Pls 7, 17–18), Gamla (Cassuto 2016), Dor (Stern 2000) and 'En Gedi (Shamir 2007). The truncated-pyramid loom weight is the dominant type found at Maresha (Israel's Judean Foothills), comprising more than one-half of the total for the Hellenistic period (Stern and Shamir 2019; Shamir forthcoming). Masada yielded 47 identifiable pyramid loom weights. This is the dominant type found at the site, accounting for 50% of the total sample (Shamir 1994). Similar loom weights were also uncovered at Transjordanian sites, such as Pella, in contemporary strata (Table 13.2; McNicoll et al. 1982: 75, Pl. 15b) and Tell Mazar (Boertien 2013:80).

Mass and Thickness There are two factors that influence the choice of loom weights: mass and thickness. The mass determines how many threads can be attached to a loom weight. The mass provides equal tension to each warp thread to which a loom weight is tied. If the weight is too light and does not create sufficient tension, it is difficult for the weaver to manage the warp threads during weaving. The thickness determines thread density (Andersson Strand 2012; Meo 2015; Olofsson et al. 2015; Ferrara and Meo 2017: 120; Ulanowska 2020: 147). Thick warp threads require more tension than thin warp threads (Mårtensson et al. 2009: 378). Additionally, linen warp threads require higher tension than wool warp threads (Andersson 1999: 20; Nelson 2016). The average weight (g) of loom weights is greater at Tell Iẓṭabba than at other Hellenistic and Roman sites (Table 13.2). It is possible that some other sites mentioned in Table 13.2 produced wool textiles. The Bet She'an region, however, was famous for its linen production.

The thickness of loom weights can be the same as its diameter, depending on the loom weight's shape. The former dictates the nature of the fabric produced (Mårtensson et al. 2009; Andersson Strand 2010). In addition to creating tension in the warp threads, loom weights serve to organize the warp threads during weaving. For a plain weave, two rows of loom weights equal in length to the width of the finished fabric are used. The warp threads are tied to the weights alternating between rows (i.e. the first warp thread is tied to the front row of loom weights and the second warp thread is tied to the back row of loom weights). This holds the warp threads in place and ensures that the width of the fabric is the same as that of the rows of loom weights. If the rows of loom weights are narrower than the width of the fabric, it becomes progressively narrower during weaving. Alternatively, if the

Table 13.2 Average weight (in g) of pyramidal loom weights from Hellenistic and Roman sites in Israel

Site	Number of loom weights	Weight (g)	Average weight (g)	Reference
Gamla	57	140 and 320	230 ± 54	Cassuto 2016
Maresha	313	64–479	240.8 ± 113.7	Shamir forthcoming
Maresha Complex 169		86.9–535.5	153.5	Stern and Shamir 2019
Masada	47	91.4–590	194.9 ± 109.2	Shamir 1994
Dor			100	Stern 2000: 194–195, Fig. 130
Akko	8	115–187	148 ± 28	Shamir 2016: 91–92
‘En Gedi		91–590	195 ± 109	Shamir 2007
Tell Iẓṭabba (Lichtenberger-Tal excavations)	3	220–310	253.3 ± 55	G. Mazor and W. Atrash, personal communication
Tell Iẓṭabba (Mazor-Atrash excavations)	300	214.3–234.1		

rows of loom weights are wider than the fabric, the fabric becomes wider during weaving (Mårtensson et al. 2009; Nelson 2016).

Mårtensson et al. (2009: 396) state: “If a thread requires 10 g warp tension, the weaver must attach c. 18 warp threads to each loom weight of 177 g (Troy). The textile will be dense with very thin threads; if the warp thread requires 20 g tension, the weaver can only attach c. nine warp threads to each loom weight and the textile will be coarser”. The loom weights from Tell Iẓṭabba weigh 214–310 g (from both excavations), on average 250 g (Table 13.2), while in the Tell Iẓṭabba textile, there are 15 warp threads per cm and at Wadi ed-Dâliyeh caves (late fourth century BCE) the linen textiles have 13–19 warp threads per cm (Crowfoot 1974:67–68). This means that the set-up for this textile, probably manufactured at Beth She’an, would require 12 threads per loom weight (i.e. 250 g divided by 20 g per warp thread).

Mazor and Atrash excavations at Tel Iẓṭabba yielded four loom weights concentrations (100, 80, 80, 40), located in different rooms, two in one house and two in another, but all together with burned remains of looms. In two cases, the width of the loom could be determined by its burned remains, 0.80 m and 0.90–1.00 m wide respectively. It would be reasonable to assume that the third and fourth looms were no different in width (G. Mazor and W. Atrash personal communication).

These loom weights, from both excavations, are similar in weight to the 169 gypsum loom weights found at Tel ‘Amal (near Tell Iẓṭabba). These date from the tenth century BCE and were probably used to produce linen fabric. They range in weight from 84.4 to 518.7 g, with an average of 242.9 ± 80.3 g; 69 of these loom weights (48%) weigh between 200 and 300 g (Shamir 2013).

In the tenth century BCE stratum at Tell el-Hamma in the Jordan Valley linen was also produced as attested by a spindle wrapped with linen threads. A total 158 unfired clay doughnut-shaped loom weights discovered weigh with an average 286.4 ± 81.7 g (Shamir 1996: 142).

13.6.1 Number of Loom Weights per Loom

At Beth She'an, most of the 109 loom weights dated to eighth century BCE were found in two concentrations along the eastern wall of the central hall of Building 28636. In the southernmost concentration, the weights extended along 1.5 m, probably indicating the width of the loom. The loom weights were apparently arranged in two rows. Their range weight is 273.1 ± 82.2 (Shamir 2006: 481). The charred beam (made of olive wood) found nearby may have belonged to the loom.

At Tel 'Amal, tenth century BCE, two-meter-long parallel rows of loom weights were found *in situ* (Levy and Edelstein 1972: 331–335, 342). The exact provenance of 116 of the 171 loom weights could not be determined as a result of a recording problem. Their range weight is 242.9 ± 80.3 (Shamir 2013).

13.7 Conclusions

The textile fragments from Seleucid Tell Iẓṭabba provide us with a unique insight into the economy of the Seleucid settlement. The way in which the textile was produced shows that local techniques of linen textile production were used: s-spun, tabby weave with slightly more warp threads than weft threads. The textile is identical to the ones from the Qumran Caves including the bleaching.

Furthermore, the textile tools found at the site, loom weights and spindle whorls, were suitable to produce a textile of similar technical qualities. In addition, the significant numbers of linseeds found at archaeological sites in the area, from as early as the Bronze Age, as well as in the Hellenistic context at Tell Iẓṭabba (Orendi et al. 2021), and in conjunction with historical evidence for the cultivation and production of high-quality linens at the time, provide the contextual setting in which it is highly likely that our textile was manufactured.

Although it has to be assumed that a large part of the population in this newly founded Seleucid settlement was non-indigenous, their technical knowledge, as indicated by s-twist and warp/weft proportions, was firmly rooted in the local Near Eastern tradition. This textile (most likely produced locally and in a technique that was used over a long period of time in the region) may attest the exchange of both products and technical know-how between the local and the new Seleucid settlers. This proposition challenges an idea of an isolated colonial Hellenized-type settlement where the new settlers separated themselves from the indigenous population. Instead, it promotes the opposite view, suggesting an assimilation of knowledge with cultural exchange of daily used products, where everyday objects were exchanged between ethnic groups. A similar observation was made when we analysed the archaeobotanic remains from the German-Israeli Tell Iẓṭabba Excavation Project. Our results illustrate a strong continuity of locally used crops, underlining that the new settlers interacted closely with the local population.

The textile fragment is also the evidence for the recycling economy of the Seleucid settlement. It has been demonstrated that the bleached textile is likely to have originally been part of a garment. Although the reuse of textile cloth is a widespread phenomenon in the pre-industrial world, it can rarely be traced step by step in the Hellenistic period because it is rarely preserved in the archaeological record. Therefore, any example that is documented adds to our knowledge of ancient recycling practices. In our case, the garment became a purse that enclosed the coins.

Textiles were too costly to throw away. When a garment had reached a state where patching and mending were no longer possible, it was cut into pieces and either used as patches, or was remade into another garment or repurposed. Due to the cost of new clothes, there would have been a flourishing secondary use clothing industry (Mannering 2000: 15; Grömer 2017).

References

Ancient Sources

Codex Theodosianus

Theodosiani libri XVI cum constitutionibus Sirmondianis et leges novellae ad Theodosianum pertinentes. Mommsen CMT, Meyer PM (ed.) (1905). Weidemann, Berlin.

Diocletian Edictum de pretiis rerum venalium

Diocletian Edict of maximum prices. Graser ER (ed., transl.). In: Frank T (1940). *An economic survey of Ancient Rome* V. The Johns Hopkins University Press, Baltimore.

Diokletians Preisedikt. Lauffer S (ed., transl.) (1971). Texte und Kommentare 5. De Gruyter, Berlin.

Expositio totius mundi et gentium

Expositio totius mundi et gentium. In: M Stern (ed.) (1980). *Greek and Latin Authors on Jews and Judaism*, Vol. II: CXXXIV. Israel Academy of Sciences and Humanities, Jerusalem, p 495–498.

Pausanias

Pausanias, *Description of Greece* vol. V. Wycherley L (ed., transl.) (1935). The Loeb Classical Library. Harvard University Press, Cambridge.

Rabbinic Literature

yAZ. Talmud Yerushalmi, ᶜAbodah Zarah.
yQid. Talmud Yerushalmi, Qiddušin.

Modern Works

Abbo S, Zezak I, Lev-Yadun S, Shamir O, Fridman T, Gopher A (2014) Harvesting wild flax in the Galilee, Israel and extracting fibers bearing on Near Eastern plant domestication. *Israel Journal of Plant Sciences* 62:52–64.

Andersson Strand E (1999) *The common thread: Textile production during the Late Iron Age–Viking Age*. Institute of Archaeology Report Series 67. University of Lund, Lund.

Andersson Strand E (2010) The basics of textile tools and textile technology: From fibre to fabric. In: Michel C, Nosch M-L (eds) *Textile terminologies in the Ancient Near East and Mediterranean from the third to the first millennia BC*. Ancient Textiles Series 8. Oxbow Books, Oxford and Philadelphia, p 10–22.

Andersson Strand E (2012) From spindle whorls and loom weights to fabrics in the Bronze Age Aegean and Eastern Mediterranean. In: Nosch M-L, Laffineur R (eds) *KOSMOS. Jewellery, adornment and textiles in the Aegean Bronze Age. Proceedings of the 13th international Aegean conference, Copenhagen, 21–26 April 2010*. Aegaeum 33. Peeters, Leuven and Liège, p 207–214.

Avi-Yonah M (1962) Scythopolis. *Israel Exploration Journal* 12:123–134.

Boertien JH (2013) *Unravelling the fabric: Textile production in Iron Age Transjordan*. Unpublished dissertation, University of Groningen.

Brandl B (1993) Clay, bone, metal and stone objects. In: Finkelstein I, Bunimovitz S, Lederman Z (eds) *Shiloh: The archaeology of a biblical site*. Tel Aviv University Institute of Archaeology Monograph Series 10. Tel Aviv University Institute of Archaeology, Tel Aviv, p 223–262.

Cassuto D (2016) Textile production implements. In: Syon D (ed.) *Gamla. The Shmarya Gutman excavations 1976–1989*, III.2. Israel Antiquities Authority Reports 56. Israel Antiquities Authority, Jerusalem, p 261–282.

Conder CR, Kitchener HH (1882) *The survey of Western Palestine*. Memoirs 2. Palestine Exploration Fund, London.

Crowfoot GM (1957) Spindle whorls and loom weights. In: Crowfoot JW, Crowfoot GM, Kenyon KM (eds) *Samaria-Sebaste III: The objects from Samaria*. Palestine Exploration Fund, London, p 398–402.

Crowfoot E (1960) Textiles, matting and basketry. In: Kenyon K (ed.) *Excavations at Jericho I: The tombs excavated in 1952–1954*. British School of Archaeology in Jerusalem, London, p 519–526.

Crowfoot E (1974) Textiles. In: Lapp PW, Lapp NL (eds) *Discoveries in the Wâdi ed-Dâliyeh*. Annual of the American Schools of Oriental Research 41. American Schools of Oriental Research, Cambridge Mass., p 60–77.

Dothan M, Freedman DN (1967) *Ashdod 1: The first season of excavations 1962*. ʻAtiqot 7 (English Series). Antiquities and Museums in the Ministry of Education and Culture, Jerusalem.

Ebeling P, Edrey M, Harpak T, Lichtenberger A, Tal O (2020) Field report on the 2019 German-Israeli Tell Iẓṭabba excavation project (Beth She'an), Israel. *Zeitschrift des Deutschen Palästina-Vereins* 136(1):176–190.

Ebeling P, Edrey M, Harpak T, Lichtenberger A, Tal O (2021) Field report on the 2020 German-Israeli Tell Iẓṭabba excavation project (Beth She'an), Israel. *Zeitschrift des Deutschen Palästina-Vereins* 137(1):60–74.

Elgavish J (1968) *Archaeological Excavations at Shikmona, Field Report No. 1: The Levels of the Persian Period, Seasons 1963–1965*. The City Museum of Ancient Art, Haifa (in Hebrew).

Ferrara B, Meo F (2017) Loom weights in sacred contexts: The square building of the Heraion near the Sele River. In: Brøns C, Nosch M-L (eds) *Textiles and cult in the Ancient Mediterranean*. Ancient Textiles Series 31. Oxbow Books, Oxford and Philadelphia, p 112–125.

Finkielsztejn G (1999) More evidence on John Hyrcanus I's conquests: Lead weights and Rhodian amphora stamps. *Bulletin of the Anglo-Israel Archaeological Society* 16:33–63.

Flohr M (2013) *The world of the fullo: work, economy and society in Roman Italy*. Oxford Studies on the Roman economy. Oxford University Press, Oxford.

Forbes RJ (1956) *Studies in ancient technology* IV. Brill, Leiden.

Grömer K (2017) Recycling of textiles in historic contexts in Europe. Case studies from 1500 BC till 1500 AD. In: Miloglav I, Kudelić A, Balen J (eds) *Recikliraj, ideje iz prošlosti*. Arheološki Muzej u Zagrebu, Filozofski Fakultet Sveucilišta u Zagrebu, Institut zu Arheologiju, Zagreb, 75–98.

Hall R (1986) *Egyptian textiles*. Shire Egyptology, London.

Herbert SC (1979) Tel Anafa 1978: Preliminary report. *Bulletin of the American Schools of Oriental Research* 234:67–83.

Holland L (2009) *Weights and weight-like objects from Ceasarea Maritima*. Lionel Holland, Hadera.

Kislev ME, Simchoni O, Melamed Y, Maroz L (2009) Botanical remains. Food and industrial crops. In: Panitz-Cohen N, Mazar A (eds) *Excavations at Tel Beth-Shean 1989-1996. III. The 13th–11th century BCE strata in Areas N and S*. The Israel Exploration Society, Jerusalem, p 764–771.

Kislev ME, Simchoni O, Melamed Y, Maroz L (2011) Flax seed production: Evidence from the Early Iron Age site of Tel Beth-Shean, Israel and from written sources. *Vegetation History and Archaeobotany* 20:579–584.

Kol-Yaakov S (2000) Various objects from the Hellenistic, Roman, and Byzantine periods. In: Hirschfeld Y (ed.) *Ramat Hanadiv excavations: final report of the 1984–1998 seasons*. Israel Exploration Society, Jerusalem, p 473–503.

Kool R, Berman A, Shamir O, Tepper Y (2011) A late tenth-century Fatimid coin purse from Beth She'an. *'Atiqot* 67:32–41.

Landau Y, Tzaferis V (1979) Tel Iṣṭabah, Beth Shean: The excavations and Hellenistic jar handles. *Israel Exploration Journal* 29:152–159.

Larson KA, Erdman KM (2018). Tools for textile manufacture. In: Berlin AM, Herbert SC (eds) *Tel Anafa* II, iii. Kelsey Museum Fieldwork Series, Ann Arbor, p 145–210.

Levy S, Edelstein G (1972) Cinq annẽes de fouilles a Tel 'Amal. *Revue Biblique* 79:325–67.

Lichtenberger A (2008) Artemis and Zeus Olympios in Roman Gerasa and Seleucid religious policy. In: Kaizer T (ed.) *The variety of local religious life in the Near East in the Hellenistic and Roman periods*. Brill, Leiden and Boston, p 133–153.

Lichtenberger A (2020) Decapolis. *Brill encyclopedia of Early Christianity online*. https://doi.org/10.1163/2589-7993_EECO_SIM_00000878, accessed 27 Nov 2020.

Lichtenberger A, Meyer C, Tal O (2020) Magnetic prospecting at Nysa-Scythopolis (Tell Iẓṭabba, Beth She'an, Israel): Deciphering urban planning at a newly founded Hellenistic town of the Decapolis. *Bulletin of the Anglo-Israel Archaeological Society* 38:45–70.

Lichtenberger A, Tal O; with an appendix by Shamir O (2020) A hoard of Alexander II Zabinas coins from Tell Iẓṭabba (Bet She'an), Israel. *Israel Numismatic Research* 15:45–59.

Mannering U (2000) The Roman tradition of weaving and sewing: A guide to function? *Archaeological Textiles Newsletter* 30:12–15.

Mårtensson L, Nosch M-L, Andersson Strand E (2009) Shape of things: Understanding a loom weight. *Oxford Journal of Archaeology* 28:373–398.

Mazar A (2019) Weaving in Iron Age Tel Reḥov and the Jordan Valley. *Journal of Eastern Mediterranean Archaeology and Heritage Studies* 7:118–137.

Mazar A (2020) Objects related to textile production. In: Mazar A and Panitz-Cohen N (eds) *Tel Reḥov: A Bronze and Iron Age City in the Beth-Shean Valley. Vol. V: Various Objects and Natural-Science Studies*. Qedem 63. Institute of Archaeology, Hebrew University of Jerusalem, Jerusalem, p 81–129.

Mazor G (2008) Beth-Shean: the Hellenistic to Early Islamic periods. The Israel Antiquities Authority excavations. In: Stern E (ed.) *The new encyclopedia of archaeological excavations in the Holy Land* V. Israel Exploration Society, Jerusalem, p 1623–1636.

Mazor G, Atrash W, Finkielsztejn, G (2018) *Bet She'an IV. Hellenistic Nysa-Scythopolis: the amphora stamps and sealings from Tel Iẓṭabba*. Israel Antiquities Authority Reports 62. Israel Antiquities Authority, Jerusalem.

Mazor G, Bar-Nathan R (1998) The Bet She'an excavation project – 1992–1994. Antiquities Authority expedition. *Excavations and Surveys in Israel* 17:7–36.

McNicoll A, Smith RH, Hennessy B (1982). *Pella in Jordan 1: An interim report on the joint University of Sydney and the College of Wooster excavations at Pella 1979–1981*, 2 vols. Australian National Gallery, Canberra.

Meo F (2015) New archaeological data for the understanding of weaving in Herakleia, southern Basilicata, Italy. In: Harlow M, Nosch M-L (eds) *Greek and Roman textiles and dress: An interdisciplinary anthology*. Ancient Textile Series 19. Oxbow Books, Oxford and Philadelphia, p 236–259.

Nelson TJ (2016) *Artifactual evidence for the role of the warp-weighted loom: The transformation of textile production in the Iron Age Levant*. Unpublished dissertation, Stony Brook University.

Olofsson L, Andersson Strand E, Nosch M-L (2015) Experimental testing of Bronze Age textile tools. In: Andersson Strand E, Nosch M-L (eds) *Tools, textiles and contexts: Textile production in the Aegean and Eastern Mediterranean Bronze Age*. Ancient Textiles Series 21. Oxbow Books, Oxford and Philadelphia, p 75–100.

Orendi A (2020) Flax cultivation in the Southern Levant and its development during the Bronze and Iron Age. *Quaternary International* 545:63–72.

Orendi A, Lichtenberger A, Tal O (2021) Food in a colonial setting: The flora assemblage of a short-lived Seleucid-founded site in the Near East. *Vegetation History and Archaeobotany*. https://doi.org/10.1007/s00334-020-00820-z

Papadopoulou M (2016) Wool and the city: Wool and linen textile trade in Hellenistic Egypt. In: Droß-Krüpe K, Nosch M-L (eds) *Textiles, trade and theories: From the Ancient Near East to the Mediterranean*. Kārum – Emporion – Forum 2. Ugarit Verlag, Münster, p 193–210.

Reisner GA, Fisher CS, Lyon DG (1924) *Harvard excavations at Samaria: 1908–1910*, 2 vols. Harvard University Press, Cambridge Mass.

Schick T (1995) A 10,000-year-old comb from Wadi Murrabba'at in the Judean desert. *'Atiqot* 27:199–202.

Shamir O (1992) A twelfth century BCE linen textile fragment from Beth Shean. *Archaeological Textiles Newsletter* 14:6–8.

Shamir O (1994) Loomweights from Masada. In: Aviram J, Foerster G, Netzer E (eds) *Masada* IV. The Hebrew University of Jerusalem and the Israel Antiquities Authority, Jerusalem, p 265–282.

Shamir O (1996) Loomweights and whorls. In: Ariel DT (ed.) *Excavations at the City of David 1978–85, directed by Y. Shiloh*, Vol. IV. Qedem 35. Institute of Archaeology, Hebrew University of Jerusalem, Jerusalem, p 135–170.

Shamir O (1999) Textiles, basketry, and cordage from 'En Raḥel. *'Atiqot* 38:91–124.

Shamir O (2006) Objects associated with the weaving industry (Beth Shean, Area P). In: Mazar A (ed.) *Excavations at Tel Beth-Shean 1989-1996 I. From the Late Bronze Age IIB to the Medieval period*. Israel Exploration Society, Jerusalem, p 474–483.

Shamir O (2007) Loom weights from ‘En-Gedi. In: Stern E (ed.) *En-Gedi Excavations, I: Conducted by B. Mazar and I. Dunayevsky, Final report (1961–1965)*. Israel Exploration Society, Jerusalem, p 381–390.

Shamir O (2009) A linen textile fragment. In: Panitz-Cohen N, Mazar A (ed.) *Excavations at Tel Beth-Shean 1989-1996, III. The 13th–11th century BCE strata in Areas N and S*. Israel Exploration Society, Jerusalem, p 608.

Shamir O (2013) Tel ‘Amal: Loomweights. *Hadashot Arkeologiyot — excavations and surveys in Israel* 125:1–8.

Shamir O (2014) Textiles, basketry and other organic artifacts of the chalcolithic period in the Southern Levant. In: Sebbane M, Misch-Brandel O, Master DM (eds) *Masters of fire: Copper Age art from Israel. Catalogue of Exhibition*. Institute for the Study of the Ancient World at New York University, Princeton and New York, p 139–152.

Shamir O (2015) Textiles from the Chalcolithic period, Early and Middle Bronze Age in the southern Levant – the continuation of splicing. *Archaeological Textiles Review* 57:12–25.

Shamir O (2016) Akko, the courthouse site: loomweights and whorls. In: Hartal M, Syon D, Stern E (eds) *‘Akko* II. Israel Antiquities Authority Reports 60. Israel Antiquities Authority, Jerusalem, p 91–95.

Shamir O (2020) Continuity and discontinuity from the Neolithic period to the Chalcolithic period linen textiles production in the Southern Levant. With an appendix by A. Rast-Eicher. In: Schier W, Pollock S (eds) *The competition of fibres. Early textile production in western Asia and south-east and Central Europe (10,000–500 BCE). International workshop, Berlin 8–10 March 2017*. Ancient Textiles Series 36. Oxbow Books, Oxford and Philadelphia, p 27–37.

Shamir (forthcoming) Loomweights and Whorls from Maresha. *‘Atiqot*.

Shamir O, Schick T (2019) Chalcolithic to Medieval textiles from the Judean desert caves survey. In: Peleg-Barkat O, Ashkenazi J, Leibner U, Aviam M, Talgam R (eds) *Between sea and desert: On kings, nomads, cities and monks. Essays in honor of Joseph Patrich*. Kinneret Academic College and Ostracon, Tzemach, p 195–224.

Shamir O, Sukenik N (2011) Qumran textiles and the garments of Qumran’s inhabitants. *Dead Sea Discoveries* 18:206–225.

Simchoni O, Kislev ME (2012) Food and fodder in Early Bronze Age strata in Area M. In: Mazar A (ed.) *Excavations at Tel Beth Shean 1989–1996, IV. The 4th and 3rd millennia BCE. The Beth Shean-valley archaeological project*. Israel Exploration Society, Jerusalem, p 422–429.

Simchoni O, Kislev ME, Melamed Y (2007) Beth-Shean as a trade center for crops in the Bronze Age: Botanical and entomological evidence. In: Mazar A, Mullins RA (eds) *Excavations at Tel Beth Shean 1989–1996, II. The Middle and Late Bronze Age strata in Area R. The Beth Shean-valley archaeological project*. Israel Exploration Society, Jerusalem, p 702–715.

Shimoni C, Shamir O (1994) Gesher Haziv-textile remains on coins. *‘Atiqot* 25:97.

Stern E (2000) *Dor. The ruler of the seas: Nineteen years of excavations at the Israelite-Phoenician harbor town on the Carmel coast*. Israel Exploration Society, Jerusalem.

Stern I, Shamir O (2019) Loomweights and whorls. In: Stern I (ed.) *Maresha subterranean complex 169, seasons 2000–2016, final report*. Annual of the Nelson Glueck School of Biblical Archaeology 12. Hebrew Union College Press, Jerusalem, p 392–404.

Sukenik N, Shamir O, Rottoli M, Belis M (2019) Textiles and strings from Qumran cave 11. In: Fidanzio M, Humbert J-B (eds) *Khirbet Qumrân and Aïn Feshkha IVA: Qumrân Cave 11Q archaeology and new scroll fragments*. Novum Testamentum et Orbis Antiquus – Series Archaeologica 8. Vandenhoeck and Ruprecht, Fribourg, p 97–118.

Tal O (2011) ‘Hellenistic foundations’ in Palestine. In: Grabbe LI, Lipschits O (eds) *Judah between East and West: The transition from Persian to Greek rule (ca. 400–200 BCE)*. Library of Second Temple studies 75. T and T Clark, New York and London, p 242–254.

Ulanowska A (2020) Looms in spaces. The warp-weighted loom in the Aegean Bronze Age and the modern experience of weaving. *Bulletin de l’association pour la promotion des recherches sur l’Âge du Bronze* 18:146–152.

Weiss E, Zohary D (2011) The Neolithic southwest Asian founder crops: Their biology and archaeobotany. *Current Anthropology* 52:237–254.

Wild JP (1970) Notes on clothing. In: Erim KT, Reynolds J (eds) The copy of Diocletian's edict on maximum prices from Aphrodisias in Caria. *Journal of Roman Studies* 50:120–141.

Yahalom-Mack N (2007) The textile industry. In: Mazar A and Mullins RA (eds) *Excavations at Tel Beth-Shean 1989–1996*. Vol. II: *The Middle and Late Bronze Age Strata in Area R*. Israel Exploration Society, Jerusalem, p 661–669.

Zohary D, Hopf M, Weiss E (2012) *Domestication of plants in the old world: The origin and spread of domesticated plants in southwest Asia, Europe, and the Mediterranean basin*. Oxford University Press, Oxford.

Zori N (1962) An archaeological survey of the Beth-Shean valley. In: *The Beth Shean valley: The 17th archaeological convention*. Israel Exploration Society, Jerusalem, p 135–198 (in Hebrew).

Orit Shamir, PhD, is the Head of the Department of Museums and Exhibits of the Israel Antiquities Authority. Her area of specialization is ancient textiles, loom weights and spindle whorls from the Neolithic to the medieval period in Israel. Her publications are accessible, i.al., at antiquities.academia.edu/OritShamir.

Achim Lichtenberger, PhD, is Professor of Classical Archaeology and Director of the Archaeological Museum at Münster University, Germany. He is co-director of the German-Israeli Tell Iẓṭabba Excavation Project, the Armenian-German Artaxata Project and the Danish-German Jerash Northwest Quarter Project.

Oren Tal, PhD, is Professor of Classical Archaeology at the Jacob M. Alkow Department of Archaeology and Ancient Near Eastern Cultures of Tel Aviv University, Israel. He is the director of the Apollonia-Arsuf Excavation Project and co-director of the German-Israeli Tell Iẓṭabba excavation Project.

Chapter 14
Unravelling the Threads of the Nubian Openworks. New Inquiries on a Unique Textile Tradition from Meroitic Sudan (c. 350 BCE–350 CE)

Elsa Yvanez and Ulrikka Mokdad

Abstract Around 100 BCE–350 CE, in present-day Sudan, the Kushite people of the Meroitic kingdom assembled long strands of cotton threads to create lattice works of elaborate design at the bottom of clothing and furnishing textiles. These openwork borders concluded the weave of large fabrics woven on the warp-weighted loom. Their ubiquity in the textile corpus of the Meroitic period makes these openworks a classic fixture of Meroitic weaving practices, reflected through iconography in other media as well.

In 1984 and 1998, the renowned Nubian textiles experts Elisabeth Crowfoot and Nettie K. Adams studied the Meroitic openwork technique developed at Qasr Ibrim and compared it to other openworks from the Bronze Age site of Kerma. Highlighting their common decorative vocabulary, the authors proposed to see the two techniques as different manifestations of the same craft tradition. Twenty years later, the present authors re-examined this hypothesis using an interdisciplinary approach based on the meeting of textile studies with experimental archaeology. Bringing heads and hands together, this project and the present article provided the opportunity to reconstruct and understand the manufacture and history of this remarkable technique.

Keywords Sudan and Nubia · Kerma · Meroitic · Craft · Textile · Openwork technique · Experimental archaeology

E. Yvanez (✉)
Polish Center of Mediterranean Archaeology, University of Warsaw, Warsaw, Poland

U. Mokdad
Centre for Textile Research, University of Copenhagen and Expert Weaver, Copenhagen, Denmark

A. Ulanowska et al. (eds.), *Ancient Textile Production from an Interdisciplinary Perspective*, Interdisciplinary Contributions to Archaeology,
https://doi.org/10.1007/978-3-030-92170-5_14

14.1 Introduction

Along the course of the Nile, in today's Sudanese and Egyptian Nubia, the inhabitants of the Meroitic kingdom (c. 350 BCE–350 CE) developed a rich textile tradition, moulding artistic influences from Pharaonic and Hellenistic Egypt to the local Kushite heritage.[1] The distinctive Meroitic textiles are characterized by the very common use of cotton fibres, the weaving of simple or extended tabby fabrics on the warp-weighted loom, frequent blue ornaments in either tapestry or embroidery techniques, and complex decorative finishings along the lower edge of textiles. A particularly striking example of these ornamental features is a singular form of openwork border – an open lattice measuring between 1 and 5 cm high – preceding a heavy row of tasselled fringes. To our knowledge, this openwork technique is unique to Meroitic Sudan and Nubia. It has a particularly strong visual impact, illustrated in both iconography and preserved textiles of the period. It seems to find its origin in the creative use of the warp lengths available at the bottom of textiles woven on a warp-weighted loom, and is strongly associated with the frequent fringes and tassels characteristic of the Meroitic ornamental vocabulary.

Because of their widespread presence in the archaeological textile record, Meroitic openwork borders have attracted the attention of the three pioneers of Nubian textile research. Grace Crowfoot reported their presence as early as 1934 (Griffith and Crowfoot 1934: 10) and gave a first drawing of the fine specimen recovered in the elite cemetery of Meroe (Dunham 1963: 143; 147). Her daughter Elisabeth Crowfoot then produced an in-depth analysis of the technique based on the tens of examples she observed at Qasr Ibrim (Crowfoot 1984). Nettie K. Adams later described similar textiles and iconographic representations in her discussion on Meroitic dress (Adams 1989). Struck by their resemblance with Kerma pot-nets, she devoted an entire article to the comparison of the two techniques, stressing the unicity of this "unique decorative tradition" (Adams 1998: 17).

The present article aims to build on this scholarship, first by providing an up-to-date overview of the corpus. Going beyond textile analyses alone, we chose to adopt an interdisciplinary approach merging different sources: textile artefacts, iconography, archaeological data, and experimental testing. Combining expertise in Meroitic textile archaeology (Yvanez) with weaving and Late Antique art history (Mokdad), the two authors joined heads and hands together to re-examine these remarkable openwork borders, questioning past hypotheses and attempting to better assess their place in the ancient Sudanese textile tradition.

[1]The present study is part of the TexMeroe project (Elsa Yvanez), which aims to produce an in-depth knowledge of several techniques specific to Meroitic textiles as well as understand the socio-economic context of their production. Marie Skłodowska-Curie fellowship n°743420. This project is supported by the European Commission, Horizon 2020 Research and Innovation program.

14.2 The Meroitic Openworks: Presentation

14.2.1 General Overview

Our investigations in the Meroitic corpus have revealed a group of 77 textiles with openwork borders, to which more specimens will no doubt continue to be added as studies in this area progress.[2] So far, our records show examples from Meroe, Sai island, Aksha, Gebel Adda, Qustul-Ballana, Karanog and Qasr Ibrim (Fig. 14.1, Table 14.1). The overwhelming majority of the corpus comes from Nubia, especially from Qasr Ibrim. This geographical concentration does not seem to indicate a local weaving technique or taste, since one specimen and many iconographic representations are also known from Central Sudan. It matches the general repartition map of Meroitic textile finds, reflecting the dry climatic conditions of Nubia and the good preservation of organic material in this region.

All recorded openworks are created on a fabric predominantly consisting of cotton, with both a warp and a weft made of S-spun cotton fibres. However, a few specimens incorporate wool threads in the openwork section, as extra wefts to create the pattern.

The preserved sections of openwork borders tend to be better preserved than the fabric they originally belonged to. Fragments often show large portions of the openwork and the fringes, but very few remains of the woven fabric itself, making the identification of their original use difficult. We see two reasons why that is the case: (1) the openwork borders are very densely made and tightly assembled, which makes them strong and more resistant to time and decay, and (2) the openworks were often torn off and re-used (see below). A few better-preserved textiles and iconographic representations show that this type of borders was used for both clothing and furnishing items, making open lattices and long fringes a ubiquitous element of the Meroitic textile landscape.

14.2.2 Definition

The openwork borders form a small portion of a woven textile, generally located at the bottom edge of a fabric or, less often, added along the upper edge. Despite its use of both warps and wefts, this portion is technically non-woven but *wrapped* by hand. The progress of supplementary wefts through the warp is "repeatedly broken, [their] movements is alternately forward and back, [creating] a progressive encircling, or *wrapping*, of warp-by weft-elements" (Emery 2009: 214, our emphasis). In the case of the Meroitic openworks, the supplementary wefts tightly turn around groups of several warp threads. Successive passages of weft-wrapping

[2] A PhD about the Qasr Ibrim textile collection curated at the Bolton Museum (U.K.) is currently in preparation by Sarah Hitchens at Liverpool University.

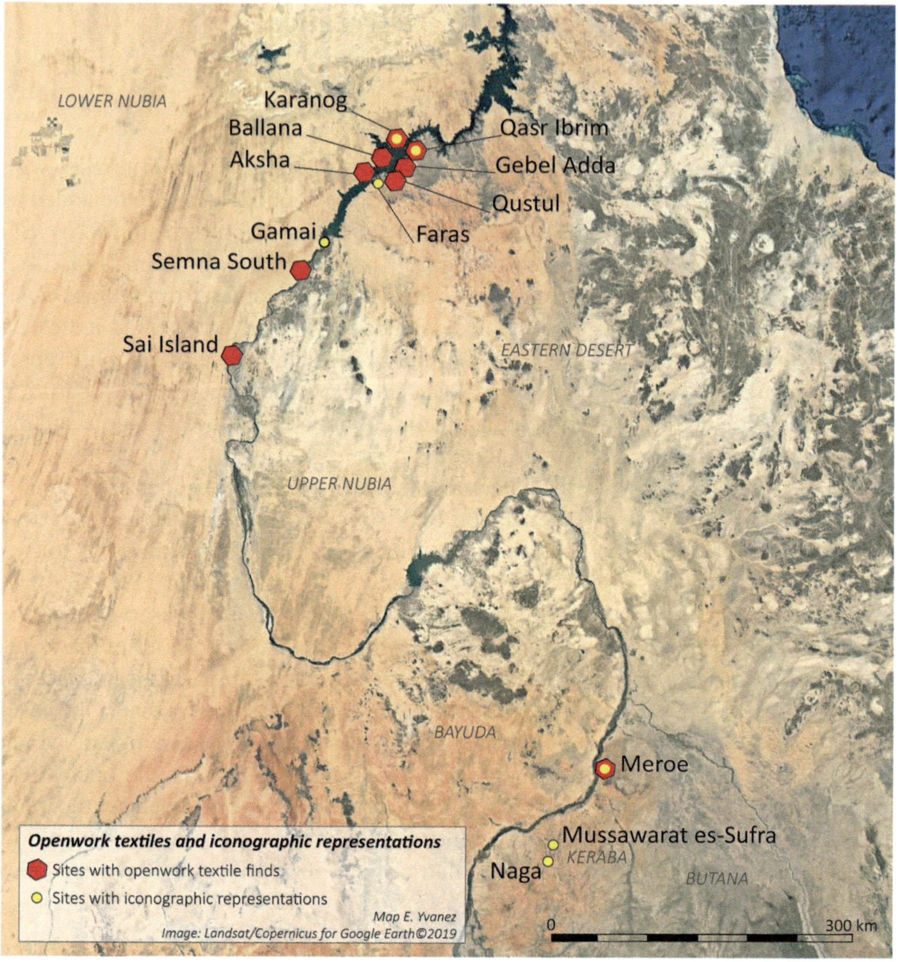

Fig. 14.1 Map of Meroitic Sudan, showing the location of openwork textile finds and iconographic representations. (Map by E. Yvanez, background image Landsat/Copernicus for Google Earth 2019)

Table 14.1 Repartition of openwork borders per sites, with general indications on fibre types (these numbers are mere indications, as the corpus is growing along ongoing excavations and collection reassessment)

Site	N° of recorded specimen	Fibres	Blue threads
Meroe	1	Cotton	
Sai island	2	Cotton (+ wool)	
Aksha	1	Cotton (+ wool)	
Gebel Adda	5	Cotton (+ wool)	X
Qustul-Ballana	7	Cotton (+ wool)	
Karanog	7	Cotton	
Qasr Ibrim	53	Cotton (+ wool)	X

over the same warp groups create tight vertical or diagonal bundles of threads, leaving large empty spaces in the fabric's structure. Consecutive grouping and re-grouping of the warps lead to an "openwork patterning" exhibiting rows of geometric shapes arranged according to a "vertical ordering of the openings" (Emery 2009: 85).

14.2.3 *Description of Archaeological Examples*

This type of border is prevalent in the cotton fabrics characteristic of the Meroitic textile tradition. These textiles were usually woven in plain tabby weave (simple or extended), with reinforced selvedges. They are predominantly white or beige in the natural colour of cotton. Decorations were common, encompassing horizontal bands of a plain contrasting colour or tapestry patterns, overwhelmingly created with dark and light blue threads. Tone-on-tone patterning was also used, especially incorporating bundles of supplementary wefts in the weave, adding a three dimensional and textural effect along the horizontal borders (cf. Figure 14.2). After weaving, the large rectangle of fabric was generally used without tailoring of any kind, as wrap-around garments or furnishings.

Fig. 14.2 Cotton 'sash' with supplementary weft bundles, openwork lattice and fringed tassels, from the Nubian site of Karanog. (E7511E, photo by E. Yvanez, with the courtesy of Penn Museum)

Before removing the finished textile from the warp-weighted loom,[3] the bottom end of the ground weave could be completed by yet another type of decoration: the openwork and fringed border. The weave itself would first be secured by several rows of twinning or a flat braid (see below). A portion of openwork would then be created over an area covering between 1 and 5 cm of the warp. It consists of a lattice-like structure where the threads form different geometric patterns alternating with empty spaces. The patterns involve an ever-changing combination of vertical hatches, lozenges, and cruciform shapes (Fig. 14.2), sometimes enhanced by the use of extra wrapping threads of a different colour – generally in dark blue wool or cotton – forming another layer of bi-colour geometric shapes (Fig. 14.3).

Once the openwork was completed, another braid or rows of twinning would keep all the threads together and be used to attach a dense row of tasselled fringes, usually quite bushy and long. Among the 77 recorded specimens, 46 fragments show the presence of fringes: always made of s2Z threads, they are preserved up to a 9–13 cm length. Alternatively, the fringed tassels can be replaced by small 'bobbles', particularly on blue and white specimens where the blue threads would be used to form the exterior.

This technique is generally observed at the bottom of large woven fabrics, but a smaller number of specimens also seem to function as stand-alone pieces: the openwork is created directly under the starting border, and then used to anchor a series of very long, fringed tassels. The main feature of these short textiles is their fringes

Fig. 14.3 Openwork border from Qasr Ibrim: bicolour lattice pattern showing the use of natural-color cotton threads and dark blue wool threads. (BM 71880, photo by E. Yvanez ©Trustees of the British Museum)

[3] The discovery of many clay loom weights is Central Sudan and Nubia, found as single units or large groups, attest the widespread use of the warp-weighted loom during the Meroitic period, e.g. Yvanez and Woźniak 2019, Fig. 14.6b.

and the resulting fabric could have been used as ornaments around the body or around a small structure or object (see below).

As the example in Fig. 14.2 shows, archaeological finds of openworks have a very matted and almost felted appearance, due to the many years of deposition in the ground. They are also fragmented, desiccated, and friable. Their condition makes observing and understanding their structure particularly difficult.

14.3 Manufacturing Techniques

Textile scholars of the past had access to a large number of textiles, often during excavations, and managed to build a first-hand knowledge of the openworks by unravelling – "dissecting" – several specimens (Crowfoot 1984: 11). In contrast, modern observation standards in museums do not allow us to look 'inside' the openworks. We therefore chose a different approach, uniting archaeology and weaving experiments, to try to understand the manufacturing technique and thereby question past hypotheses.

14.3.1 Our Approach

As defined by the *Oxford Companion to Archaeology* (Asher Silverman 2012, vol 1: 564), "experimental archaeology refers to the application of experimental methods in data collection and description, interpretation, and explanation of the archaeological record". This definition emphasizes the positivist approach that cemented the discipline, based on "laboratory (replicable) experiments", "the careful control of variables", "an hypothetico-deductive process", and the use of "authentic materials" (Outram 2008).[4] In textile research especially, experimental archaeology encompasses three different approaches based on first-hand knowledge production: ethnographic studies, experience archaeology, and true experimental archaeology (Andersson Strand 2009). Often used to test the function and efficiency of specific textile tools, experimental protocols have been developed by the Centre for Textile Research (Andersson Strand and Nosch 2015). Guidelines include the selection of similar raw material, the use of precise tool replicas, and the reproducibility of the experiment by at least two experienced craftspeople.

As such, our study does not meet the criteria of 'true' experimental textile archaeology. We did not manipulate specific variables, and we used modern looms (albeit of basic construction) and modern raw materials, as we were not able to procure hand-spun cotton thread resembling Meroitic cotton. The absence of living

---

[4] For further reference on the methodology of experimental archaeology, see e.g. Ascher 1961; Comis 2010; and O'Sullivan and Souyoudzoglou-Haywood 2019.

hand spinning and weaving traditions[5] in Sudan also precluded modern ethnographic observations. In her 1931 *Methods of Hand Spinning in Egypt and the Sudan,* Grace Crowfoot did record local cotton spinning, but the warp-weighted loom had been long dis-used and such lattice structure was not known (or reported) in the local textile production.

Moreover, the experiments were not geared towards the testing of a tool's function but focussed on exploring a textile-making technique. Our approach was therefore slightly more subjective, as it rested upon gestures and the weaver's experience, two parameters difficult to objectively assess from archaeological remains alone. Our investigation still achieved two of the five types of archaeological experiments: they led to a "scale construction that test a hypothetical design for a structure" in the framework of a general "investigation into how things were achieved in the past" (as defined by Reynolds and Outram, summarized in Outram 2008: 3).

Despite these discrepancies, we believe in the usefulness of our experiments and in their sharing the same heuristic approach as more rigorous protocols. Our approach combined a deep knowledge of Meroitic textile production in general, the study of numerous preserved openwork borders in museums, expert skills in weaving crafts, and experienced manipulation of the warp-weighted loom.

14.3.2 Textile Experimental Reconstruction

The recreation of ancient textiles is a search for possible answers to specifically designed questions, which requires close co-operation between researchers coming from the different fields of textile archaeology and craft (Sarri and Mokdad 2019). The answers found can never be exact. However, even though the answers we find may seem speculative, they form useful stepping stones that will carry future research to a better understanding of ancient artefacts and craftspeople. In other words, recreating ancient textiles can be regarded as a large puzzle where our efforts represent but a single piece.

Craft knowledge and academic knowledge are two very different things (Ciszuk 2007). The art of weaving is, more than anything, *tacit knowledge*. Through observation and experience, the weaver gains an understanding of how structure, colour and design work in combination to produce a distinct surface. During weaving, the weaver's body and mind must stay concentrated and work together as one unit while the same small movements are repeated over and over again. An important thing for a weaver is a keen perception. Mistakes must be discovered and corrected immediately, otherwise one will have to unravel a large piece of cloth. Unravelling takes a

[5] Traditional weavers are still known in the town of Shendi, although their number is declining rapidly. They belong to the Coptic community that settled in the region at the beginning of the twentieth century and use the foot-treadle loom. Interviews conducted by E. Yvanez in 2009 indicated the possible survival of cotton hand-spinning in a small and remote community of women, but their location proved impossible to trace at that time.

lot of time that could more usefully be spent on weaving. As a weaver, one must also have a sense in one's hands for how much the weft should be tightened and beaten. If the weaver tightens the weft yarn too much, the edges of the cloth will begin to shrink. If the weaver gives too much weft yarn to the warp, the cloth will start to bulge. The sense for exactly how to place the weft yarn and how much it should be tightened is gained with the weaver's experience. It cannot be acquired only intellectually, e.g. from reading.

Joining heads and hands in a fruitful collaboration, the present study adopted a decisive "knowledge-based" approach compatible with a more "experiential" element (Outram 2008: 4–5; Souyoudzoglou-Haywood and O'Sullivan 2019: 2). Basing the project on our study of many Meroitic openwork borders, we decided to test Elisabeth Crowfoot's reconstruction by directly experiencing the technique and recreating similar openwork lattices.

14.3.3 Creating Openworks: Elisabeth Crowfoot's Reconstruction

Working at Qasr Ibrim, Elisabeth Crowfoot "dissected" several openwork borders,[6] which enabled her to follow the path of both the warp threads and the wrapping threads. From these examples, she reconstructed the following *chaîne opératoire*:

1. After weaving the main part of the fabric, the weaver untied the warp threads from the loom weights and Z-plied them. The threads would then be assembled in "a braid in which the warps themselves are used to produce a narrow entwined band, from which they re-emerged at regular intervals" (Crowfoot 1984: 11);
2. Once re-attached to the loom weights as a single shed, the plied warps were then used as a base to anchor the openwork pattern. The weaver used a finer Z-plied yarn as wrapping thread, first attaching it to the selvedge thread at the lower corner of the intended openwork portion, leaving the long end hanging for later use. Working from the bottom to the top, the weaver wrapped this extra thread around groups of warps, separating and re-grouping the warps depending in the intended pattern. Once at the top, the wrapping thread was passed through the braid and went down again, completing another portion of the lattice. The end was left hanging;
3. The technique was repeated until all the wrap threads were wrapped according to a variety of geometric patterns;
4. Once again detached from the loom weights, all the hanging threads were assembled in a finishing braid, from which the warps re-emerged to form a cord supporting the fringed tassels. The threads used for the wrapping were knotted to form the core of the tassels and secure the extra threads of the fringes.

[6] "all these borders so far dissected", Crowfoot 1984: 11.

Elisabeth Crowfoot produced a series of highly useful drawings (Crowfoot 1984: Fig. 1–2) to support her reconstruction and help in the understanding of this complex-looking technique. While her description explains the appearance of the borders and the completion of the bushy fringes, it seemed to the present authors very complicated indeed. Our hesitation mainly resided in the fact that the weaver would have had to untie the loom weights not once but twice, *before* the completion of the piece. The loom weights would be untied a first time to create a flat braid with the fabric still attached on the loom but not in tension. The weaver would then reattach the warp threads to the weights, create the openwork, and then detach the weights again to make the finishing braid and the fringed tassels. As any weaver using a warp-weighted loom would know, it is very difficult to keep the warp threads in order once they are not under tension, let alone repeatedly retie them to the weights in perfect position. It is also very hard to create an orderly flat braid without optimal tension.

To us, the many manipulations of the loom weights implied by Elisabeth Crowfoot's technical reconstruction did not seem to fit the working patterns or solutions normally sought by weavers used to work on this type of loom. We therefore wanted to test if it was possible to re-create an openwork border while keeping the warps attached to the loom weights until the end. Our approach does not negate Elisabeth Crowfoot's reconstruction. It is an addition, trying to explore the roots of the wrapped openwork tradition from a weaver's perspective. If this technique was created to make use of the warp-lengths left hanging after a fabric is woven on the warp-weighted loom, then surely it should be possible to use it while following the basic principles of this tool.

14.3.4 'Thinking with My Hands': A Weaver's Experiments

Several trial openwork patterns were made in order to investigate whether Crowfoot's hypothesis on the threads' course inside the openwork borders could serve as a model for a recreation. After having read the article and looked carefully at the step-by-step drawings in "Openwork fringes from Qasr Ibrim", we suspected that the Meroitic weavers may have had an easier and faster way of carrying out their decorative repertoire of openwork borders than the one suggested by Crowfoot in her article (Crowfoot 1984). To test this interpretation, we decided to try out some of the patterns shown on the drawings.

The first border experiments were performed on a tapestry frame loom (Fig. 14.4).[7] The warp yarn chosen for our experiments is a very strong one, normally used for both macramé and as warp for rag rugs. The deliberate choice of warp material allowed us the freedom to manipulate the warp threads without the risk of breaking them during work so that they could be pulled out of their natural

[7] 55 cm Mirrix Zach loom.

Fig. 14.4 Experiment n°1. Recreation of a simple diamond-shape lattice on a tapestry frame loom: archaeological example from Sai 8-B-32.B, B2-38. (Photo by E. Yvanez ©Sai Island Archaeological Mission), U. Mokdad setting up the loom and final result. (Photos by E. Yvanez)

vertical position and come together in groups. The relatively thick warp yarn also gave us quick results in our investigations, and the patterns appeared clearly visible on the small samples. A warp was set up with an undyed 9-stranded cotton macramé string at a density of 5 ends per centimetre. A few centimetres of weft faced tabby were woven with a weft of undyed three-stranded cotton macramé string in order to build a piece of woven cloth as a basis for the openwork experiments.

As mentioned above, the openwork borders themselves are not woven. The intricate lacelike patterns are produced by wrapping thinner threads, in this case the three-stranded macramé cotton weft, tightly around groups of two or four warp threads thereby covering the warp completely. We discovered that the wrappings looked more regular and were easier to create if the supplementary weft threads were always wrapped in the same direction instead of wrapping them in two different directions as shown in Crowfoot's article (1984, Fig. 2e). In order to make the patterns look regular, it was also essential to count the number of times each supplementary weft thread was wrapped around the warps. We also discovered that if the thin wefts were only wrapped around the grouped warped threads where they were supposed to form a cross, the result did not look like the actual model. However, if the two thin supplementary wefts were put together and two knots were tied – one after the other – the grouped warp threads could in this way be pulled out of their vertical position and result in a nice triangle, i.e. the first half of the cross.

These rather promising results of our first openwork border experiments convinced us to try out similar patterns on the warp-weighted loom.[8] A warp of single-ply cotton threads was set up with loom weights and heddles, and a few centimetres of weft faced tabby were woven, just as we did on the tapestry frame loom. However, instead of removing the loom weights before creating the starting braid, we removed the heddles and brought all the loom weights together in front of a lower beam, thereby removing the natural shed. Having the warp threads suspended from the upper beam, still taut because of the loom weights but without sheds, created the

[8] Modern recreation of a warp-weighted loom set up at the Centre for Textile Research with a set of truncated pyramidal loom weights made of terracotta (c. 250–275 g each). The same frame – with different loom weights – is illustrated in Öhrman 2017: 27.

perfect opportunity to make two rows of weft-twining on top of each other. Thus, we could produce nice starting- and ending braids with a supplementary weft of the thick warp yarn without having to remove any loom weights from the warp. It was then possible to create the openwork pattern without difficulty, in the same way as we did on the frame loom.

A step-by-step hypothesis (Fig. 14.5):

1. Weave on a warp-weighted loom;
2. Bring the two sheds together: remove the heddles, bring the two sheds in front of a low bar;
3. Interlace weft threads on a few rows of compact 2-strand 'countered' and 'reversed' weft-twining, grouping the warps in small groups;
4. Install long threads for wrapping: a long end, passed through the twining every two warp groups, repeated along the entire width of the fabric;
5. Wrap the extra thread, each length around one warp group. Group the warps and wrap them together to form the middle of the cross (X) pattern. To tighten this central part, the extra threads can be knotted. Separate the warp groups and wrap them separately. Repeat to the end, using the other lengths of thread as more ground is covered;
6. Leave all the threads hanging and secure everything with several rows of twining (or detach the weights and braid a flat finishing braid);
7. Detach the loom weights;
8. Form and attach the fringed tassels.

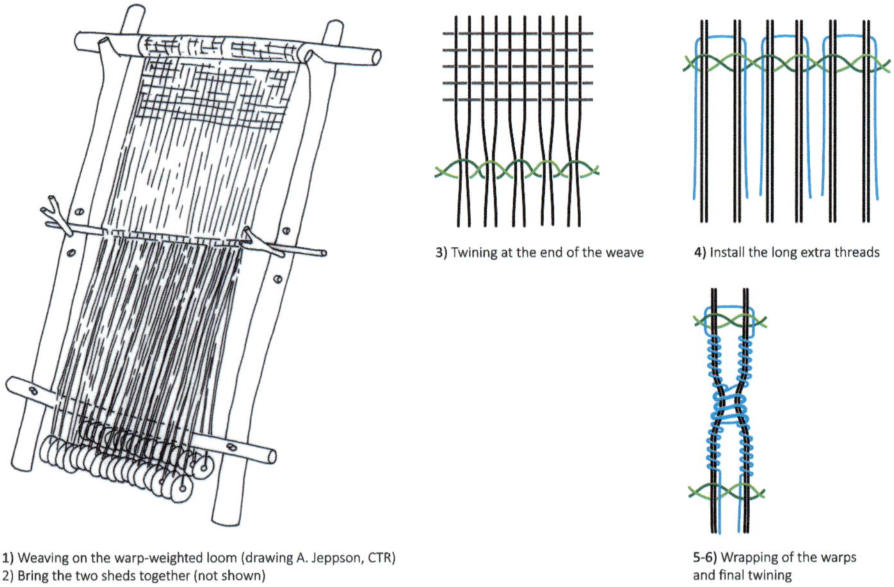

Fig. 14.5 A step-by-step hypothesis for the manufacture of wrapped openwork borders on the warp-weighted loom. (Drawing of the loom by A. Jeppson CTR, technical drawings by E. Yvanez)

This hypothesis offers a much simpler and straightforward execution, which complies with the usual manipulations of the warp-weighted loom. It could have been used to create a basic form of openwork borders, particularly the finer ones at the edge of lighter fabrics. It does not account, however, for the plying of the warp threads or the thick finishing braids sometimes used to anchor particularly heavy rows of fringed tassels. Yet, it seems to be supported by several unfinished pieces found at Qasr Ibrim, interpreted as training examples (Crowfoot 1984: 12; Adams 1987: 96).

Making these samples have shown us that it is in fact possible to create such well-structured patterns by counting threads and repeating the same graphic modules. Our exercise with the simple patterns of openwork have also made us realize that creating the lacelike borders is easier than we first believed. Our second experiment demonstrated that, if we use twining instead of a flat braid, making wrapped openworks may not have necessarily involved the complicated removal and reinstallation of loom weights. Their creation could have been much straight-forward and indeed well suited to the warp-weighted loom. Their stunning appearance results more from the careful repetition of detailed but simple gestures than from heavy manipulations of the fabric and loom set-up.

Fond of elaborate tapestry weaving, the Meroitic weaver would have been used to manipulating long lengths of extra threads. Therefore, this process seems well-adapted to both the weaving tool – the warp-weighted loom – and the weaving 'brain' of the craftspeople.

14.4 Functions and Contexts

As mentioned in our general description of the openwork borders, their fragmentary state and the lack of completely preserved pieces – with both ground fabric and border – hinder the identification of their original use. We need to broaden our study beyond individual pieces, cross-referencing data between textiles and iconographic representations, to see a more coherent picture emerge, in which geometric lattice patterns and long fringes formed a very important motif in the Meroitic textile repertoire used in both dress practices and furnishings.

On the one hand, reliefs, paintings and engravings were commissioned by the royal family and its affluent subordinates in Meroe and throughout the kingdom's regional capitals. These documents mostly depicted their patrons during their official duties or in highly codified funerary scenes. On the other hand, most of the textiles have been recovered from the rich graves of civil, military, and religious administrators. From an urban setting, even the plethora of textiles from the town of Qasr Ibrim is strongly associated to a small population of local officials regulating the activities of this important pilgrimage centre and trading post (W.Y. Adams and N.K. Adams 2013). Our knowledge is therefore very biased in favour of the elite part of the Meroitic society, as we remain blind to other population groups more discreet in the archaeological record (Yvanez 2018). The nature of the sources forces us to limit our discussion to royal and elite contexts, but it is our impression that lattices, fringes and tassels were wide-spread in other Sudanese clothing traditions as well, possibly over a long-time frame and interpreted in other media such as animal skin, feathers, and other ornaments.

Supplementing the evidence from textiles, we were able to assemble a group of c. 50 iconographic documents representing possible openworks. In most cases, it is much easier to identify the fringed tassels, with their small circular tops and long vertical hatches, than it is to spot the openwork borders themselves. It seems that they are indicated by the presence of one or several horizontal lines preceding the tassels. Short vertical hatches or zigzags sometimes complete the drawing, giving a more faithful rendition of a lattice. Such representations were notably recognised at Meroe, Mussawarat es-Sufra, Naga, Faras, Gammai, Karanog and Qasr Ibrim (cf. map in Fig. 14.1), mostly on garments. They are depicted on temple reliefs and paintings, on funerary reliefs in the royal pyramids, on painted private funerary stelae, on *ba*-statues representing the deceased, and on engraved bronze bowls.

14.4.1 Clothing

14.4.1.1 The Royal Sash

Of the thousands of Meroitic textiles inventoried, only one single fabric can be securely identified as belonging to a member of the royal family.[9] For this reason, iconography is our primary source for the study of royal dress and, despite the lack of any openwork borders or fringed garment found in royal context, many figures illustrate their primordial importance in the 3-piece costume used in enthronement ceremonies (Török 1990). In these scenes, the king and the reigning queen wear a long tunic covered by a coat, fastened by a long-tasselled cord. This outfit is completed by a fringed sash, worn on the right shoulder, across the chest, and around the left side of the body.

As time progressed, the fringes became longer and longer, until they covered the entire front. On several representations, such as on the Lion Temple at Naga (Fig. 14.6a), the sovereign is shown with his or her right arm passing through the fringes to hold diverse attributes (Gamer-Wallert 1983). These long tassels appear to be attached to a band, c. 5–10 cm high, decorated with geometric patterns (circles, check patterns, chevrons) and, sometimes, the face of the divine ram Amun. It is possible that this band was created with yarn, in an openwork technique, which would have formed a soft and flexible base for the tassels and added beads or precious ornaments. Fascinating glimpses in the ancient Meroitic polychromy, rare fragments of painted wall décor show that these fringed sashes were made of red threads. Small fragments found in the Taharqa temple at Qasr Ibrim display red

[9] Boston Museum of Fine Arts 23.812.

Fig. 14.6 The function of openwork borders as seen through iconographic documents: (**a**) Queen Amanitore at Naga Lion temple, after Gamer-Wallert 1983, Pl. 7a.; (**b**) Painted stelae from Karanog, Cairo museum JE40229, after Wenig 1978, no.°127, 206; (**c**) Standard of god Apedemak at Naga Lion temple, after Gamer-Wallert 1983, Pl. 4a. (Figure by E. Yvanez)

tassels, attached together by white, blue and red threads and secured to a red band (Pyke 2007). Circular yellow (gold?) beads are inserted in between the tassels.

The elaborate construction shown in iconography, predominantly red, is far from the archaeological textiles. It stems from earlier models of royal garb, which exhibited a red sash made of fabric and passed across the chest (e.g. worn by the Napatan king Tanoutamon in his funerary chapel at El-Kurru, Török 1990: 177). The Meroitic sovereigns seem to have reused older clothing items in their own ceremonial dress, using Pharaonic elements as well as their own Kushite heritage from the Napatan period. They, however, added several elements of their own textile and ornamental traditions. Beside the small tapestry patterns, common on their tunics and coats as well as on preserved fabrics, they also integrated the tasselled fringes and, most probably, the openwork borders. In this specific 3-piece costume, the fringed sash was imbued with a rich symbolic meaning. As a last layer of protection, it was laid over the enthronement garments, after each item was given to the sovereign by the dynastic gods Amun and Apedemak.

14.4.1.2 Wrapping-around Skirts

Moving away from such powerful garments, the next category of clothing making use of openwork borders is the wrap-around garments – i.e. skirts – made of a large rectangular piece of fabric and finished, along the bottom, by a portion of openwork. Two fragmented specimens, made of cotton, have been recorded at Qasr Ibrim (Adams 1998: 25; W.Y. Adams and N.K. Adams 2013: Pl. 41a-b). They both exhibit use-wear and a thin openwork border, c. 2 cm high, forming a pattern of lozenges.

The borders are not finished by fringes but by a flat braid. A similar piece was also discovered in a female grave at Ballana (Mayer-Thurman and Williams 1979: 61). Found in the cemetery of Karanog, a stunning painted stela shows how such a textile could have been worn in female clothing (Fig. 14.6b). As the main figure, a lady is wearing a long white skirt formed by two layers of textiles, with the upper layer closing on her left hip. This fabric is decorated by a large swastika figure (in tapestry?) and a heavy-looking border at the bottom, drawn with a series of parallel horizontal lines and small hatches (the openwork?) and vertical lines representing the fringes. In Meroitic art, many female statues are shown dressed in a similar, but usually plain long skirts (e.g. *Sudan, Royaumes sur le Nil*: 290). It is probable that details, such as openworks and other decorative elements, were painted on rather than carved and have long since disappeared.

14.4.1.3 Sashes, Belts and 'Aprons'

Openwork borders and fringes also seem to have been commonly used to decorate the bottom of garments worn around the waist. It is especially the case in male clothing, which often includes a sash or a belt, knotted on the front, with the two long sides hanging down the legs. The bottom extremities, often flaring so as to create a sort of triangular apron-like garment, can be decorated with fringes represented by a row of vertical lines (Yvanez 2018: Fig. 7). Such iconographic representations find an interesting parallel in a cotton textile from Karanog, here illustrated in Fig. 14.2. Preserved as two fragments, the piece is comprised of the two ends of a long but narrow fabric, 26 cm wide. The main weave is a balanced cotton tabby, with nine picks of supplementary wefts along the top and bottom edge and an elaborate portion of openwork, 3 cm high, on both borders. It is finished by a row of thin but numerous fringes, preserved up to 12 cm long. Attached around the waist, such a textile would have formed wide panels in front of the legs, with long fringes on the shins, moving in a stunning fashion together with the wearer's movements.

Either placed at the bottom of a skirt or front apron, it is easy to imagine the effect created by such heavy ornaments. Their presence would have greatly influenced the range of movements and the posture of the wearer, as well as confer a high visual impact to an otherwise simple garment. Openworks and fringes, in helping to create highly recognisable outfits, without a doubt played an important role in setting the Meroitic elite aside from the rest of the population.

14.4.2 Furnishing

To our knowledge, there exist six representations of textiles in Meroitic art that do not depict clothing, and five of them are directly related to a cult setting. Interestingly, all five present fringes and what can be interpreted as an openwork border. The first type of object is a border with long fringes used to decorate the base of divine insignias (Fig. 14.6c). It is known by three reliefs from the Lion temples of Mussawarat es-Sufra (third century BCE) and Naga (first century CE), showing sceptres with figures of the gods Apedemak and Sebiumeker standing on a platform lined with a fringed border (Gamer-Wallert 1983, Pl. 4a; Török 1987: Fig. 28–29). Belonging to the same genre of objects, another scene shows a large figure of a ram – the god Amun – standing on a litter heavily decorated with several layers of fringed fabric, festooning the litter's sides and front. The check-pattern surmounting the fringes very clearly resembles bands of openwork. Part of a religious procession, this scene is engraved on a bronze bowl found in a grave at Gammai and used for ritual libations during the funeral (*Sudan Royaumes sur le Nil*: 286). The last representation is a graffito, recently rediscovered on a column shaft of the funerary temple K. 1500 at El-Kurru (Emberling and Davis 2019: 161). The graffito shows a rectangular fabric (or possibly a mat) with a patterned central field, fringes at the bottom and a check-pattern border and corner tassels at the top. In such a setting, this drawing recalls the wall hangings and altar cloths decorating cultic buildings in the Late Antique Mediterranean basin (De Moor and Fluck 2009).

An extraordinary find comes to supplement the iconographic documents: a set of six large textiles decorated with openwork borders and/or tapestry, discovered in the destruction layers of the Isis temple at Qasr Ibrim, dated to c. 525–575 CE (Adams 1987, 2006). They were found scattered through the building together with a series of textile miniatures – small but complete woven pieces with intact borders and selvedges – many of which included expert or beginner openwork patterns. Torn off in many pieces, three large fragmentary fabrics were preserved up to 85 × 115 cm. They are made of fine tabby cloth, with dark or lighter blue threads. One of the best-preserved pieces shows a narrow band of openwork along one side (Adams 1987: 87; 2006: Fig. 2). Three other textiles bear a large rectangular tapestry insert depicting a range of geometric patterns and Isiac symbols, made of very fine tapestry in blue and white colours and framed in a dark blue border. Once again, one specimen has a preserved portion of openwork lattice (Adams 1987: 91; 2006: 203). Identified by Nettie K. Adams as curtains, wall decors, and liturgical covers (for altars, for example), these high-quality fabrics are among the finest examples of the Meroitic textile arts. Mended in antiquity and reused until the very last decades of the Kushite rule in Qasr Ibrim, they all exhibit a juxtaposition of tapestry figures, the colour blue, and openwork borders.

In addition to clothing and furnishing items, several fragments of openwork borders, either finished by fringes or embroidered cords and bobbles, seem to have been created as independent pieces, without any woven section. It is difficult to ascertain their original function, but it is likely that they were used as ornaments on the body, around limbs for example, or in association with important objects or pieces of furniture in the household or the temple.

14.4.3 Reuse

Whatever their original function and use, the openwork borders were clearly prized by their owners. A high proportion of the preserved specimen exhibit different stages of reuse and recycling. The first type of reuse strategy, attested by five fragments found at Qasr Ibrim,[10] was to sew an openwork border to a different woven fabric (as seen in Fig. 14.3). The borders were teared off their original fabric or

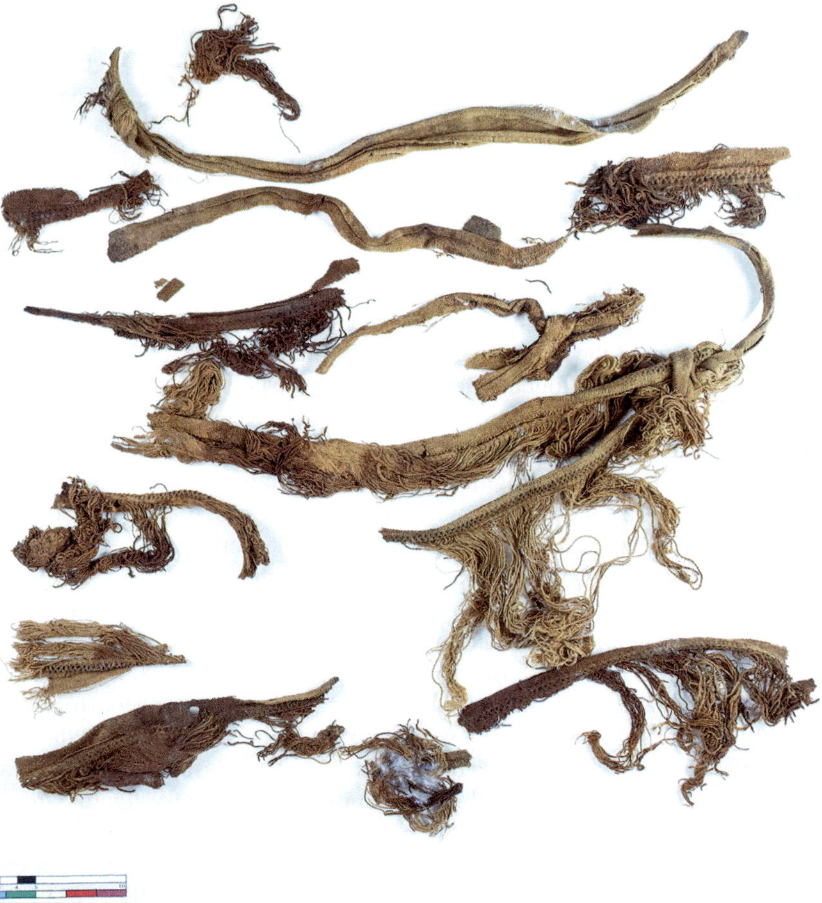

Fig. 14.7 Example of funerary reuse of fringed openwork borders from Gebel Adda: long decorative 'ribbons' teared off the fabric and tying the shroud. (ROM 973-24-2895, photo by E. Yvanez, with the courtesy of the Royal Ontario Museum)

[10] Textiles stored in the British Museum and Bolton Museum, unpublished. The British Museum objects are visible on their online object collection database (BM 71880 and BM 72292).

former independent pieces, and reattached to a supposedly new woven fabric. The seam can be discreet, made with a simple flat seam on the textile's reverse using a s2Z thread, or much coarser with a flat or rolled hem maintained with overcast irregular stitches. This type of reuse would have prolonged the active use of the textile and border for a much longer period.

The second strategy was to tear the openwork section off, together with the fringed tassels, along the warp or the weft. These long 'ribbons' could then be used as fastenings to securely attach the shroud around the deceased's body. Two sets of such fastenings were found knotted among the skeletal remains in graves at Gebel Adda (Fig. 14.7) and Sai. Other textile parts could be reused this way, especially tearing off a band along the reinforced selvedges. The openwork portions, being so tightly wrapped, are strong as well as decorative, fulfilling a practical use while conferring an aesthetic quality to the body wrappings. Finally, it is important to note that, outside of Qasr Ibrim, all openworks have been found in a funerary setting, where the items of clothing they originally decorated went through a last phase of reuse, enveloping the body in large and ornate shrouds.

14.5 Conclusion: A Unique Textile Tradition?

We hope to have demonstrated that the Meroitic openwork borders were very much linked to the use of the warp-weighted loom: presented with the long lengths of threads left hanging at the bottom of a fabric, the weavers made use of them in a very creative and ornamental way. Our experiment has confirmed the talent of Meroitic weavers, who seem to have adopted the warp-weighted loom around the same time as cotton fibres (Yvanez 2015), and quickly developed a unique textile tradition exploiting the qualities of both. Intended for elite consumption, these textiles certainly had a striking appearance, especially when enhanced by openwork borders and fringes. However, their manufacture does not seem to have required a very high level of expertise. This new textile data finds interesting parallels in data coming from the field, where the distribution of textile tools does not seem to indicate the existence of dedicated workshops but rather the use of multi-functional sectors for craft activities in public places or domestic quarters.[11] In the absence of written sources, a better understanding of textile techniques coupled with our growing knowledge on textile activities is the only way to reach the artisan behind the artefact. However, much work remains to be done before archaeology can securely identify the degree of craft specialisation and organisation of the ancient Sudanese textile production.

The fringed borders became an intrinsic part of the Meroitic weaving arts, finding their place in cotton textiles alongside the stripes, tapestry bands and

[11] The analysis of textile tools and their distribution is ongoing on several sites (Yvanez in preparation).

supplementary weft bundles that decorated the edges of fabrics. Weavers explored a local repertoire of lattice work, drawing from older Kushite traditions such as the Kerma pot-nets or openwork leather garments, dated to c. 1700–1500 BCE (Adams 1998; Reisner 1923: Pl. 64–65). Instead of creating stand-alone structures, like the Kerma nets, they linked the wrapping technique to their weaving practice and tremendously developed it to integrate two of the key patterns of Meroitic art: geometric repetitive modules and hanging ornaments.

In that sense, openwork and fringed borders link the textile arts to the rest of the contemporary artistic production. Across media, artefacts share the same decorative repertoire, where tassels, pompoms, fringes and other hanging or dangling elements were especially prominent. Fringes made of fibres (threads or plant material) or leather were a common element in clothing: they were used at the edge of skirt-like garments or forming an independent band used as a girdle (Bates and Dunham 1927: Pl. 67.1). Fringed textiles, most probably based on an openwork band, were also used as furnishing for cultic objects or the temple. Tassels and pompoms made of threads, leather strips, or feathers have also been found attached to bags and quivers, as well as in connection with animal trappings such as harnesses (e.g. Dunham 1963: 227). Even jewellery displays hanging ornaments, e.g. the gold pendants in the shape of cowries arranged in a row at the bottom of earrings or broaches (*Soudan, Royaumes sur le Nil*: 310–315). As such, openworks, fringes, and tassels were an essential part of the decorative landscape of the Meroitic kingdom, interpreted in different ways according to the social context and medium, and closely associated with the clothing of both humans and their animals.

References

Adams NK (1987) Textile remains from a late temple in Egyptian Nubia. *Ars Textrina* 8:85–124.

Adams NK (1989) Meroitic high fashions: Examples from art and archaeology. *Meroitica* 10:747–755.

Adams NK (1998) Ancient lace of Nubia and the Sudan: A unique decorative tradition. *Cahiers de recherche de l'Institut de papyrologie et d'égyptologie de Lille* 17/3:17–25.

Adams NK (2006) Ideologies in conflict: The textiles from the Isis temple at Qasr Ibrim. In: Schrenk S (ed.) *Textiles in situ: Their find spots in Egypt and neighbouring countries in the 1st millennium CE.* Riggisberger Berichte 13. Abegg-Stiftung, Riggisberg, p 197–210.

Adams WY, Adams NK (2013) *Qasr Ibrim: The Ballaña phase.* EES excavation memoir 104. Egypt Exploration Society, London.

Andersson Strand E (2009) Experimental textile archaeology. In: Andersson Strand E, Gleba M, Mannering U, Munkholt C, Ringgaard M (eds) *North European symposium for archaeological textiles X.* NESAT X. Ancient Textiles Series 5. Oxbow Books, Oxford, p 1–3.

Andersson Strand E, Nosch M-L (eds) (2015) *Tools, textiles and contexts: Textile production in the Aegean and Eastern Mediterranean Bronze Age*. Ancient Textile Series 21. Oxbow Books, Oxford and Philadelphia.

Ascher R (1961) Experimental archaeology 1. *American Anthropologist* 63:793–816. https://doi.org/10.1525/aa.1961.63.4.02a00070.

Asher Silverman, N (2012) *The Oxford companion to archaeology* (second edition). Oxford University Press, Oxford.

Bates O, Dunham D (1927) *Excavation at Gammaï, varia Africana* IV. Harvard African Studies 8. Peabody Museum of Harvard University, Cambridge.

Ciszuk M (2007) The academic craftsman – a discussion on knowledge of craft in textile research. In: Gillis C, Nosh M-L (eds) *Ancient textiles: Production, craft and society: proceedings of the 1st international conference on ancient textiles*. Ancient Textile Series 1. Oxbow Books, Oxford, p 13–15.

Comis L (2010) Experimental archaeology: Methodology and new perspectives in archaeological open air museums. *EuroREA* 7:9–12.

Crowfoot E (1984) Openwork fringes from Qasr Ibrim. *Meroitic NewsLetter* 23:10–16.

Crowfoot GM (1931) *Methods of hand spinning in Egypt and the Sudan*. Bankfield Museum Note n°12. F. King & Sons Ltd., Halifax.

De Moor A, Fluck C (eds) (2009) *Clothing the house. Furnishing textiles of the 1st millennium AD from Egypt and neighbouring countries*. Tielt, Lannoo.

Dunham D (1963) *The West and South cemeteries at Meroe*, *Royal Cemeteries of Kush* 5. Museum of Fine Arts, Boston.

Emery I (2009, first edition 1966) *The primary structures of fabrics*. The Textile Museum, Washington, New York.

Emberling G, Davis S (2019) *Graffiti as devotion. Along the Nile and beyond*. Kelsey Museum Publication 16. University of Michigan, Ann Harbor.

Gamer-Wallert I (1983) *Der Löwentempel von Naqa in der Butana (Sudan). III. Die Reliefs*. Beihefte zum Tübinger Atlas des Vorderen Orients 48/3. Reichert, Wiesbaden.

Griffith FLL, Crowfoot GM (1934) On the early use of cotton in the Nile Valley. *Journal of Egyptian Archaeology* 20:5–12.

Mayer-Thurman C, Williams BB (1979) *Ancient textiles from Nubia. Meroitic, X-group, and Christian fabrics from Ballana and Qustul*. Art Institute of Chicago, Chicago.

O'Sullivan A, Souyoudzoglou-Haywood C (2019) Defining experimental archaeology: Making, understanding, storytelling? In: Souyoudzoglou-Haywood C, O'Sullivan A (eds) *Experimental archaeology: making, understanding, storytelling*. Archaeopress, Oxford, p 1–4.

Öhrman M (2017) A warped version: Manipulating Roman looms for metaphorical effect – Potamius of Lisbon's Epistula de substantia 5–9. *Humanitas* 71:51–70. https://doi.org/10.14195/2183-1718_71_3.

Outram AK (2008) Introduction to experimental archaeology. *World Archaeology* 40(1):1–6. https://doi.org/10.1080/00438240801889456.

Pyke G (2007) The fragmentary wall plaster. In: Rose P, *The Meroitic temple complex at Qasr Ibrim*. Egypt Exploration Society, London, p 44–68.

Reisner GA (1923) *Excavations at Kerma, Part IV–V.* Harvard African Studies 6. Peabody Museum of Harvard University, Cambridge Mass.

Sarri K, Mokdad U (2019) Recreating Neolithic textiles: An exercise on woven patterns. In: Souyoudzoglou-Haywood C, O'Sullivan A (eds) *Experimental archaeology. Making, understanding, storytelling*. Archaeopress, Oxford, p 83–92.

Souyoudzoglou-Haywood C, O'Sullivan A (eds) (2019) *Experimental archaeology. Making, understanding, storytelling*. Archaeopress, Oxford.

Török L (1987) *The royal crowns of Kush. A study in the Middle Nile Valley regalia and iconography in the 1st millennia BC and AD*. BAR S338. BAR Publishing, Oxford.

Török L (1990) The costume of the ruler in Meroe: Remarks on its origin and significance. *Archéologie du Nil Moyen* 4:151–202.

Soudan, Royaumes sur le Nil (1997) Catalogue published on the occasion of the exhibition at the Institut du Monde Arabe (Paris). Flammarion, Paris.

Wenig S (1978) *Africa in Antiquity II. The arts of Ancient Nubia and the Sudan. The catalogue*. Brooklyn Museum, Brooklyn.

Yvanez E (2015) De fil en aiguille. Aspects de l'artisanat textile méroïtique. *Égypte, Afrique et Orient* 78:63–66.

Yvanez E (2018) Clothing the elite? Patterns of textile production and consumption in ancient Sudan and Nubia. In: Ulanowska A, Siennicka M, Grupa M (eds) *Dynamics and organisation of textile production in past societies in Europe and the Mediterranean.* Fasciculi Archaeologiae Historicae XXXI:81–92.

Yvanez E, Woźniak MM (2019) Cotton in ancient Sudan and Nubia: Archaeological sources and historical implications. In: Bouchaud C, Yvanez E (eds) *Cotton in the Old World. Domestication, cultivation, uses and exchange. Revue d'ethnoécologie* 15. Museum d'Histoire Naturelle, Paris. https://doi.org/10.4000/ethnoecologie.4429.

Elsa Yvanez is an archaeologist specialized in the textile production of ancient Sudan and Nubia, in the *chaine opératoire* and economic significance of spinning and weaving, as well as in the use of textiles for clothing and burial. This research was part of her activities as a Marie Skłodowska-Curie fellow at the Centre for Textile Research (UCPH), Copenhagen, Denmark, in the TexMeroe project (European Union's Horizon 2020 research and innovation programme, grant agreement n°743420). www.texmeroe.com, @texmeroe.

Ulrikka Mokdad is an art historian specialised in late antique Egyptian textiles. She is also a tapestry weaver who has exhibited her artworks in Denmark and abroad since 1997. In 2019, she digitalized the Margrethe Hald Archive (https://ctr.hum.ku.dk/nyhedsliste/margrethe-halds-archive-2/) at the Centre for Textile Research in Copenhagen, Denmark, where she is currently working as a research assistant.

Chapter 15
Interdisciplinary Methods and New Perspectives on Inscribed Textiles: A Case Study of Christian 'Tiraz'

Julia Galliker and Helga Rösel-Mautendorfer

Abstract World collections include an important group of textiles with inscriptions in Greek and Coptic dating from the Late Antique and early Islamic periods in Egypt. Among these we find a group of cloths with elaborated inscriptions that represent a Christian version of '*tiraz*'. The evidence indicates that extensive inscriptions were embroidered on funerary cloths as a means to convey prayers and blessings for a loved one.

Through this paper, we demonstrate how interdisciplinary research methods are necessary to interpret these complex artefacts. Our goal is to share work in progress through a case study of an unpublished funerary shroud in the Kelsey Museum of Archaeology (KEL 88017a,b). Research methods include textual analysis, technical textile analysis and experimental archaeology. The evidence suggests that this artefact was the product of a professional industry that produced luxury funerary shrouds for the Christian market. The authors are currently working to create an interdisciplinary international project to analyse and publish a corpus of inscribed textiles.

Keywords Coptic · Greek · Egypt · Technical analysis · Experimental archaeology · Epigraphy

J. Galliker (✉)
University of Michigan, Ann Arbor, MI, USA
e-mail: jlgallik@umich.edu

H. Rösel-Mautendorfer
Independent Scholar, Vienna, Austria
e-mail: helgo@roesel.at

A. Ulanowska et al. (eds.), *Ancient Textile Production from an Interdisciplinary Perspective*, Interdisciplinary Contributions to Archaeology,
https://doi.org/10.1007/978-3-030-92170-5_15

15.1 Introduction

World collections include an important group of textiles with inscriptions in Greek and Coptic dating from the Late Antique and early Islamic periods in Egypt. These artefacts contain important cultural and documentary evidence in the form of moveable objects. After the Arab conquests of Egypt in the mid-seventh century CE, markings on cloth signified expanded official control of the textile industry. '*Tiraz*' originally referred to a type of embroidered band (Golombek and Gervers 1977; Salim 1997; Mackie 2015). Over time the term evolved to become a more generic designation for messages on cloth. Among the textiles recovered from Upper and Middle Egyptian burials, we find evidence of a distinctive Christian form of *tiraz*. These remains represent a synthesis of Antique and Arabic religious cultures in Egypt under Muslim rule as a means to convey prayers and blessings. Funerary textile fragments with Christian inscriptions contain extensive social and documentary information about cultural practices including religious beliefs, personal relationships and commerce.

The two goals of this paper are to present this little-known group of textiles to a broader audience and to demonstrate how interdisciplinary research methods are necessary to interpret these complex artefacts. This paper presents a work in progress to share research to date and to show the direction of and scope for a larger project to study and publish a corpus of inscribed textiles. Our current methodology is based on non-invasive analytical techniques that bring textual scholars, historians, technical textile analysts and experimental archaeology practitioners together. We demonstrate our methods through a case study of a large inscribed fragment at the University of Michigan's Kelsey Museum of Archaeology.

15.2 Inscribed Coptic Textiles

In the late nineteenth and early twentieth centuries, non-scientific excavations in Egypt led to the large-scale transfer of textile remains, papyri and other artefacts to the antiquities market (Hagen and Ryholt 2016). Fluck et al. (2000: 125–131, Thomas 2007: 141–145) describe the problems associated with reconstructing find spot information many decades after the materials were shipped to Europe and elsewhere. Most remains were found in Late Antique and early Islamic cemeteries adjacent to settlements in the Nile Valley. Desert conditions inhibited decomposition and preserved many grave goods in a desiccated state. By the time Egyptian authorities enforced export controls in the mid-twentieth century, hundreds of thousands of textiles had been accessioned into collections around the world.

While imprecise, the collective body of Late Roman, Late Antique and early Islamic cloths is conventionally identified as Coptic. In the past century, dozens of museum catalogues have been published to present these artefacts to the general public. One of the central challenges facing scholars studying this material is how to find significance within a population of poorly documented remains that have been entirely divorced from their contexts. Research strategies include classifying fragments according to characteristics such as weave structure, expensive fibres, iconography or categorising remains by their function. Analysis of funerary textiles with embroidered inscriptions offers a promising research approach to help situate this group of textiles within their social context.

The large majority of inscribed textiles recovered from Egypt are written in Arabic script. *Tiraz* has a long history of scholarship as an important source of documentary and dating information. The volume co-authored by Kühnel and Bellinger (1952) provides a guide to the research value of cross-disciplinary collaboration to interrogate the material fully. Publications by other researchers including Mackie (1989) and Helmecke (2006a, b) are important contributions that help situate textiles within their appropriate historical context.

While far fewer in number, Coptic-inscribed textiles were recognised early in the history of antiquities excavations. In 1891, Robert Forrer, a dealer and excavator, published drawings and a photograph of a Coptic inscribed textile (Pls VIII, X). Early scholarship includes an article published in 1923by Gaselee who analysed Greek and Coptic textile inscriptions in the Victoria and Albert Museum. While noted by various researchers during the past century, work by Cäcilia Fluck (1996, 1997, 2006) has created a foundation for Greek and Coptic inscribed textile research through various publications.

15.3 Cross-Collection Survey

One of the first challenges in pursuing research on Greek and Coptic inscribed textiles is to identify fragments scattered among museums around the globe. Cross-collection research is essential because the incidence of these inscriptions is low. Many major textile collections own just one or two fragments. Web-based catalogues have made some collections more accessible, but many others are not documented or published.

To date, we have identified over 250 inscribed fragments in 40 European and North American collections. There are certainly more to be found as we continue our work. Our research protocol excludes inscriptions that are painted or written on cloth. Froschauer (2006) published some examples of ink inscriptions on linen cloth from the Österreichische Nationalbibliothek collection (see his Figs. 24, 25). Our research focus is on the interaction between the cloth medium and inscription content. Inscriptions in this context are defined as textile materials worked into the body of a cloth for the purpose of representing script, either in readable or pseudo form.

One of the great surprises of our cross-collection research was to find that the Kelsey Museum of Archaeology at the University of Michigan contains the largest known collection of Greek and Coptic inscribed fragments with 29 fragments. The collection was formed in the 1930s–1950s as a textile analogue to the university's papyrus library, the largest in North America. After the 1960s, research priorities changed, and the collection was largely forgotten.

15.4 Inscription Typologies

Inscribed textiles fall into distinct categories based on characteristics including the form and content of the inscription, iconography, materials, production method and usage evidence. Figure 15.1 provides an example of some of the types of inscriptions included in the corpus.

A group of textiles known as *gammadia* incorporate what appears to be a single Greek letter. These letter forms are typically read as *eta, iota* or *gamma*. Both the meaning and function of these signs are obscure and have been interpreted by various scholars (Nauerth 1986; Bénazeth 2006: 116–118; Szymaszek 2014: 189; 2017). *Gammadia* are typically formed by a contrasting colour of wool tapestry inserted into a linen tunic or monastic stole. They appear on Late Roman garments and representations up to the fifth century CE.

Fine tapestry wall hangings are among the most widely known group of inscribed textiles. These colourful luxury hangings portray mythological or Christian scenes. According to Durand (2006, 2009: 159–164) text adjacent to key figures serves as an interpretational guide to support a narrative message. In contrast, Nauerth (2006: 97) presents a convincing argument that some inscriptions 'wander' and lose their relationship to adjacent figural scenes. Examples of representations accompanied by text include the Judgement of Paris (Harvard 1975.41.32), the Dionysos hanging fragment (Museum of Fine Arts Boston 1973.290), Hestia Polyolbos (Dumbarton Oaks Museum Washington BZ.1929.1) and the Icon of Virgin and Child (Cleveland Museum of Art 1967.144). Similar to these tapestry wall hangings is a small group of resist dye hangings with similar themes and use of text (Fluck 2006: 152; Schrenk 2002).

Fig. 15.1 Examples of Greek and Coptic inscribed textiles. From the left: Kelsey 94405, British Museum 1904,0706.41, Kelsey 94470. (Photo credits: authors, British Museum)

In the period following the Islamic conquests, inscribed tunic ornaments underwent significant alterations that continued into the tenth century CE. Consistent with papyrological evidence, Coptic supplanted Greek as the language of cloth inscriptions. Durand (2006: 87) observed that over time, letter forms became more ornamental, suggesting the influence of Kufic calligraphy. Readable inscriptions indicate changes in Coptic transcription and include errors and copying mistakes (Fluck 2006: 158). On many such cloths, the vertical and horizontal arrangement of text hinders legibility. Rather than legible writing, some tunic ornaments include pseudo text or repeating letter patterns that suggest an apotropaic purpose.

A second trend evident among Christians in post-conquest Egypt is the use of text for onomastic purposes. A small group of weft-faced compound weave figured silks dated to the seventh–ninth centuries CE are identified by the names Joseph or Zacharias (Thomas 2012: 107–114). In some cases, the same scene is portrayed, but with a different name given. Researchers including Fluck (2006: 161) and Durand (2009: 158) suggest that the name was used to identify a specific workshop. Technical analysis of these textiles supports this theory. Surviving fragments show that the inscribed names were clearly part of the 'programmed' pattern sequence tied-up on the drawloom. On one half of the textile, the letters appear in correct letter order, but are reversed at the midpoint because the pattern unit is mirrored through horizontal symmetry (Galliker 2014: 250–251).

15.5 Christian *Tiraz*

When *tiraz* emerged in Arab culture in the eighth century, we see three developments in Coptic textile inscriptions that suggest assimilation of Islamic customs. First, the purpose of Christian inscriptions evolved from labelling represented figures to blessings and supplications. Second, there is a marked reduction in representation in favour of decorative and ornamental designs. Third, there is a shift away from patterning through tapestry weave in favour of mixed techniques including extensive use of embroidery.

Tiraz-like inscriptions appear on the remains of various garments and cloths. For many fragments, the original purpose of the textile is unknown because dealers cut away damaged portions leaving only intact areas with decorations. Some short inscriptions include names such as "Kristophoros" (Kelsey 94469) and "Saint Jouob, the Lord" (Thomas Jefferson Univ. 1974.52.3a, b). Others contain short appeals. Examples include "my brother Kire" (Kelsey 93914) "hagios" (Cooper Hewitt Museum New York 1902-1-128) and "Hosanna Maria" (Textile Museum of Canada Toronto T88.00220).

Among surviving Christian *tiraz* fragments, there is a group that is remarkable for their size and elaborate inscriptions. Some of these textiles are complex constructions that integrate different techniques and materials in ornate compositions. To date, only two radiocarbon dated fragments of this type have been published. In the Phoebus Foundation Antwerp, KTN 711 dates to 1029–1219 CE, 95.4%

probability (De Moor et al. 2006; De Moor et al. 2008: 204–205). Van der Vliet (2006: 52–55) read and analysed the text on this fragment. The Metropolitan Museum fragment (MMA 90.5.877) has an earlier radiocarbon date between 810 and 1010 CE, 95.4% probability. While interesting, these results demonstrate that the number of cloths that have been scientifically dated are too few to assign the group as a whole to a specific time period (Fluck 2006: 183–185).

Excavation reports from explorations from an extensive cemetery site at Naqlun in the Fayum provide an important perspective on inscribed textiles in collections. Godlewski (2002) describes how graves dating to the Fatimid and Mamluk periods yielded extensive textile finds. A series of analyses by Czaja-Szewczak (2002: 182; 2004; 2011: 413) shows that textile remains included funerary textiles as well as clothing and cloth goods from daily life. Finds include many examples of ornately decorated cloths with Arabic inscriptions as well as a few examples in Coptic. Notably, one disturbed grave excavated in 2000 contained remains of two Coptic inscribed wool textiles. A decorated funerary tunic (Nd.00.083) is inscribed with text from Psalm 46.2–3 (Czaja-Szewczak 2003: 159–163, Figs. 1–6). A large woollen shawl or shroud (Nd.00.082) inscribed with four extensive lines of is considered to be pseudo text (Czaja-Szewczak 2004: 209–210, Fig. 9).

Within the corpus collected during the survey phase of this project, we identified 45 textiles that appear to be funerary shrouds and contain extensive text. The Kelsey Museum collection is especially noteworthy because 13 fragments include elaborate Coptic inscriptions. While publications typically refer to these textiles as shawls or veils, our analysis suggests that these cloths were used principally as body shrouds for the following reasons:

- The size and shape of these textiles is very large. The dimensions of an intact piece in the Phoebus Foundation collection (KTN 711) are 220.5 × 105 cm (De Moor et al. 2008);
- This group of cloths is constructed from high twist wool yarn singles with an open weave structure. While ideal for draping around a body, these characteristics would have been impractical in use with snags and pulls in the loose yarns. Apart from one fragment with a patch carefully matched to the ground weave (Museum of Applied Arts in Vienna T10618), we see no signs of wear, damage or structural deformation that would occur with use;
- Similarly, microscopic inspection of yarns does not reveal any signs of wear such as piling or abrasion. Yarns in the ground weave are very intact in areas that are not stained by body fluids;
- Surviving evidence indicates that the cloths were constructed for funerary purposes. All fragments in this group are extensively decorated and inscribed near the two ends at the starting and finishing borders. The long centre sections are plain and unembellished. Staining and destroyed sections in the remains of the cloths show that the body was placed in their centre section with the intact ends of the cloths placed on top of the body.

15.5.1 Case Study – Kelsey Museum 88018a, b

Given the significance of these objects to scholarship, it is important to apply research methods that fully explore the content of this material. In keeping with our goal of presenting interdisciplinary research in progress, we have included an example of our current work on an unpublished Coptic inscribed funerary shroud (KEL 88017a,b) at the Kelsey Museum of Archaeology at the University of Michigan.

The photograph in Fig. 15.2 shows the two surviving fragments placed side by side. The ‘b’ piece is at the bottom of the photo; the ‘a’ portion is on the top. These remains are the two ends of a large shroud. Both the starting and finishing borders survive with fragments measuring about 80 cm in height. The selvedges are intact which show that the finished width of the cloth is 150 cm. Extensive staining from body fluids indicates that the shroud was used in a burial context.

Figure 15.3 illustrates our interdisciplinary methodology. Research is separated into two categories based on the need for physical sampling through invasive analysis. To date our work on KEL 88017a,b is focused mainly on non-invasive analysis techniques.

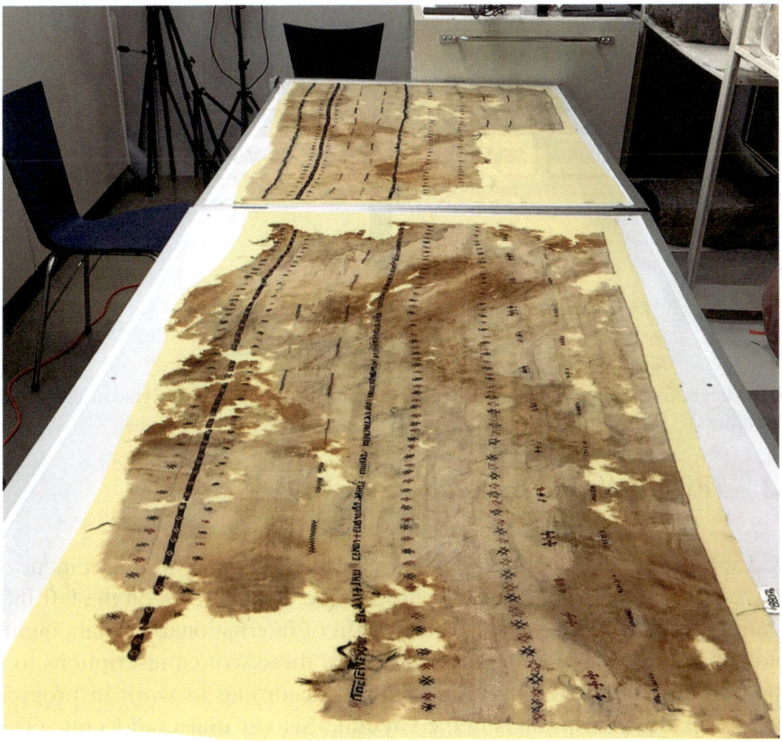

Fig. 15.2 Kelsey 88017 **b** fragment in the lower part of the image; the **a** fragment is at the top. (Photo credit: authors)

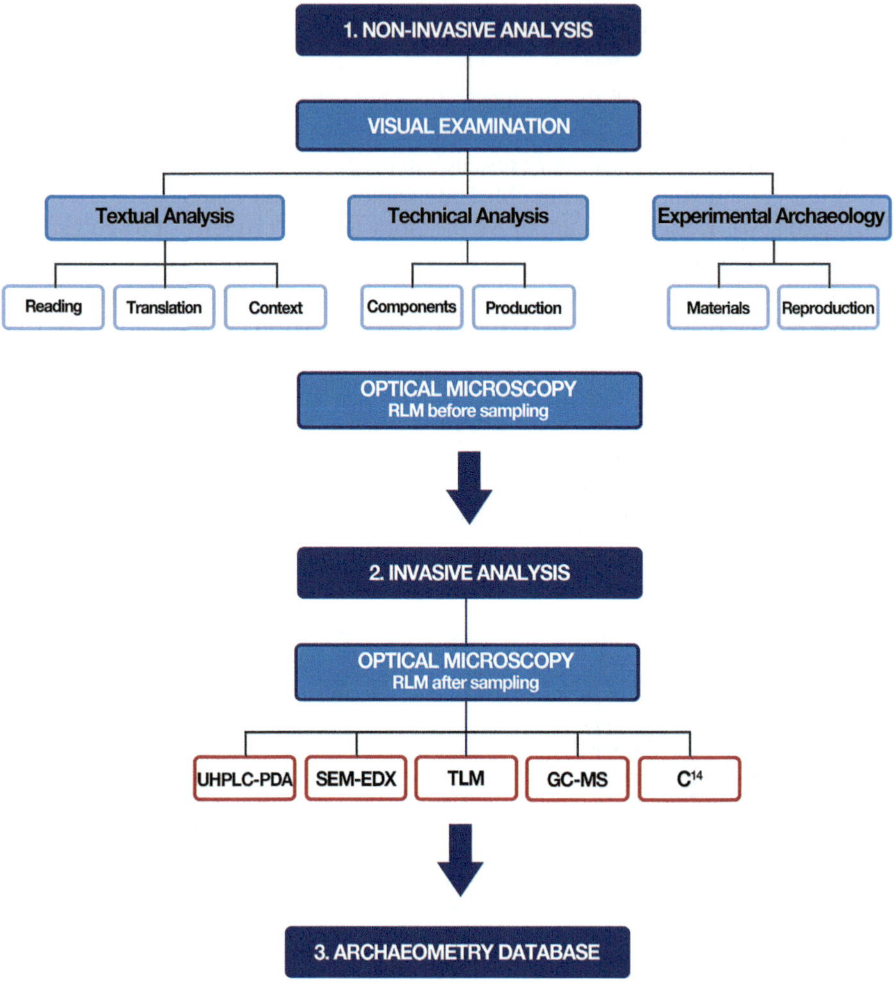

Fig. 15.3 Diagram of the interdisciplinary research model applied to investigate inscribed textiles. (Credit: authors)

15.5.2 Textual Analysis: Reading and Translation

In Fig. 15.4, the 'a' and 'b' fragments are oriented to show the placement of the decorative program at the ends of the shroud. The black arrows on the left indicate the location of the three lines of text. Our team of international scholars includes a Coptologist to study the unique palaeography of these written inscriptions to interpret their form, content and cultural context. According to work in progress by Suzana Hodak, the inscription is in the Fayumic-Sahidic dialect of Coptic. The text is readable, but there are several errors in inscription such as debased or reversed letter forms.

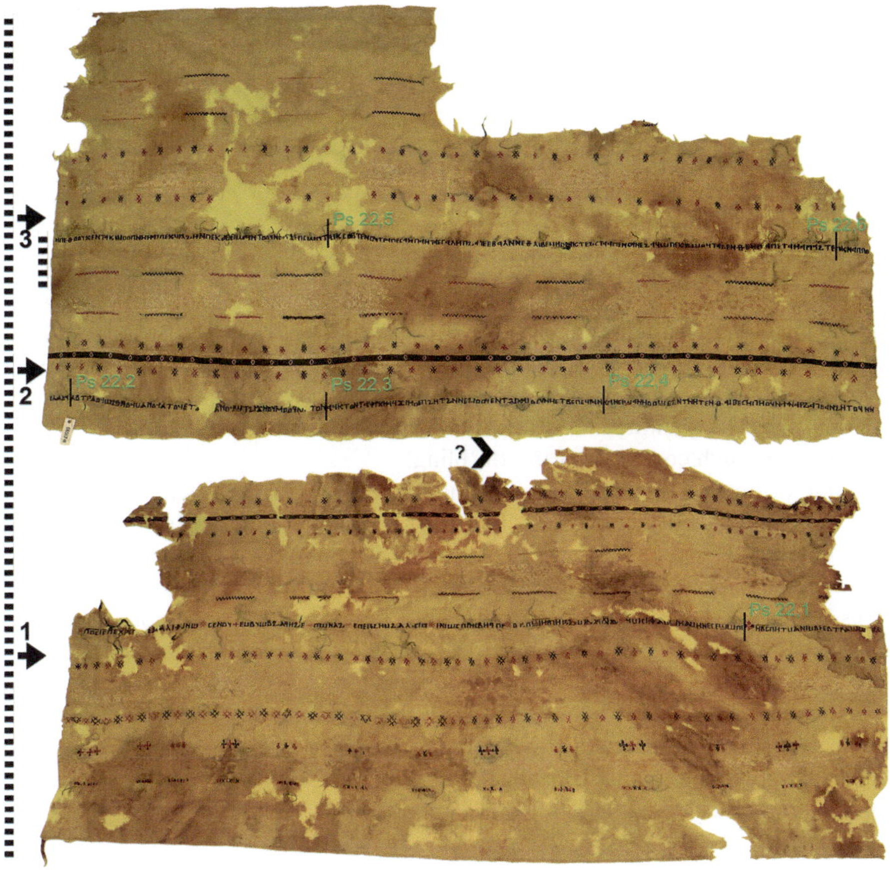

Fig. 15.4 Kelsey 88017 **b** at the bottom, **a** fragment above. The black arrows on the left indicate lines of text. (Photo credit: authors)

Beginning on the left, the 'b' fragment (below) contains a personal prayer with the following preliminary translation: "Lord, Jesus Christ, our true God, bless, help and protect the life of our servant Sisinios, the priest, son of your servant, the deacon Houmisi, the carpenter. Amen. It shall be."

The green line on the right side of line 1 on the 'b' fragment indicates the start of Psalm 22 of the Septuagint. As shown by the notations in green, the psalm extends over two lines on the 'a' fragment (above).

Some of the letter forms seen on this textile are unusual even within the corpus of similar fragments. At some places in the inscription, a few letters appear in stylised form. In addition, certain letters are written or embellished with red thread as compared to the blackish-blue colour used for most of the inscription. One explanation is that the stylised presentation of letters might be influenced by Kufic writing. More likely, the embellished characters reflect the influence of Coptic manuscripts which used similar colours and letter forms. A question for further research is to analyse manuscripts in the same dialect as the textile to detect conventions for stylised and embellished letter forms.

15.5.3 Technical Textile Analysis

The basic structure used to weave this textile is an open balanced tabby with approximately 14 warps and wefts per cm. However, there is considerable variation in warp and weft spacing throughout the cloth. While the ground weave structure is plain, the shroud ends are extensively embellished. As shown in Fig. 15.4, the ends have a symmetrical design program with embroidered ornaments and weft stripes in a contrasting yarn. The quantity and type of decorations indicates the cloth was designed to impress through conspicuous use of 'bling'.

While difficult to see in the photos, the 'a' and 'b' fragments each contain five registers of brocaded white silk included for decorative effect. Magnified images show that the weavers chose an unusual brocading method. Brocade is formed by introducing a supplementary weft between two or more ground wefts to lock pattern yarns into place. Brocaded designs are mainly determined by colour or texture and the arrangement and length of floats.

On this textile, brocading wefts were inserted into the same shed as the tabby ground weave. This choice is puzzling because the short float lengths reduce the glossy appearance of the silk yarns. Future reconstruction experiments will explore this method of brocading to determine if the structure was influenced by loom technology or other technical factors.

15.5.4 Experimental Reproduction

Among the technical attributes involved in constructing this shroud, the embroidery techniques used for decoration and inscription are distinctive. Analysis and reproductions by experienced practitioners provide extensive information about embroidery techniques.

As noted by Vogelsang-Eastwood (2016: 58–60), embroidered textiles are rare among textiles attributed to Late Antiquity before the Arab conquests. A few finds from Mons Claudianus, dating to 100–120 CE, are decorated with stem stitches and couching. A textile from Abu Sha'ar, dated to the fourth century, is decorated with chain stitching. A textile from Deir Abu Metta in the Dakhla oasis, dated to the fifth century CE, was embroidered with alternating colour chain stitches imitating twining.

Recent scholarship concerning embroidery in the Late Antique period includes papers by Fluck (2011) and Droß-Krüpe and Schieck (2014). Mérat (2013: 127–139) published an analysis of seven radiocarbon-dated embroidered textiles in the Louvre collection. Two tapestries embroidered with satin stitches are dated to the fifth–sixth centuries CE. The remaining embroideries were produced much later in the thirteenth–fourteenth centuries.

Embroidery stitches on KEL 88017a,b show a number of interesting conventions. For all embroidered forms, the starting and finishing stitches for decorations were at the same spot. The two ends of the embroidery yarn were twisted together and left uncut. These long embroidery yarns are visible through the open weave on the right side. No knots were used to secure embroidered decorations. Surviving uncut ends do not show signs of wear as would be expected if the cloths were used in life.

Letter forms show particular technical adaptations to the loose weave structure. To prevent the ground fabric from becoming stretched or tensioned, each letter was worked individually, with no stitches carried from one letter to the next. The artisan started all letters at the top, stitched downward to the bottom of the letter, then up again. As shown in Fig. 15.5, this sequence resulted in crossings on the front side and parallel stitches on the reverse side of the cloth. Similarly, the same conventions were applied to other embroidered decorations. While the heights of each letter vary at the top, the bottom of all letters share the same weft for each line of the inscription.

Close analysis of individual letters on this piece results in new insights about the planning and construction of stitched inscriptions. Figure 15.6 is a composite of individual *nu* letters from the 'a' and 'b' fragments. We see a high degree of consistency in the letter forms as shown by the intersection of the diagonal and vertical lines as indicated by the red arrows. The slightly indented join above the bottom of the letter reflects a specific stylistic choice.

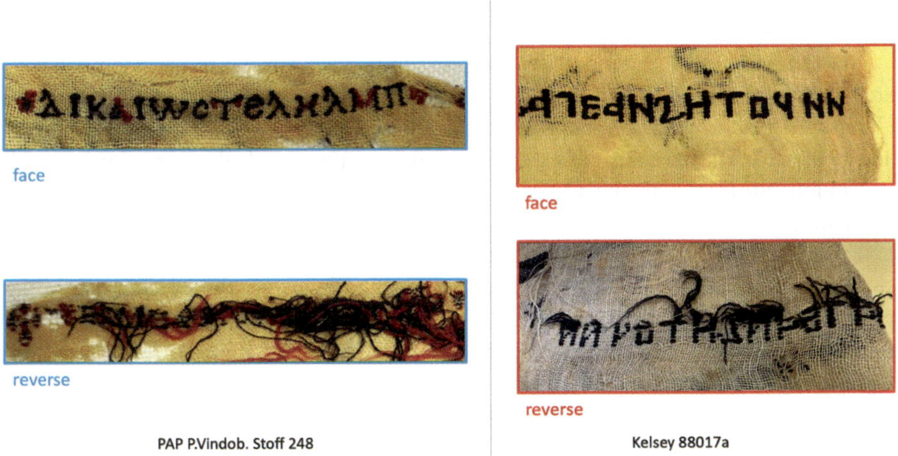

Fig. 15.5 Comparison of the letter forms and embroidery stitches. The right side of the image shows the face and reverse sides of text from Kelsey 88017a; on the left is the face and reverse of text from P. Vindob. Stoff 248 in the ÖNB collection. (Photo credit: authors)

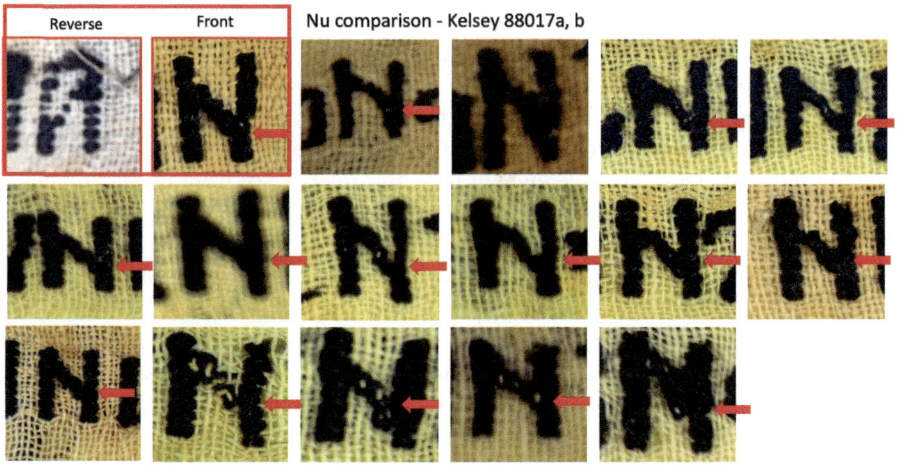

Fig. 15.6 Composite image comparing various *nu* letter forms embroidered on 88017a,b. An image of the reverse side of the letter *nu* appears in the upper left. Red arrows indicate a stylist convention applied to most verticals on the right member. (Photo credit: authors)

Most letters are consistency spaced with 3 or 4 warps in-between. There are also examples of kerning, in which spacing is adjusted for certain letters to give a more pleasing effect. Our analysis shows that letter spacing was planned before the cloth was embroidered.

We also compared stylistic features for other frequently used letters such at the *tau*. Comparing individual letters, there are differences in the top member of the *tau*. Analysis of the stitches shows that a slight dip in some of the horizontal strokes was caused by improper alignment of the stitches that would not occur if written with a stylus. These examples show that analysis and documentation of embroidery stitches provide an aid to recovering damaged text.

Superficially at least, KEL 88017 appears to be very similar to fragments in several other collections. Several textiles share common features including materials, elaborated Coptic inscriptions, use of particular colours, similar decorative forms and brocaded silk registers. Figure 15.5 compares parts of Kelsey inscription with P. Vindob. Stoff 248 in the Österreichische Nationalbibliothek (ÖNB) collection in Vienna.

From the reverse side, we see that embroidery techniques and conventions are similar on both textiles. Side by side comparison shows a marked difference in the palaeography between the two fragments. The Vienna fragment on the left has a more ornate style with ligatures on the letter forms. Despite these different lettering styles, both inscriptions share common stitching conventions that were adapted to the cloth medium to create inscriptions that were intended to be seen and read.

15.5.5 Invasive Analysis

As shown in Fig. 15.3, our research integrates detailed technical analysis with objective testing methods. These include dyestuff analysis with UHPLC-PDA, fibre analysis with TLM, SEM-EDX and radiocarbon dating. One area of particular interest pertains to dye techniques to achieve strong dark colours.

Since no invasive analysis has been performed on the Kelsey fragment, we are presenting results from the comparison sample in Vienna. Both textiles share the same colour scheme with black and red letters (Palme and Zdiarsky 2012: 105). Detailed analysis studies on P. Vindob. Stoff 248 show that the black embroidery yarn was dyed with woad or indigo. An unknown dye source was added to obtain the deep black colour. The red yarn was dyed with dye made from lac. Notably, this dye stuff is not seen in Egyptian textiles prior to the Arab conquests. The presence of this substance is consistent with changing patterns in Indian Ocean trade.

15.6 Conclusions

Our work to identify Greek and Coptic inscribed textiles scattered among world collections allowed us to segment the population and identify changing practices over time. Inscriptions in Late Antiquity labelled key figures and were adapted to support Christian iconography. Over time, the labelling function evolved to validate and intensify the representational message with text. Christian use of text for religious meaning merged with the messaging role of Islamic inscription to create a distinctive form of Christian *tiraz*.

Analysis of 45 textiles with elaborate Coptic inscriptions shows that highly embellished, luxury funerary shrouds were produced in professional workshops roughly between the tenth and twelfth centuries. The case study of KEL 88017a,b demonstrates the value of interdisciplinary research to integrate textual, technical and experimental approaches. The inscription combined personal appeals with quotations from the Coptic psalter. The message asserted religious identity and invoked God's protection.

Analysis of the *chaîne opératoire* provides valuable insight to detect workshop and craft conventions. Careful analysis of embroidery stitches indicates labour specialisation and planning to customise funerary shrouds according to the wishes of grieving families. We see that letter forms were adapted to the medium and production method. Comparison of letter forms shows consistent stylistic forms, but variations also occurred due to the medium of production.

This case study has produced some interesting findings but has also created a number of questions which have stimulated new perspectives. We are currently working to prepare a grant proposal to finance a full-scale project including 14C, fibre and dye analysis to publish a cross collection corpus of Christian *tiraz*.

References

Bénazeth D (2006) Textiles avec inscriptions du premier millénaire, conservés au musée du Louvre (Département des antiquités égyptiennes) In: Fluck C, Helmecke G (eds) *Textile messages: Inscribed fabrics from Roman to Abbasid Egypt*. Brill, Leiden, p 115–130.

Czaja-Szewczak B (2002) Burial tunics from Naqlun. *Polish Archaeology in the Mediterranean* 14:177–184.

Czaja-Szewczak B (2003) Naqlun 2003 From scraps to tunic. *Polish Archaeology in the Mediterranean* 15:159–164.

Czaja-Szewczak B (2004) Textiles from Naqlun, 2004. *Polish Archaeology in the Mediterranean* 16:203–210.

Czaja-Szewczak B (2011) Funerary textiles from the Medieval cemetery of Naqlun. *Polish Archaeology in the Mediterranean* 20:413–420.

De Moor A, Verhecken-Lammens C, Van Strydonck M (2006) Relevance and irrelevance of radiocarbon dating of inscribed textiles. In: Fluck C, Helmecke G (eds) *Textile messages: Inscribed fabrics from Roman to Abbasid Egypt*. Brill, Leiden, p 223–232.

De Moor A, Verhecken-Lammens C, Verhecken A (2008) *3500 years of textile art*. Lannoo, Tielt.

Droß-Krüpe K, Paetz gen. Schieck A (2014) Unravelling the tangled threads of ancient embroidery – a compilation of written sources and archaeologically preserved textiles. In: Harlow M, Nosch M-L (eds) *Greek and Roman textiles and dress: An interdisciplinary anthology*. Ancient Textiles Series 19. Oxbow, Oxford and Philadelphia, p 207–235.

Durand M (2006) Vers une pseudo-épigraphie textile en langue copte: Diogène, Panopé, Thétis. In: Fluck C, Helmecke G (eds) *Textile messages: Inscribed fabrics from Roman to Abbasid Egypt*. Brill, Leiden, p 83–94.

Durand M (2009) Inscribed fabrics from Egypt: a study in Greek and Coptic textile epigraphy. *Journal of Coptic Studies* 11:157–180.

Fluck C (1996) Koptische Textilien mit Inschriften in Berlin (I.). *Bulletin de la société d'archéologie copte* 35: 161–172.

Fluck C (1997) Koptische Textilien mit Inschriften in Berlin (II.). *Bulletin de la société d'archéologie copte* 36:59–70.

Fluck C (2006) „Denkt liebevoll an mich ..." Textilien mit Inschriften im Museum für Byzantinische Kunst, Berlin. In: Fluck C, Helmecke G (eds) *Textile messages: Inscribed fabrics from Roman to Abbasid Egypt*. Brill, Leiden, p 151–172.

Fluck C (2011) Bestickte Kleider aus dem spätantiken Ägypten. In: Eaton-Krauss M, Fluck C, Loon van GJM (eds) *Egypt 1350 BC–AD 1800: Art historical and archaeological studies for Gawdat Gabra*. Reichert Verlag, Wiesbaden, p 75–86.

Fluck C, Helmecke G (eds) (2006) *Textile messages: Inscribed fabrics from Roman to Abbasid Egypt*. Brill, Leiden.

Fluck C, Linscheid P, Merz S (eds) (2000) *Textilien aus Ägypten: Teil 1: Textilien aus dem Vorbesitz von Theodor Graf, Carl Schmidt und dem Ägyptischen Museum Berlin*. Reichert, Wiesbaden.

Forrer R (1891) *Die Graeber- und Textilfunde von Achmim-Panopolis*. E. Birkhäuser, Strassburg.

Froschauer H (2006) Koptische Textilien mit Inschriften in der Papyrussammlung der Österreichischen Nationalbibliothek. In: Fluck C, Helmecke G (eds) *Textile messages: Inscribed fabrics from Roman to Abbasid Egypt*. Brill, Leiden, p 131–150.

Galliker J (2014) *Middle Byzantine silk in context: Integrating the textual and material evidence*. Unpublished dissertation, University of Birmingham Pages.

Gaselee S (1923) Lettered Egyptian textiles in the Victoria and Albert Museum, *Archaeologia* 73:73–84.

Godlewski W (2002) Naqlun excavations, 2002. *Polish Archaeology in the Mediterranean* 14:163–171.

Golombek L, Gervers V (1977) Tiraz fabrics in the Royal Ontario Museum. In: Gervers V (ed.) *Studies in textile history: In memory of Harold B. Burnham*. Royal Ontario Museum, Toronto, p 82–125.

Hagen F, Ryholt K (2016) *The antiquities trade in Egypt 1880–1930: The H.O. Lange papers*. Det Kongelige Danske Videnskabernes Selskab, Copenhagen.
Helmecke G (2006a) Ein ägyptischer Tiraz auf den 'Abbasidenkalifen al-Mu'tamid 'ala-llah. In: Fluck C, Helmecke G (eds) *Textile messages: Inscribed fabrics from Roman to Abbasid Egypt*. Brill, Leiden, p 233–238.
Helmecke G (2006b) Tiraz-Inschriften im Berliner Museum für Islamische Kunst. In: Fluck C, Helmecke G (eds) *Textile messages: Inscribed fabrics from Roman to Abbasid Egypt*. Brill, Leiden, p 173–192.
Kühnel E, Bellinger L (1952) *Catalogue of dated tiraz fabrics*. Textile Museum, Washington, DC.
Mackie L (1989) Textiles. In: Kubiak W, Scanlon GT (eds) *Fustat expedition final report*, Vol. 2. Eisenbrauns, Winona Lake, IN, p 81–97.
Mackie L W (2015) *Symbols of power: Luxury textiles from islamic lands, 7th to 20th century*. Cleveland Museum of Arts, New Haven.
Mérat A (2013) Étude technique et iconographique d'un ensemble de broderies égyptiennes antiques conservés au musée du Louvre. In: De Moor A, Fluck C, Linscheid P (eds) *Drawing the threads together: Textiles and footwear of the 1st millennium A.D. from Egypt: Proceedings of the 7th conference of the research group Textiles from the Nile Valley, Antwerp, 7–9 October 2011*. Lannoo, Tielt, p 127–139.
Nauerth C (1986) Zur Herkunft der sogenannten Gammadia In: Feld O, Bockelberg S, Peschlow U (eds) *Studien zur spätantiken und byzantinischen Kunst, 3. Friedrich Wilhelm Deichmann gewidmet*. R. Habelt, Bonn, p 113–119.
Nauerth C (2006) David oder Achill? In: Fluck C, Helmecke G (eds) *Textile messages: Inscribed fabrics from Roman to Abbasid Egypt*. Brill, Leiden, p 95–101.
Palme B, Zdiarsky A (2012) *Gewebte Geschichte: Stoffe und Papyri aus dem spätantiken Ägypten*. Phoibos Verlag, Wien.
Salim MAM (1997) Textilien mit gewirktem und gesticktem Dekor aus Ägypten. In: Salim MAM (ed.) *Islamische Textilkunst des Mittelalters: Aktuelle Probleme*. Abegg-Stiftung, Riggisberg, p 21–107.
Schrenk S (2002) Die „topographischen" Friese auf den Behangfragmenten mit Danielszene und Petrusszene in Berlin und M. Stein, Die Inschriften auf dem Daniel- und dem Petrus-Stoff in Berlin. In: Hoheisel K, Hutter M, Klein WW, Vollmer U (eds) *Hairesis: Festschrift für Karl Hoheisel zum 65. Geburtstag*. Aschendorff, Münster, p 72–98.
Szymaszek M (2014) The distribution of textiles with 'Greek letter' signs in the Roman world: The case of the so-called *gammadia*. In: Droß-Krüpe K (ed.) *Textile trade and distribution in Antiquity*. Harrassowitz Verlag, Wiesbaden, p 189–197.
Szymaszek M (2017) The textile term *gammadia*. In: Gaspa S, Michel C, Nosch M-L (eds) *Textile terminologies from the Orient to the Mediterranean and Europe 1000 BC–AD 1000*. Zea E-Books, Lincoln, NE, p 483–491.
Thomas TK (2007) Coptic and byzantine textiles found in Egypt: Corpora, collections, and scholarly perspectives. In: Bagnall RS (ed.) *Egypt in the Byzantine world, 300–700*. Cambridge University Press, Cambridge, p 137–162.
Thomas TK (2012) Silk. In: Evans HC, Ratliff B (eds) *Byzantium and islam: Age of transition, 7th–9th century*. Metropolitan Museum of Art, New York, p 148–159.
Van der Vliet J (2006) "In a robe of gold": Status, magic and politics on inscribed christian textiles from Egypt. In: Fluck C, Helmecke G (eds) *Textile messages: Inscribed fabrics from Roman to Abbasid Egypt*. Brill, Leiden, p 23–68.
Vogelsang-Eastwood G (2016) *Encyclopedia of embroidery from the Arab world*. Bloomsbury, London.

Julia Galliker is a postdoctoral researcher studying textiles at the Kelsey Museum of Archaeology, University of Michigan, US. She completed her PhD in history at the University of Birmingham, UK. Her dissertation is a parallel investigation of textual and material evidence for figured silks attributed to Mediterranean workshops between AD 600 and 1200. She co-founded the spiral textile project to bridge the gap between academic research and the practical expertise of handweavers around the world. She also holds an MBA in corporate finance and MA degrees in history and international relations.

Helga Rösel-Mautendorfer studied Celtic studies at the University of Vienna. For her master's degree she conducted research on textiles from Hallstatt. She has participated in various academic projects including: "HallTexFWF – Dyeing techniques of the prehistoric textiles from the salt mine of Hallstatt" (University for Applied Arts Vienna, Department of Archaeometry) and the "HERA Project CinBA – Creativity and Craft Production in Middle and Late Bronze Age Europe" (Natural History Museum Vienna). She has published various articles on archaeological textiles and experimental archaeology. In researching prehistoric and late antique textiles, her particular interests are dyeing techniques, experimental archaeology and reconstruction of ancient artefacts.

Chapter 16
Weaving Experiments with the Rigid Heddle: Woven-in Metal Spirals from Siksälä and 'Auleja' Technique

Ronja Lau

Abstract Throughout history, various band weaving techniques were used to create ribbons, bands or girdles that comprise a specific group in the archaeological evidence of textile artefacts. Quantitatively, tablet weaving is among the main techniques, but there is also evidence for other weaving techniques. For example, rigid heddles were used to create different weave structures and patterns, from simple repp or tabby weaves to complex geometric designs. This contribution examines weaving on a rigid heddle using an experimental approach, with textile finds from the site of Siksälä in Estonia dating from 800 to 1475 AD serving as the basis for archaeological weaving experiments. In addition, a weaving technique called 'Auleja' is used to recreate ribbons with geometric colour design. From as early as the tenth century AD, narrow bands play a central role in the textile tradition around the Baltic Sea and a wider area in Scandinavia. Written sources and folk beliefs passed on in oral history tradition, as well as ethnographic comparisons shed light on the production and also on symbolisms connected with the bands and their patterns.

Keywords Estonian Iron Age · Siksälä · Rigid heddle · Weaving techniques · Weaving experiment · 'Auleja' · Ethnography

16.1 Introduction

The Baltic countries and Scandinavia (Estonia, Latvia, Lithuania, Sweden, Finland and Norway) have always been regions with a rich textile tradition. Due to the local chronological system within the modern borders of Estonia and ongoing Iron Age Tradition, the Early Middle Ages date to 1200/1220 AD till fourteenth century AD

R. Lau (✉)
Ruhr-University Bochum, Bochum, Germany

A. Ulanowska et al. (eds.), *Ancient Textile Production from an Interdisciplinary Perspective*, Interdisciplinary Contributions to Archaeology,
https://doi.org/10.1007/978-3-030-92170-5_16

(Sperling 2014: 21). Iron Age burial practises can be observed for an even longer period of time (Valk and Laul 2014: 236).

The site Siksälä in south-eastern Estonia provides the frame of reference for the experiments due to its rich textile finds. Two Siksälä cemeteries are placed on top of two hills. The first, Kalmetemägi, contains several burials from the late Iron Age to the Middle Ages (800–1475 AD). On the second hill, Kirikumägi, only burials dated to the sixteenth to the eighteenth century were found (Valk and Laul 2014: 234). Due to a large number of metal objects in the graves and wet conditions provided by the surrounding two lakes, the organic preservation is relatively good. Most of the textile finds were identified as clothing mostly made of wool. The lack of plant fibres could be explained by the preservation properties of the local sediment. The majority of the textile finds is dyed in different bright colours, such as red and blue (Rammo and Matsin 2014: 352). In four early Medieval/late Iron Age burials, different kinds of ribbons are preserved.

In grave 143 with a male burial dated to c. 1300–1350 AD, several fragments of narrow bicoloured bands have been found, in the following referred to as 143, no. 11 (Fig. 16.1, Valk et al. 2014: 152). The bands were wrapped around the man's belt, so no functional interpretation can be made for them. The narrow bands appear to be in repp weave, which could indicate the use of a rigid heddle or heddle rods and shed sticks. The red and yellow dyed fragments are made of Z2s spun thread and can be separated into two original ribbons. The first band is made with three stripes containing a red one in the middle. The second ribbon has a reversed pattern with the yellow stripe in the centre.

Grave 243, containing the body of a 9–14-year-old girl, has yielded various kinds of jewellery and to which textile fragments are attached. These parts belong to a head gear (Fig. 16.2) and should drop down the back of a woman or girl. The items with the reference 243, no. 1, are made of narrow ribbons and decorated with bronze spirals, which form a geometrical pattern at the end. The ribbons are in most cases red and blue with an approximate length of 30 cm (Valk et al. 2014: 256).

Simple repp ribbons (Siksälä grave 143, find no. 11) and a ribbon decorated with bronze spirals (Siksälä grave 243, no. 1) were an inspiration for the discussed weaving experiment.

16.2 Evidence for the Rigid Heddle and Rigid Heddle Bands in Northern Europe

To set these finds into a wider context, the archaeological evidence for tools for band weaving, iconography of weaving and actual repp ribbons is discussed with a reference to a wider time-span.

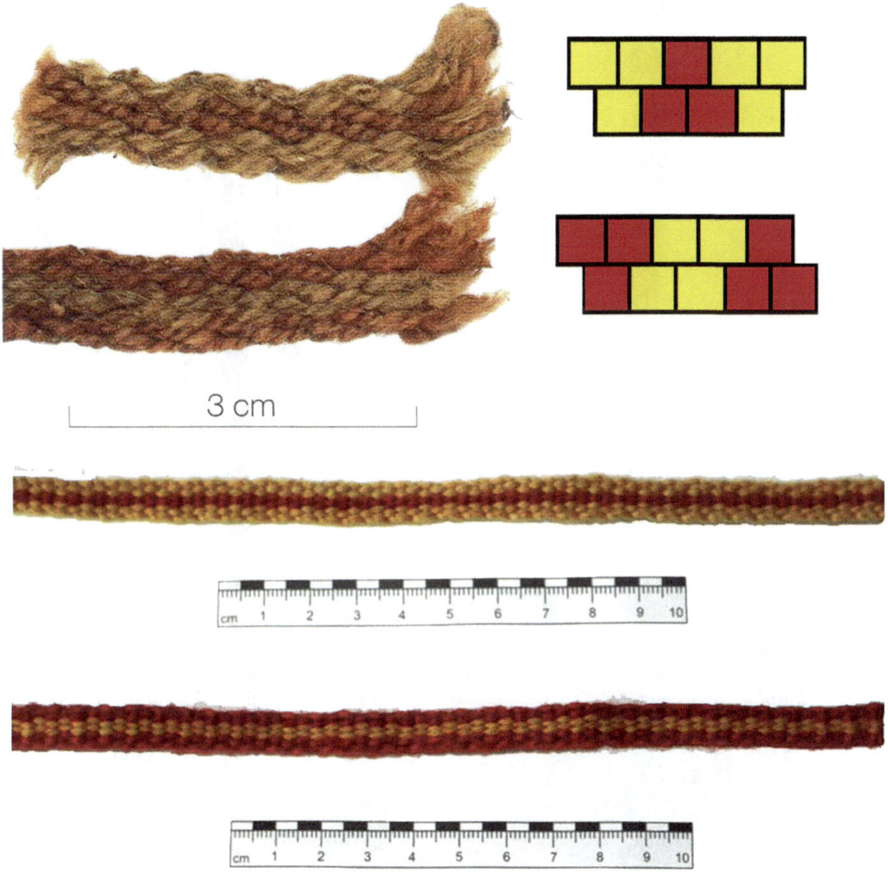

Fig. 16.1 Upper left: textile finds from Siksälä grave 143 no. 11, after Valk and Laul 2014: 152; Upper right: schematic warp alignment of the two ribbons from grave 143 no. 11; Bottom: reconstructed ribbons. (Reconstructions and photos by R. Lau)

16.2.1 Tools

A rigid heddle consists of alternating holes and slots. The warp is threaded through holes and slots in alternation and thus creates two sheds when the heddle is moved up and down. A shuttle or a different tool for holding the weft can be used to insert the weft through the sheds. It is possible to weave narrow bands with different colours and patterns in a nearly unlimited way (Dixon 2012: 8).

There are only a few archaeological finds of actual rigid heddles in Europe (Anderson 2003: 34; Foulkes 2011: 45). One reason could be that they might have been made of wood, and organic material usually disintegrates. Therefore, small wooden fragments of a potential rigid heddles found in the archaeological record might be difficult to identify.

Fig. 16.2 Left: reconstructed head piece from grave 243 no. 1. (Reconstruction and photo by R. Lau); Right: original head piece fragment from Siksälä grave 243 no. 1, after Valk and Laul 2014: 256

Roman period rigid heddles made of bone and wood are described by S. J. Foulkes (2011). According to her, there are four different variants of bands that could have been made with Roman rigid heddles: (1) a plain weave tabby band; (2) a warp dominant weave, with the weft pulled quite tightly; (3) a warp faced or repp band appearance weave, to be produced by pulling the warp; (4) tubular weaves (Foulkes 2011: 44).

The earliest evidence for a rigid heddle in northern Europe is the heddle from South Shields from northern England, dated to the Roman Iron Age (Wild 1988: 38). It is fragmented and made out of bone and decorated with bronze parts. Two other rigid heddle finds from northern Europe are from the Medieval period site Bryggen in Norway; dates in the period between the twelfth to fourteenth centuries

AD are suggested for both. One is manufactured from wood and the other is made of elk antler. Both are quite small and well preserved (Øye 1988: 80).

Until the fourteenth century AD, there are few traces of additional tools connected to rigid heddle weaving in the archaeological evidence in northern Europe. If shuttles were used, they are again difficult to identify, like finds identified as possible shuttles among the finds from Bryggen. Here, a total of eight narrow wooden objects with concave ends were found in house contexts. These objects are between 15.2 and 21.5 cm long and, since they are associated with textile production, they could have been used as shuttles. These objects date to between 1170 and 1413 AD. A specific feature of these finds are the holes. However, recent shuttles have no hole, which speaks against the above interpretation (Øye 1988: 81).

16.2.2 Iconographic Sources

Iconographic evidence for rigid heddles can be detected in the sixteenth century AD. The so-called weaving boxes from Cologne (Fig. 16.3) depict a small band loom standing on a table with the weaver pulling and pushing a rigid heddle up and down (Van Groningen 1520–1540). Small portable frames for rigid heddle weaving were traditionally used in Norway (Torgenrud 2014: 46). There are no older traces of weaving boxes or frames that helped warping the rigid heddle, but it is possible to stretch the warp directly between two fixed objects like trees, fences or posts.

16.2.3 Finds of Excavated Repp Bands

The earliest textile finds of narrow ribbons in northern Europe are dated to the Bronze Age and were found in oak coffin graves, for example in Stræ rup Mark, Langvad, Baunehøj, Guldhøj, Hølling and Schöningstedt (Bender Jørgensen 1986: 289). These ribbons were mainly used as part of clothing. To date, Iron Age finds of repp weave ribbons in northern Europe are rare: one item was found in the Vaalermoor bog (Neumann et al. 2011: 213).

From London and Dublin, we know of some settlement finds of different ribbons dated to the tenth and eleventh century AD (Crowfoot et al. 1992: 142). Beside some fragments without context, one such 10 cm long ribbon was sewn onto a bag, probably for decorative reasons. The rigid heddle ribbons appear again in the fourteenth century in a higher number than before (Crowfoot et al. 1992: 142).

Fig. 16.3 Weaving boxes from Cologne, after Van Groningen 1520–1540

16.2.4 Ethnographic Comparisons

Different countries around the Baltic Sea developed a strong textile tradition lasting into modern times. Estonia, Latvia, Norway and Finland are such examples with a rich textile heritage. To document this, since the year 1869, the Riga Latvian Society has collected different dress components, including rigid heddle woven ribbons (Karlsone and Vaska 2009: 285). Weaving ribbons served various purposes not only in terms of practical use, but also meeting symbolic and ritual needs present within the specific cultural mindset. In rural areas in Scandinavia, narrow ribbons were used at least from the seventeenth century for both practical and decorative reasons. Belts, hair ribbons, braces or carry belts are just a few examples of their practical use (Torgenrud 2014: 13). Until the 1930s, belts were used for carrying children during the day. A popular belief documented for the nineteenth century and beginning of the twentieth century was that these belts had magical powers protecting wearer and children from evil forces (Torgenrud 2014: 30; Rand 2019: 7). Historically, textile traditions were transmitted from mother to daughter. Most of those traditional ribbons were made of white linen and red wool, creating a powerful colour contrast (Torgenrud 2014: 18).

16.3 The Experiment – Methods and Objectives

The archaeological textile finds provide no information as to whether these textiles were made with the rigid heddle. Ribbons or bands could have been woven using different tools, such as heddle rods and shed sticks, or on a warp weighted loom (Foulkes 2011: 45). The aim of the experiment described below is to evaluate if bands from Siksälä grave 143 no. 11 and grave 243 no. 1, could have been made using the same tool – a rigid heddle, presumably known in this period – and how much technical leeway is possible. Specifically, the experiment is focused on reconstructing specific techniques and finds, i.e. the bi-coloured bands from grave 143, the technique of in-weaving metal spirals as in the textile from the grave 243 and investigating pattern structure and appearance of 'Auleja' woven ribbons from the same time period. The bands from Siksälä were chosen for their excellent preservation and their geographical and chronological classification to c. thirteenth century AD. Although there are no 'Auleja' bands found at this site, chronology and the rigid heddle are connecting elements. Alongside the technical question on how the excavated and 'Auleja' bands were created, the experiment was also vital for visualising a colour scheme, patterns and features such as bronze spirals.

The experiment is divided into three parts answering different questions about the rigid heddle. In the first part, weaving two simple bands with the rigid heddle aims at recreating the original visual appearance of ribbons 143 no. 11. The second part explores how efficient the rigid heddle is for inserting bronze spirals into the band. The third part aims at producing a band in 'Auleja' technique on the basis of ethnographic knowledge and modern descriptions of Estonian and Latvian archaeological bands from the eleventh to thirteenth century AD woven in this technique (Karlsone and Vaska 2009).

For all three parts of the experiment, wool threads (machine-spun yarn) were used together with a handmade wooden rigid heddle and a shuttle. Thus, the experimental reconstructions of textiles from the two graves 143 and 243 from Siksälä are categorised as B-standard reconstructions, according to I. Demant's definitions (Demant 2009: 145). As classified by Demant, the use of machine spun yarn, but plant dyed and hand-woven fabrics, can create an idea of how the fabric could have appeared (Demant 2009: 146).

16.3.1 Weaving Experiment – Part One: Siksälä Bands from Grave 143 No. 11

The visual appearance of repp ribbons woven on rigid heddle depends mostly on how tightly the weft is pulled during weaving. The weft itself is only visible as a small 'dot' on the side of the repp band. In the basic weaving technique, both the front and the back of the band are identical in structure and appearance. To thread the rigid heddle, the yarn has to be threaded through the holes and the slots. The number of warp threads in holes and slots is alternating from even to uneven (Torgenrud 2014: 99). For the reconstruction of the ribbons 143 no. 11, nine threads were threaded through the holes and 10 warp threads through of the slots (Fig. 16.1).

Material analyses of textiles in Siksälä point to the use of sheep wool in most of the fabrics (Rammo and Matsin 2014: 352). Plant dyed wool threads where also used for the experimental weaving. Different common dyes could be used for dyeing yellow, especially *reseda luteola, genista tinctoria, tanacetum vulgare*, birch leaves or yellow onions (Schweppe 1992: 177). For the experiment on reconstructing the bands 143 no. 11, threads were dyed using madder (red) and onion (yellow).

The weaving process itself is not difficult but consists of lifting or pushing the rigid heddle down to open the different sheds. To achieve the desired visual appearance of the pattern in the final product, preparing the warp is essential. As the warp threads were very few (Fig. 16.1), preparing the warp was an easy task. Altogether, weaving of the two ribbons 143 no. 11 of approximately 20 cm length each, required only about 20 minutes. The number of warp threads in the original textile bands 143 no. 11 and their reconstruction fits astonishing well with the number of slots and holes in the two rigid heddles from Bryggen Norway (Øye 1988: 80), which must have been used for weaving narrow ribbons. Due to the Baltic Sea connection, a similar weaving tradition in Norway, Scandinavia and Estonia is possible (Laul and Valk 2007: 79).

16.3.2 Weaving Experiment – Part Two: Siksälä Head Piece from Grave 243 No. 1

The original artefact from grave 243 displays a repp structure ground weave decorated with bronze spirals (Fig. 16.2). In total, the ribbon (without the head band), has a length of about 30 cm and a width of about 1 cm (Valk et al. 2014: 256). As identified in the published images, nine warp threads were threaded in the holes of the rigid heddle and 10 threads into the slots. According to the colour of the original textile, threads dyed in indigo (*indigofera tinctoria*) and madder were used. Before the actual experiments started, some test weaving was done.

The main experiment consisted of following steps: making bronze/brass spirals and preparing the warp. Due to the difficult process of making bronze wire, machine-made wire was used and 48 spirals were produced for this project. Crafting them by hand, bending the wire around narrow metal nails, took about one hour. To ensure an efficient workflow, long spirals where made, split and cut according to the needed length.

As the original band is of three red and blue vertical stripes, the spirals were pushed over each section of red or blue warp threads. The weft was led through either the left or the right spiral. Otherwise, the weft would be visible on the outside of the spiral. Threading the weft through the spiral, together with the warp, makes it invisible. Sliding the spirals onto the warp threads while threading was impossible, since the rigid heddle would be a barrier, preventing the spirals from passing through its slots or holes. Therefore, to slide the spirals, the warp threads were pulled out of the rigid heddle. This was laborious and a very slow process. Alternatively, these bands could have been braided by hand as suggested by Laul and Valk (2007: 65). Also, heddle rods could have been used with the spirals slid onto the warp threads before weaving. It is possible that the spirals could be pulled from behind, through string heddles, towards the weaver when necessary. This assumption requires further experimentation.

One side of the head piece was made in an hour, plus the one hour required for spiral production. Two pieces were woven for the experiment, doubling the time, so that the entire task required four hours of calculated work. In retrospect, the whole process took considerably longer, because the time for figuring out how to proceed and then correcting the mistakes were not included. The upper head band is only a C-standard reconstruction according to Demant (2009: 144). In conclusion, using the rigid heddle during weaving this band was more of a disadvantage than an advantage. It slowed the weaving down, which leads to the conclusion that the ribbons of the head piece are more likely to have been woven without a rigid heddle.

16.3.3 Weaving Experiment – Part Three: 'Auleja' Technique

The rigid heddle can not only be used for weaving ribbons with repp or tabby structure, but also to create geometric and multi-coloured patterns, which are traditionally called the 'Auleja' technique. For weaving an 'Auleja' ribbon, the first step is to choose a background colour and a pattern colour. Then, the pattern threads are put into every third hole or slot of the rigid heddle (Fig. 16.4). The rest of slots and holes are threaded with the background warps. An example with a red pattern-thread is shown in Fig. 16.4. The pattern warp threads can be pushed down if the pattern demands that they should not appear on the front, and pulled up, if they should be seen. Therefore, due to the manipulation of the warp, the pattern will be visible as flotation (Fig. 16.5) of the warp threads (Karlsone and Vaska 2009: 306).

Fig. 16.4 Top: reconstructed bicolour ribbon in 'Auleja' technique; Bottom: warp alignment for the 'Auleja'/pick-up technique. (Reconstructions and photos by R. Lau)

The earliest evidence for the 'Auleja' technique can be traced back around the tenth century AD in the Baltics,[1] where most of ribbons made in this technique are found in burials from Latvia, Lithuania, Belarussia and parts of Russia (Karlsone and Vaska 2009: 306). It is said that they were mostly bicolour in blue and red, or red and yellow (Karlsone and Vaska 2009: 308). Wool and linen were the basic raw materials for creating such ribbons. Due to the geometric design concept, the patterns should repeat themselves, be symmetrical and therefore pleasing to the eye. This balance is one of the important things to achieve. The pattern differs on the front and back side of the ribbon: as the threads are pushed down on the front side, they create the floats at the back side and vice-versa. The creativity is only limited by the number of pattern threads related to the construction of the rigid heddle (Karlsone and Vaska 2009: 318).

The experimental ribbon was designed with the help of the visual descriptions from *Art of the Balts* (Karlsone and Vaska 2009: 306) in an attempt to create a balanced pattern. In order to better understand the technique and principles of patterning, two contrasting colours (red and blue) were chosen, as known from the ribbons from the eleventh to thirteenth century (Karlsone and Vaska 2009: 308). Aligning the warp threads can be seen in Fig. 16.4. The red warp threads will be the pattern threads and are alternating evenly (Fig. 16.4). While weaving, the pattern threads can be manipulated by hand. In this manner, the backside of the ribbon forms the exact negative of the front pattern. During the weaving process, it is possible to see what the next row will look like (Fig. 16.5). In this way, the weaver controls whether he or she has picked up or pushed down the right threads.

To weave a c. 10 cm section of band in 'Auleja' technique took about 20 minutes. Especially for techniques like this, experience is important for the weaving process.

[1] The oldest 'Auleja' or pick-up woven band from Northern Europe is from the site Jõuga in eastern Estonia. The fragment has been dated to the late Iron Age c. 800–1225 AD and described as tabby with repp character (Rammo 2014: 112).

Fig. 16.5 Weaving the 'Auleja' ribbon, pattern warp threads (red) chosen for a new shed. (Photo by R. Lau)

Future experiments should include comparisons in work speed efficiency on the same pattern by weavers with different levels of experience.

16.4 Discussion and Conclusions

The quantity and quality of ribbons and bands found in archaeological excavations do not represent the historical reality. Bad preservation conditions leading to few surviving finds is one reason. The state-of-the-art with publications of different standards in many languages, e.g. Russian, Ukrainian, Byelorussian, Lithuanian, Latvian, Estonian, etc. (cf. Karlsone and Vaska 2009: 306; Valk et al. 2014: 152), makes it difficult to follow up on these finds in international research. Unfortunately, so far, no *in situ* finds of rigid heddles with attached warp and woven band have been found – there is no equivalent to e.g. the find of weaving tablets in the Oseberg ship burial (Götze 1908: 488).

Therefore, the hypothesis that the bands 143 no. 11 and 243 no. 1 from Siksälä were made with a rigid heddle has to be critically evaluated. For the bands 143 no. 11, the use of this tool seems feasible. However, since it is very difficult to handle and insert the metal spirals of 243 no. 1 while weaving on the rigid heddle, more experiments are necessary to explain how these bands have been made.

Even if the rigid heddle is an expedient tool to weave ribbons 143 no. 11, this is of course only one possibility for creating similar bands. Weaving with a shed stick and heddle rod is another plausible alternative (Foulkes 2011: 41). Tablet weaving is yet another technique, which is better attested archaeologically and already suggested as the technique of production of other bands from the Siksälä graves (Matsin 2013: 77). The combinations of red and yellow or red and blue (Karlsone and Vaska 2009: 308) can be found in simple bands from 143 no. 11, as well as in more elaborate products, such as bronze spirals bands from 243 no. 1, or in tablet woven ribbons from Siksälä (Matsin 2013: 69) and also in contrasting 'Auleja' patterns.

At Siksälä, the bands have been placed in graves, thus creating a symbolic context for these finds. Although the ribbons are found in graves of both men and women, in male graves they are simple bands serving as belts (Laul and Valk 2007: 64). Only female graves, of girls as well as adult women, contain the special head pieces with attached ribbons decorated with bronze spirals. At Siksälä, 66 women were found with a head band in addition to deceased juveniles or infants (Laul and Valk 2007: 65). This dress component was part of an emerging change of the female costumes and burial traditions in the thirteenth century in this community, together with blue shawls, cowry shell necklaces and glass bead pendants, which were also common in eastern and northern Latvia (Laul and Valk 2007: 79). All these were possibly parts of a festivity dress or attire.

References

Anderson E (2003) Tools for textile production from Birka and Hedeby. In: Ambrosiani B (ed.) *Excavation in the black earth 1990–1995*. Birka Studies volume 8. Birka Project, Stockholm.

Bender Jørgensen L (1986) *Forhistoriske textiler i Skandinavien*. Det kongelige nordiske oldskrift-selskab, København.

Crowfoot E, Pritchard F, Staniland K (1992) *Textiles and clothing c.1150–c.1450*. Boydell and Brewer, London.

Demant I (2009) Principles for reconstruction of costumes and archaeological textiles. In: Alfaro Giner C, Tellenbach M, Ferrero R (eds) *Textiles y museologia*. Consejería de Cultura y Deporte, Valencia, p 143–153.

Dixon A (2012) *A brief history of the inkle. What, why, when and where*. Anne Dixon, Gasthorpe.

Foulkes SJ (2011) Roman rigid heddles: A survey. *Archaeological Texiles Newsletter* 52:41–47.

Götze A (1908) Brettchenweberei im Altertum. *Zeitschrift für Ethnologie* 40(4):481–500.

Karlsone A, Vaska B (2009) Sẽnosios Šventinių Drabužuų Puošybos Tradicijos. In: Butrimas A (ed.) *Baltu Menas. Art of the Balts*. Vilniaus dailė akademija, Vilnius, p 285–372.

Laul S, Valk H (2007) *Siksälä, a community at the frontiers*. CCC Papers 10. University of Tartu and Gotland University College, Tartu and Tallinn.

Matsin A (2013) The 13th–16th century tablet-woven bands from Estonia. In: Hopkins H (ed.) *Ancient textiles, modern science II*. Ancient Textiles Series 34. Oxbow Books, Oxford and Oakville, p 64–78.

Neumann R, Freudenberg B, Siwek M (2011) Das Vaaler Bändchen – Die Rekonstruktion eines archäologischen Kammgewebes aus Dithmarschen als Gemeinschaftsarbeit der Wollgruppe des Museumdorfes Düppel in Berlin. *Experimentelle Archäologie in Europa Bilanz* 10:213–218.

Øye I (1988) *Textile equipment and its working environment. Bryggen in Bergen c 1150–1500*. The Bryggen Papers Main Series 2. Norwegian University Press, Oslo.

Rammo R (2014) Tradition and transition: The technology and usage of plant-fibre textiles in Estonian rural areas in the 11th–17th centuries. In: Lipkin S, Vaianto K (eds) *Focus on archaeological textiles: Multidisclipinary approaches*. Monographs of the Archaeological Society of Finland 3, p 102–115.

Rammo R, Matsin A (2014) Weaving in Medieval village on the basis of Siksälä finds. In: Valk H, Laul S (eds) *Siksälä Kalme* I. Muistis Ja Ajalugu, Tartu Ülikool, Tartu, p 352–351.

Rand P (2019) *Estonian pick-up woven belts*. Estonian National Museum (ENM), Saara Kirjastus, Tartu.

Schweppe H (1992) *Handbuch der Naturfarbstoffe. Vorkommen, Verwendung, Nachweis*. Ecomed Verlagsgesellschaft, Landsberg/Lech.

Sperling U (2014) *Aspekte des Wandels in der Bronzezeit im Ostbaltikum. Die Siedlung der Asva-Gruppe in Estland.* Tallinn Estonian Academic Publishers, Tallinn.
Torgenrud H (2014) *Norwegian pick-up bandweaving*. Schiffer, Atglen.
Valk H, Laul S (2014) *Siksälä Kalme* I. Muistis Ja Ajalugu. Tartu Ülikool, Tartu.
Valk H, Ratas J, Laul S (2014) *Siksälä Kalme* II. Matuste Ja Leidude Kataloog. Tartu Ülikool, Tartu.
Van Groningen JS (1520–1540) *Van traktaat over de naaikunst* (Titlelblad), Cologne, Rijksmuseum, Amsterdam, https://artsandculture.google.com/culturalinstitute/beta/asset/titelblad-van-traktaat-over-de-naaikunst-jan-swart-van-groningen-willem-vorsterman/UQEuRhpEWPh2Eg, accessed 07 Dec 2020.
Wild JP (1988) *Textiles in archaeology*. Shire Archaeology, Aylesbury.

Ronja Lau is a doctoral research student of the Ruhr-University Bochum, Germany, examining textiles and textile tools from the salt mine complex of Dürrnberg, Austria. Her project aims to identify and analyze organic textile fragments, textile tools from the settlement and mineralized textiles from burials. In addition, she is specialized in structure from motion (SfM) technique and uses it in her work on textile tools. She takes part in projects like the Bronze Age settlement of Asva in Estonia at the University of Rostock and collaborates with U. Sperling and the Heinrich Schliemann Institute in excavations in Estonia.

Part V
Digital Tools

Chapter 17
Investigating Relations between Textile Production and Seals and Sealing Practices in Bronze Age Greece: A Presentation of the New "Textile and Seals" Project Database

Agata Ulanowska

Abstract The "Textiles and Seals" research project investigates the complex relationships between textile production and seals and sealing practices in Bronze Age Greece. These relationships can be tracked by analysing the combined evidence from seal-impressed textile tools, textile imprints on the undersides of seal-impressed clay lumps and textile production-related imagery. The main research questions posed by the project attempt to explain the use of textiles in sealing practices and the use of seals in the administration of textile production, as well as the possible meaning of iconographic references to textile production on seals. In order to facilitate these tasks, an online database was built specifically for this project through the services of the Digital Competence Centre of the University of Warsaw. Although data entry has not yet been completed, the "Textiles and Seals" database has already proved to be a powerful research tool. In this paper, the structure of the database and the possibilities of its search engines are briefly demonstrated. The paper discusses how the "Textiles and Seals" database facilitates the tracking of correlations within the examined evidence, based on a statistical search of a significant number of records and hitherto unexplored combinations of data.

Keywords Bronze Age Greece · Textile production · Sealing practices · Seals · Database

A. Ulanowska (✉)
Faculty of Archaeology, University of Warsaw, Warsaw, Poland
e-mail: a.ulanowska@uw.edu.pl

A. Ulanowska et al. (eds.), *Ancient Textile Production from an Interdisciplinary Perspective*, Interdisciplinary Contributions to Archaeology,
https://doi.org/10.1007/978-3-030-92170-5_17

17.1 Introduction

The key economic importance of textile production in Bronze Age Greece has recently been revealed through in-depth studies of the archaeological record, e.g. textile tools and workplaces (Nosch and Laffineur 2012; Andersson Strand and Nosch 2015a; Siennicka et al. 2018), as well as ethnographic and experimental data (e.g. Andersson Strand and Nosch 2015a), textual sources, specifically the Linear B tablets (e.g. Killen 1984, 2007; Burke 2010; Del Freo et al. 2010; Nosch 2012, 2014; Rougemont 2014) and textile iconography (e.g. Nosch and Laffineur 2012; Jones 2015; Shaw and Chapin 2016). Excavated textiles, limited in number due to the preservation conditions in Greece (e.g. Spantidaki and Moulherat 2012; Skals et al. 2015) and textile imprints on clay (e.g. Müller and Pini 1997; Müller 1999; Weingarten et al. 1999; Siennicka 2015–2016; Vakirtzi et al. 2018) represent the available evidence for actual textile products. However, sealing practices attested on textile tools, textile imprints preserved on seal-impressed lumps of clay, as well as textile production-related iconography on seals have not, so far, been sufficiently investigated as a potential source of knowledge about textile production and consumption in Bronze Age Greece (cf. Weingarten et al. 1999; Burke 2010; Vakirtzi et al. 2018; Karnava 2019: 502–593). Therefore, the complex nature of the relationships between textile production and seals and sealing practices in Bronze Age Greece, from the Early to Late Bronze Age (c. 2650–1200 BCE),[1] were selected as the main focus of a new research project called "Textiles and Seals".[2]

The project attempts to understand the practice of impressing seals on textile tools, specifically on loom weights, and its geo-chronological specificity in relation to other administrative practices related to textile production, such as, e.g., wool weighing (Michailidou 1990: 416; Cutler 2016: 176; Alberti 2017) and writing (e.g. Killen 1984, 2007; for the inscribed loom weight, see Karnava 2019), attested at various sites in mainland Greece, Crete and the Cyclades. It systematically explores the abundant, yet previously not sufficiently recorded evidence of textile imprints on casts of the clay sealings and their undersides, stored in the Archive of the *Corpus der minoischen und mykenischen Siegel* (hereafter, CMS) in Heidelberg (cf. Ulanowska 2020a, b). Finally, it examines the imagery of Aegean seals, specifically MBA prisms from Crete, as a potential source of iconographic references to textile production, e.g. raw materials, spinning, weaving and potential symbolic references, such as depictions of spiders (cf. Nosch and Ulanowska 2021). The main types of evidence investigated within the "Textiles and Seals" project, i.e.

---

[1] The Bronze Age in Greece is divided into three main chronological phases: Early Bronze Age (EBA 3100-2100/2050 BCE), Middle Bronze Age (MBA 2100/2050-1700/1675 BCE) and Late Bronze Age (LBA 1700/1650-1075/1050 BCE), after Manning 2010: Table 1.

[2] The research project "Textiles and Seals. Relations between Textile Production and Seals and Sealing Practices in Bronze Age Greece" has been funded by the National Science Centre, Poland, ref. no. 2017/26/D/HS3/00145, at the Institute of Archaeology, University of Warsaw, for the years 2018–2021.

seal-impressed textile tools, textile impressions and textile production-related iconography, feature different functionalities, parameters and meanings that require different manners of recording. Since this also requires scheduling of different research queries, the new "Textiles and Seals" database was designed and built online specifically for the project. This contribution discusses the overall scheme of this database, the possibilities of its search engines, as well as the first research results of specific searches on data available in the database as of October 2021. The database is made available in open access on a rolling basis, as dedicated research tasks are completed within the "Textiles and Seals" project. The modules for textile production-related iconography and textile impressions on the undersides of clay sealings have already been published in 2021,[3] while the module with seal-impressed textile tools has been published in February 2022.

17.2 The CMS Arachne and the Centre for Textile Research Databases, and Their Usefulness for the "Textiles and Seals" Project

The quantity of data available from seals and their impressions on clay is massive due to the large number of individual examples preserved and the rich iconography of seals, as well as the complexity of sealing practices whereby several seals could have been impressed on one object (Krzyszkowska 2005). Systematic and comparative recording of all the Aegean seals and sealings was the *raison d'etre* for the establishment of the CMS in 1958. Since then, CMS volumes presenting seals and sealings kept in different museums/collections worldwide have been continually published, together with the CMS *Beihefte* that discuss specific problems or specific evidence.[4] Moreover, the CMS Archive in Heidelberg preserves the unique and exquisite collection of modern casts of all seals, seal impressions and many undersides of clay sealings that have already been published in the CMS volumes, and a substantial number of the casts of objects still awaiting their publication.[5]

[3] https://textileseals.uw.edu.pl/database/, accessed 5 Oct 2021.

[4] For a general introduction and corpora of Aegean seals and sealings, see: CMS I-XIII; CMS *Beihefte* 1-10; Palaima 1990; Krzyszkowska 2005; CMS Arachne, an open access database containing images from the CMS volumes and basic descriptions of seals and sealings, https://arachne.uni-koeln.de/drupal/?q=en/node/196, accessed 6 Apr 2020. The new version of the DAI Arachne and the CMS project has been published in autumn 2021, see https://arachne.dainst.org/project/corpusminmyk, accessed 5 Oct 2021, but all further references to the CMS Arachne comprise links to its old version and search engines exclusively.

[5] For more information about the CMS Archive, see https://www.uni-heidelberg.de/fakultaeten/philosophie/zaw/cms/archive/archive.html, accessed 6 Apr 2020.

17.2.1 The CMS Arachne Database and Textile Tools

Alongside the already digitalised CMS volumes and *Beihefte* (Propylaeum eBooks),[6] large amounts of data are available in CMS Arachne – an open access online database.[7] The search engines of this database operate in two main modules: one for seal faces – each saved as an individual record – and one for seal-impressed objects (the "*surfaces containing seal impressions*"[8] search engine). The so-called "*context browser*" provides information about other objects/surfaces that were impressed with the same seal or seal face. Keywords/dictionaries for specific fields in both modules are in German, but their full lists with English translations and well-illustrated specific examples can be displayed for individual search fields and pasted in easily.

The usefulness of CMS Arachne is remarkable for any student of Aegean archaeology. However, as regards the research undertaken by the "Textiles and Seals" project, there are several limitations in using CMS Arachne as the main and only database. First, this database provides no information about textile and organic products, such as threads, cords, leather thongs, fabrics and others that were imprinted on the surfaces containing seal impressions. Secondly, the catalogued objects identified by textile archaeology as textile tools (cf. Andersson Strand and Nosch 2015a) seem to be primarily limited to those examples that bore legible seal impressions.

Textile tools can be tracked under a general keyword "*Gewicht*" in the search field "*form*", within the "*surfaces containing seal impressions*" module, with several further sub-types distinguished in relation to a specific geometric form, number of seal impressions or perforations (e.g. "*Gewicht, Pyramidenstumpf mit zwei Abdrücken, Diametralbohrung*"). However, the CMS Arachne keywords provide no method of distinguishing by type of textile tool, such as spindle whorls, loom weights or spool-shaped tools. For example, a hemi-spherical spindle whorl from Chamalevri (no. 173356 in the CMS Arachne, cf. Vlasaki 2010) and a series of spherical loom weights from Malia (e.g. nos 168570, 168572, 168573, 170093)[9] are all classified under the same keyword: "*Gewicht, Kugel, Zentralbohrung*". The measurements of individual objects described as "*Gewichte*", expressed in centimetres to two decimal places, can be found in paper or digital copies of the relevant CMS volumes, with the exception of weight, which was not recorded.

---

[6] https://books.ub.uni-heidelberg.de/propylaeum/series/info/cms?lang=en, accessed 6 Apr 2020.

[7] For the history of the CMS Arachne database, see Müller 2010; http://www.cms.adwmainz.de/cms-datenbanken-in-arachne.html, accessed 6 Apr 2020. For the recent improvements and M. Anastasiadou's manual on how to use this database, see https://www.uni-heidelberg.de/fakultaeten/philosophie/zaw/cms/databases/databasesfull.html, accessed 6 Apr 2020.

[8] Double inverted commas indicate search modules, search fields and keywords from databases other than "Textiles and Seals". The ones in the "Textiles and Seals" database are indicated by single inverted commas.

[9] All these loom weights were possibly impressed with the same seal, which was recorded as CMS II,6 213 for the impressions on the loom weights from Room 3.3 on the north-east borders of the Palace (nos 168570, 168572, 168573), and CMS II,6 203 for the impression on the loom weight from Building A in Quartier Mu.

17.2.2 CMS Arachne and Textile Production-Related Iconography of Seals

The CMS Arachne's module containing seals and seal impressions facilitates complex iconographic searches by providing a series of search fields that start from a general classification of motifs (i.e. figural or ornamental) and end with very specific identifications, such as sexual characteristics of creatures or individual signs from the Cretan hieroglyphic script. However, what has been identified as possible textile production-related motifs is only occasionally included in the CMS keywords.

Textile production-related motifs have been recognised on the basis of already existing identifications (e.g. Panagiotakopulu et al. 1997; Burke 2010: 44–47; Ulanowska 2016) and the methodology adopted in the "Textiles and Seals" project (Ulanowska 2017, 2020a; Nosch and Ulanowska 2021). They may be grouped as follows:

- potential references to raw materials, represented by: '*flax plant*', '*woolly animals*' and '*silk moth*' motifs;
- potential references to processing of fibres and spinning: '*combs*' and '*combers*' (also a potential reference to weaving), '*spindle with whorl*' and '*spinners*' motifs;
- potential references to dyestuffs and dyeing: '*murex shell*' motif;
- potential references to weaving and woven fabrics: '*loom weights*' and '*weavers*', '*warp-weighted loom*', '*rigid heddle*', '*weft-beater*', '*fabric with fringes*' and '*interlaced band*' motifs;
- potential symbolic references to textile production: '*spiders*' (Fig. 17.1).

Given the schematic character of the depictions and the limits in our understanding of Minoan imagery, none of the newly suggested interpretations can be considered definite. However, all motifs that have been recognised as related to textile production display distinct characteristics and features of functional importance for their potential real-world referents, e.g. actual plants, animals, tools and technical gestures required by textile production. They also seem to reflect the main characteristics of a textile culture specifically relevant to Bronze Age Greece, i.e. the use of wool and flax as the main raw materials, the skill of purple dyeing and the socio-economical and symbolic importance of weaving. Moreover, several iconographic comparanda for these motifs have been recognised in the arts of other cultures, specifically in the imagery of Mesopotamian glyptic from the third millennium BCE (Breniquet 2008; cf. Nosch and Ulanowska 2021).

A few of the textile production-related motifs correspond entirely (e.g. '*spiders*') or roughly (e.g. '*woolly animals*', '*silk moths*', '*murex shells*') to individual motifs and thus to keywords already distinguished by CMS. In the latter case, the CMS motifs can be termed using the same keywords, e.g. "*Ziege*" = '*goat*' in the 'Textiles and Seals' database, but the number of identified '*goat*' depictions differs due to the different methodology adopted for recognising '*woolly animals*' in the "Textiles and Seals" project (cf. Nosch and Ulanowska 2021: 85–87; Nosch et al. 2021: 39–40). Motifs described by other keywords in CMS may also correspond to textile

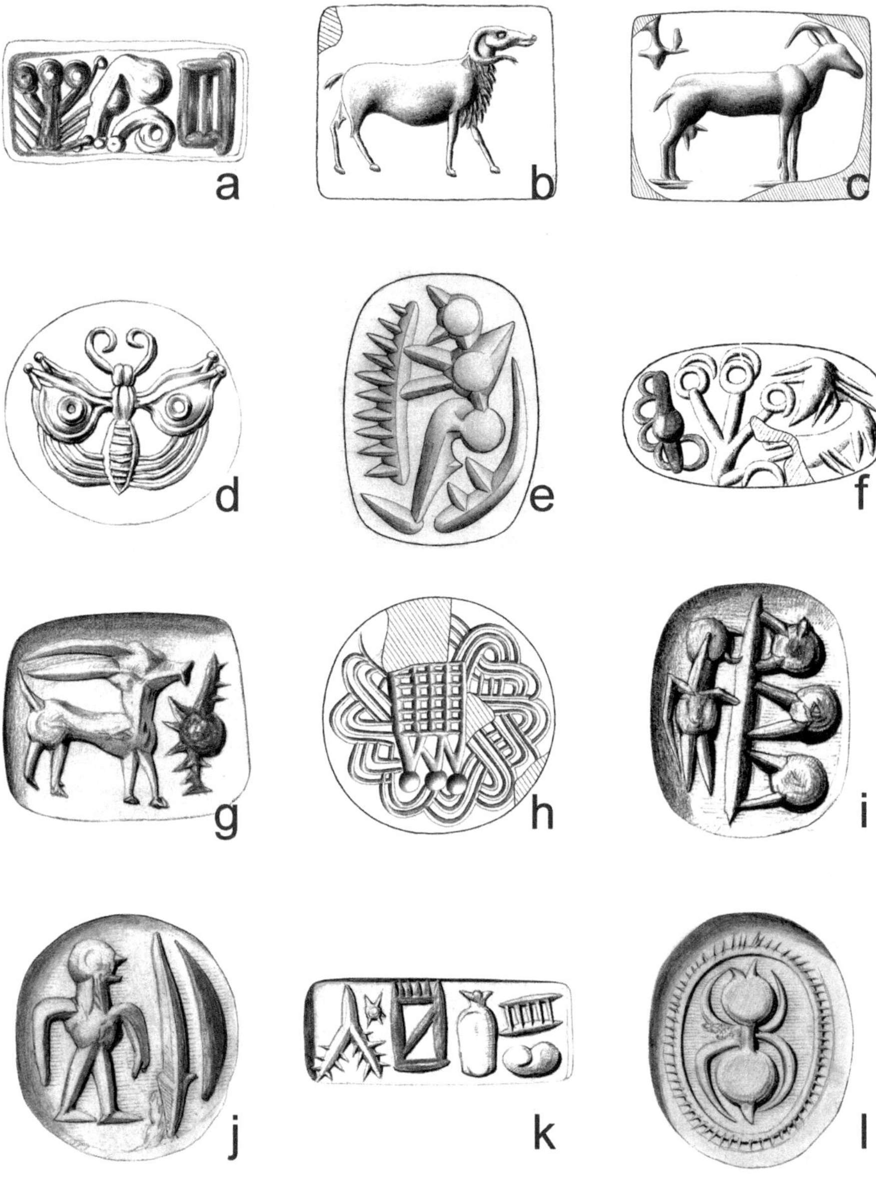

Fig. 17.1 Potential textile production-related motifs in the imagery of Aegean seals: (**a**) '*flax plant*' and '*rigid heddle*' in an inscription on a prism (CMS VII 40b/CHIC #299δ); (**b**) '*woolly animal*' – sheep (VI 177); (**c**) '*woolly animal*' – goat (II,8 378); (**d**) '*silk moth*' (VI 455); (**e**) '*comb*' (VS1A 325a); (**f**) '*spindle with whorl*' (II,8 86); (**g**) '*murex shell*' (II,2 262a); (**h**) '*warp-weighted loom*' and 'interlaced band' (II,1 64a); (**i**) '*loom weights with weaver*' (II,2 214a); (**j**) '*weft-beater*' (II,2 302a); (**k**) '*fabric with fringes*' and '*rigid heddle*' in an inscription on a prism (II,2 227/CHIC #200); (**l**) '*spider*' (II,2 224b). (Drawings after the CMS Arachne database enhanced by A. Ulanowska, courtesy of the CMS Heidelberg)

production-related identifications. For example, a motif classified as "*Stange mit Gefäßen*" corresponds roughly to the '*loom weights*' motif (cf. Burke 2010: 47; Ulanowska 2017; Militello 2018). However, there are a few depictions of "*Stange mit Gefäßen*" that are not identified as the '*loom weights*' motif in the 'Textiles and Seals' project/database (e.g. CMS XI 122b), and a few '*loom weights*' motifs that are not identified as "*Stange mit Gefäßen*" in CMS (e.g. CMS II,1 64a; Fig. 17.1.h).

Alternatively, what has been recognised as an individual textile production-related motif may not overlap with the CMS classification. For example, the '*comb*' motif has been designated in CMS as "*Tannenzweig einseitig*", which is a much wider category than combs in the 'Textiles and Seals' project/database. When the '*comb*' motif is combined with a human figure, it is classified as an "*undefinierbar*" tool in CMS (cf. Ulanowska 2020c: 69–70; Nosch and Ulanowska 2021: 87–88).

Textile production-related motifs such as '*flax*', the head of a '*woolly animal*', '*spindle with whorl*', '*rigid heddle*' and '*fabric with fringes*' correspond to graphic forms of the Cretan hieroglyphic script signs inscribed on seals (Nosch and Ulanowska 2021). Usually, each textile production-related motif corresponds graphically to a single sign of the Cretan hieroglyphic script, i.e. '*flax*' motif – to *Corpus Hieroglyphicarum Inscriptionum Cretae* (hereafter, CHIC) 031, the head of a '*woolly animal*' in profile – to CHIC 016, '*rigid heddle*' – to CHIC 038, '*fabric with fringes*' – to CHIC 041, etc. Only the '*spindle with whorl*' motif bears resemblance to multiple CHIC signs, such as CHIC 063 and 050, 062 (Nosch and Ulanowska 2021: 89–90). The motifs that are graphically abbreviated to the Cretan hieroglyphic signs may, however, also appear on non-inscribed seals, and be shown as ornaments or remain unrecognised as potential script signs, e.g., the '*flax*' on CMS III 237a; the numerous heads in profile of the '*woolly animal*'; the '*spindle with whorl*' on CMS II,2 150a (cf. Jasink 2009; the DBAS – CHS Cretan Hieroglyphic Seals database;[10] Nosch and Ulanowska 2021).

17.2.3 The CTR Database and Seal-Impressed Textile Tools from Greece

The large number of preserved textile tools provides another important source of mass information. The evidence of textile tools from Bronze Age Greece is partially recorded in the database of the Centre for Textile Research, University of Copenhagen (hereafter, CTR). The CTR database was designed in Microsoft Access in 2003, in order to record all textile tools studied within the "Tools, Textiles, Texts and Contexts" research programme (hereafter, TTTC, cf. Andersson Strand and

[10] The DBAS – CHS Cretan Hieroglyphic Seals online database, designed by M.A. Jasink and L. Bombardieri at the webpage of the Laboratory of Aegean Civilisations of the University of Florence, is a tool for searching MBA inscribed seals and tracking relations between the CHIC signs and graphic elements that were not considered script signs in CHIC, https://www.sagas.unifi.it/vp-394-dbas-chs-cretan-hieroglyphic-seals.html, accessed 6 Apr 2020.

Nosch 2015b: 145). It collects 8,725 textile tools selected from sites in Greece, Cyprus, Anatolia and the Levant, but although access to it is kindly granted on request, the CTR database is not accessible online. The CTR database with the TTTC typology of spindle whorls and loom weights created a new standard for recording vast numbers of different types of tools in a comparative way. At the same time, by recording a series of parameters such as the geometric form, weight, height, diameter/thickness, number and diameter of perforations, use-wear, etc., the CTR database allows the reconstruction of the basic functionality of textile tools and, thus, the general characteristics of textiles produced at a specific site or in a specific context (cf. Andersson Strand and Nosch 2015b).

The seal-impressed textile tools from Bronze Age Greece are occasionally recorded in this database together with other tools from archaeological sites that were investigated by the TTTC programme (cf. Andersson Strand and Nosch 2015a). However, no detailed information about the nature of the markings and stamps on textile tools is provided and, apparently, information about such markings was not always recorded.

17.3 "Textiles and Seals" Database – A General Scheme and Construction

Creating a new database was scheduled as part of the project in order to provide a specialist tool for handling and searching numerous data of different natures, characteristics and features. The database was built by the Digital Competence Centre of the University of Warsaw, the project's technological partner. The software used to construct the database is free and open source. The application tailored to the database requirements was written in Python, using the Django framework. Data is stored in a PostgreSQL database. The search and data visualisation interface was built based on Bootstrap.

The database comprises three modules reflecting the main types of the investigated evidence, such as '*seal-impressed textile tools*', '*impressions of textiles on clay*' and '*iconography of textile production*' modules. The cross-references with the data already included in CMS Arachne are provided by separate fields in each module, which comprise the CMS number of an individual seal or seal impression, the CMS keywords for its motifs, the CMS Arachne ID numbers for seal-impressed objects, as well as direct web links to the CMS Arachne records.

The first two search engines for the modules '*iconography of textile production*' and '*impression of textiles on clay*' have been published online and linked to the 'Textiles and Seals' website in March (https://data.textileseals.uw.edu.pl/iconography/search?action=search) and October 2021 (https://data.textileseals.uw.edu.pl/textile_impressions/search?impressed_textile_type=1&impressed_textile_type=13&impressed_textile_type_OR=on&action=search). The last search engine for the module with seal impressed and marked textile tools has been published in February 2022 (https://data.textileseals.uw.edu.pl/textile_tools/search) and, therefore, it is not described in detail in this chapter. Access to individual records and their editing is restricted to the collaborators of the "Textiles and Seals" project and requires a log-in. Since the database is built online, multiple users can search and edit it at the same time. The database will continue to be hosted by the University of Warsaw after the project is completed.

17.3.1 'Seal-Impressed Textile Tools' Module

At the moment, this module comprises the least number of records but gathering and processing of data is still in process. There are 121 seal-impressed and marked tools thus far recorded. Those categorised by CMS Arachne as "*Gewichte*" are represented by 47 examples. For now, the integration of data from CMS Arachne and the CTR database was completed only for three seal-impressed tools from Malia on Crete.[11] Each record entry comprises data fields that identify the specific object, its context and dating, type of tool and its metrics, the form of the impressed seal and its imagery (Table 17.1). The manner of recording data that characterise the tool and its functionality is adopted from the CTR database (cf. Andersson Strand and Nosch 2015b). Photos of an entire tool and seal-impression(s) are displayed as thumbnails

Table 17.1 Individual data fields in the '*Seal-impressed textile tools*' module

Data fields identifying the specific object	Data fields identifying the impressed seal(s)	Data fields identifying properties of the textile tool
Inv. no.	CMS no./nos	Type of tool
The CMS Arachne ID	Stylistic dating	Form
Museum	Type of seal	Fabric
Site	Motif(s) description	Surface treatment
Archaeological context	Photo of the impressed seal face(s)	Preservation in %
Context dating		Use-wear
Remarks		Weight (g)
Photo of the textile tool		Reconstructed weight (g)
		Maximum diameter (mm)
		Maximum height (mm)
		Maximum thickness (mm)
		Number of perforations
		Max. diameter of perforation (mm)
		Min. diameter of perforation (mm)

[11] I thank E. Andersson Strand for sharing this data with me. These tools are recorded in the CTR database as "*loom weights*": MAL-69 M 1662 (impressed with CMS II,6 192); MAL-69 M 2388 (impressed with CMS II,6 190); MAL-72 M 115 (impressed with CMS II,6 207).

for each record. The iconographic data in this module may be cross-referenced with the '*iconography of textile production*' module, e.g. the seal that was impressed on a textile tool may, occasionally, bear a motif identified as possibly textile production-related. This is true of the spool-shaped tool from Malia (MAL-69 M 1662) impressed ten times by a seal bearing an image of two spiders (CMS II,6 192). The search engine for this module has been published in February 2022.

17.3.2 'Impressions of Textiles on Clay' Module

Currently, this module consists of 225 records corresponding to individual silicone and plasticine casts of the undersides of lumps of clay with impressions of textiles (180 possible impressions, including uncertain identifications) as well as impressions of threads, cords, leather thongs (171 possible impressions), and mats, wickerwork and basketry (35 possible impressions). The collected data so far comes from Aghia Triada, Chania, Knossos, Malia and Phaistos on Crete, Lerna and Geraki on the Mainland, Kea and Lemnos. Fully recorded is data from Lerna (52 casts) and Phaistos (145 casts), while the comprehensive study of textile impressions on all casts in the CMS Heidelberg is to be continued.[12]

Especially interesting is comparison of textile impressions from EBA Lerna in Mainland Greece and MBA Phaistos on Crete – two sites distanced geo-chronologically and with differently organised societies, but with rich evidence for sealing. The data reflects similar sealing practices at both sites, whereby objects such as doors, chests, baskets, pots, etc. were secured by a lump of clay impressed with a seal or seals (for the sealing practices at Lerna, see: Heath 1958; Heath Wiencke 1969; Maran and Kostoula 2014; for Phaistos, see: Fiandra 1968, 1975). The imprinted threads, cords and leather thongs (the latter only at Phaistos) were used to wrap pegs/knobs/poles and rims of jars or unidentified usually convex surfaces. Textiles, mats and baskets/wickerwork were used as coverings, wrappings and products that were sealed directly. However, the origin of some textile impressions is difficult to explain, especially the ones impressed at the edges of the cast, and requires further experimental research. These might possibly be impressions of clothes worn by the seal holders or, more likely, fabrics that may have been used to wrap wet clay to keep it moist and elastic for sealing (Fig. 17.2).

Each record comprises data fields identifying the object (i.e. the CMS cast and the seal impressed lump of clay), its context and dating, the impressed seals and the impressions of textile products: threads, cords, leather thongs, textiles, etc., recorded individually with all technical details as well as pictures (Table 17.2). The technical

[12] After my study-visits to CMS Heidelberg in 2019 and 2020, the number of casts with possible textile impressions still awaiting examination is estimated to c. 1600 examples.

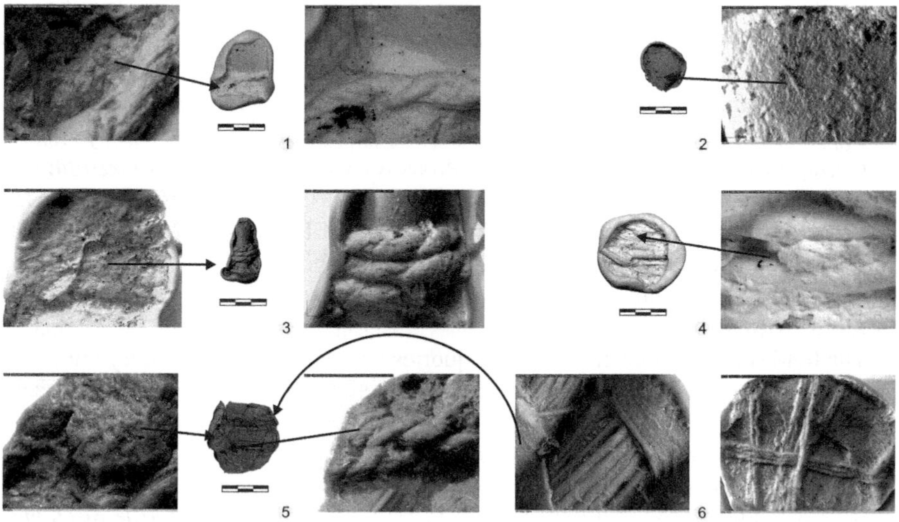

Fig. 17.2 Examples of various types of casts with textile impressions: (1) door sealing (LER 4.390), (2) jar covering (PH 764), (3) peg sealing (PH 790), (4) wickerwork covering with a possible textile imprint (LER 4.394), (5) basket or mat (PH 825o'), (6) *Päckchenplombe* (MAL 1403). Impressions of fabrics are indicated by arrows. (Casts – CMS Heidelberg, photos by A. Ulanowska)

Table 17.2 Individual data fields in the '*Impressions of textiles on clay*' module

Data fields identifying the specific object	Data fields identifying the impressed seal(s)	Data fields identifying qualities of the impressed textile product
Cast/Inv. no.	CMS no./nos	Type of impressed textile product
The CMS Arachne ID and link	Stylistic dating	Construction characteristics: plied/single twist direction twist angle thread/cord diameter=impressed width
Museum	Type of impressed seal	
Site	Motif(s) according to the CMS	
Archaeological context	Motif(s) according to Anastasiadou (2011)	
Context dating		
Remarks	Textile production-related motif(s)	Estimated no. of threads per cm
Type of the seal-impressed object	Inscriptions	Raw material
Type of the modern cast	Photo of the impressed seal face(s)	Production technique
Photo of the cast		Characteristic features
		Photo of each textile impression

characteristics of impressed threads, cords and textiles are documented according to general standards in textile archaeology, so that, whenever possible, the fields include data such as the diameter of the thread/cord, the production technique of an impressed product (e.g. spinning, drilling, weaving, basketry etc.), the direction of twist and twist angle, the number of threads per cm, and the suggested raw material group. Features reflecting the type of the sealed object (e.g., '*flat-based nodule*', '*stopper*'), binding/weave of the impressed product ('*faced tabby*', '*coiled basketry-like*'), its appearance ('*textile bulb*', '*fold-like feature*', '*wet clay-like*'), use-wear and the manner in which the sealed object was protected by cord (e.g., '*wrapped and tied*', '*cross-wrapping*') are recorded as '*characteristic features*'. The iconographic data in this module may be cross-referenced with the '*iconography of textile production*', e.g. through the '*woolly animals*' (CMS II,5 253/PH 703; CMS II,5 263/PH 705), '*spider*' (CMS V 115/ LER 4.448, LER 4.446; CMS II,5 313/PH 726) and '*interlaced band*' (CMS V 111/LER 4.358 and LER 4.348; CMS II,5 152/ PH 758 and PH 769) motifs on the impressed seals by choosing 'Yes' in the field 'Sealings with textile motifs'.

The fields of the search engine allow queries according to the inventory numbers preceded by the abbreviations for the site, CMS numbers, sites, type of impressed textile products, production techniques, raw materials and characteristic features. For search fields with several keywords/dictionaries, a scrolling list of all keywords in alphabetical order is provided. If the search word is typed, the available keyword(s) are suggested from the list. For the search fields '*type of impressed textile product*', '*raw material*', '*characteristic features*' and '*production techniques*' an inclusive disjunction OR can be chosen if the search is scheduled to encompass more data. For example, by choosing Lerna in the site field, and choosing OR for the type of impressed textile product field, all casts with impressed threads and/or cords from Lerna can be found in one search.

In all search results the requested data appears according to the following scheme: ID number in the "Textiles and Seals" database with the inventory number in brackets (this reference also serves as a link to the record), type of textile product, material, technique and construction characteristics of system one and system two, if recorded (Fig. 17.3). The latter is displayed in a row separated by 'III' graphic divisions. Minimum and maximum values for the twist angle (in grades) and diameters (in mm) are separated by '/'. If only one measurement was registered, its value is considered minimal and displays with a dash, e.g., a description of the thread/cord from the cast PH 772iB (record 1 in the database) displays as: '*Plied: Z/25-/1.74-2.38*', where '*Z*' is the twist direction of the plied cord, '*25-*' is the minimal twist angle, while '*1.74-2.38*' is the range between the maximum and minimum diameter of this thread/cord. The thumbnails are displayed for pictures of all seal impression(s) (if the same seal was impressed more than once, the thumbnails are multiplied accordingly), photos of the entire cast and each individual textile product impressed on it. Each search result can be printed or exported into an Excel.xls format, including an export of all the collected records, which is displayed in the search engine when none of the search fields is specified. This option is especially useful, since the database tools do not include the creation of charts.

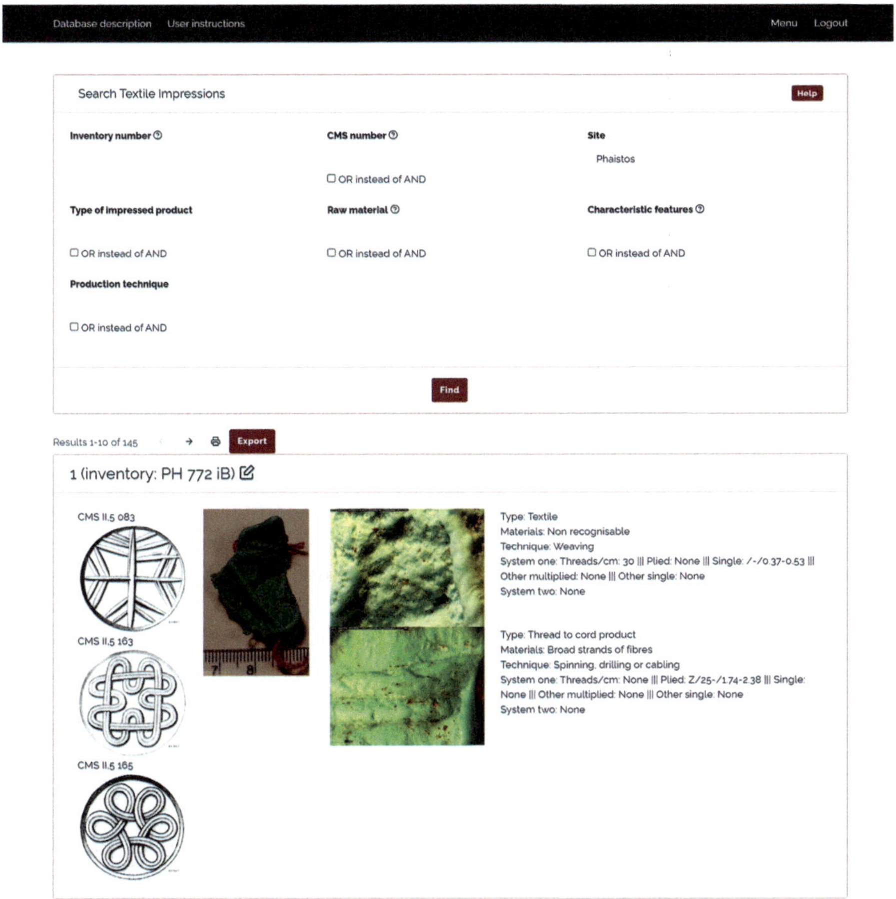

Fig. 17.3 Print screen of the search engine for the *'impressions of textile on clay'* module

17.3.3 'Iconography of Textile Production' Module

This module comprises 592 seals and impressions of seals (not seal faces as in the CMS Arachne database). The collected data predominantly documents textile production-related motifs on MBA Cretan seals and sealings (cf. Nosch and Ulanowska 2021), such as *'flax plant'*, *'combs'* and *'combers'*, *'spindle with whorl'* and *'spinners'*, *'loom weights'* and *'weavers'*, *'warp-weighted loom'*, *'rigid heddle'*, *'weft-beater'* and *'fabric with fringes'*. Motifs such as *'woolly animals'*, *'spiders'*, *'silk moths'* and *'interlaced bands'* are also present on seals dated to the earlier or later periods. The identifications which cannot be securely attested using the applied methodology are recorded as the names of motifs with an adjective *'possibly'*. Especially numerous is the category *'possibly woolly animals'* (203 seals), since distinguishing sheep and goats and other species of the Caprinae family, such as feral goats *agrimia* (a very popular motif in Aegean glyptic art), is often

Table 17.3 Individual data fields in the '*Iconography of textile production*' module

Data fields identifying the seal/seal impression	Data fields identifying the imagery of seal
CMS no	CMS face ID and Arachne link
Museum	Motif(s) according to the CMS
Site	Motif(s) according to Anastasiadou (2011)
Archaeological context	
Type of the seal-impressed object	Textile production-related motif(s)
Context dating	Elements of inscription
Stylistic dating	Inscription
Remarks	Photo of the seal face(s)

impossible.[13] The collected records come exclusively from the published evidence: the main sources are the CMS volumes and CMS Arachne, as well as CHIC, a monograph about MBA three-sided steatite prisms from Crete by M. Anastasiadou (2011) and more recent publications of seals from specific sites (e.g. Krzyszkowska 2012, 2017). As in other modules, each record comprises data fields identifying the seal or seal impression, its context and dating (contextual and stylistic), and imagery of the seal face (Table 17.3). A description of the imagery comprises references to the CMS motifs, motifs defined by M. Anastasiadou (2011), individual script signs and, if relevant, inscriptions. In the case of multi-faced seals, these data fields can be repeated for each seal face.

The fields allow for queries according to the CMS numbers, sites and stylistic dating (Fig. 17.4). The fields for iconographic searches, i.e. '*CMS motifs*', '*Anastasiadou's motifs*' and '*textile motifs*', are provided with a choice of an inclusive disjunction OR. All search fields with several keywords are provided with a scrolling list of the keywords in alphabetical order. The fields '*contains script sign*', '*sealing*', '*seal-impressed textile tool*' provide yes/no choices. In the search results, the CMS number of the seal/seal impression, which serves as a link to the record, is displayed together with thumbnail(s) of seal face(s). Therefore, if a specific textile production-related motif is recognised on one seal face of a multi-faced seal, all the other faces are displayed in such a search (Fig. 17.4). If the AND conjunction is chosen, e.g. for the '*loom weights*' and '*spider*' motifs, the search result shows all the seals with '*loom weights*' and '*spiders*' depicted on different seal faces of one seal. Each search result can be printed and there is a function for exporting data to Excel.

[13] The methodology for identification of '*woolly animals*' on seals has been developed together with the project collaborators: M.-L. Nosch, K. Żebrowska, K. Bigoraj and A. Gręzak, see Nosch et al. 2021. Features such as long spiral horns curved downwards, a short tail turned down, the head in profile with a bump, perhaps a mane, and occasionally long fleece shown at the neck or chest of the animal, are considered woolly sheep characteristics. Females of the domesticated goat, *Capra hircus* may be characterised by the specific shape of their udders and teats, occasionally rendered on seals. Other features of Caprinae, such as small beards, slightly curved horns with ridges or fluting that grow out from one spot, and short tails turned upwards may characterise all goats: domesticated, feral and wild.

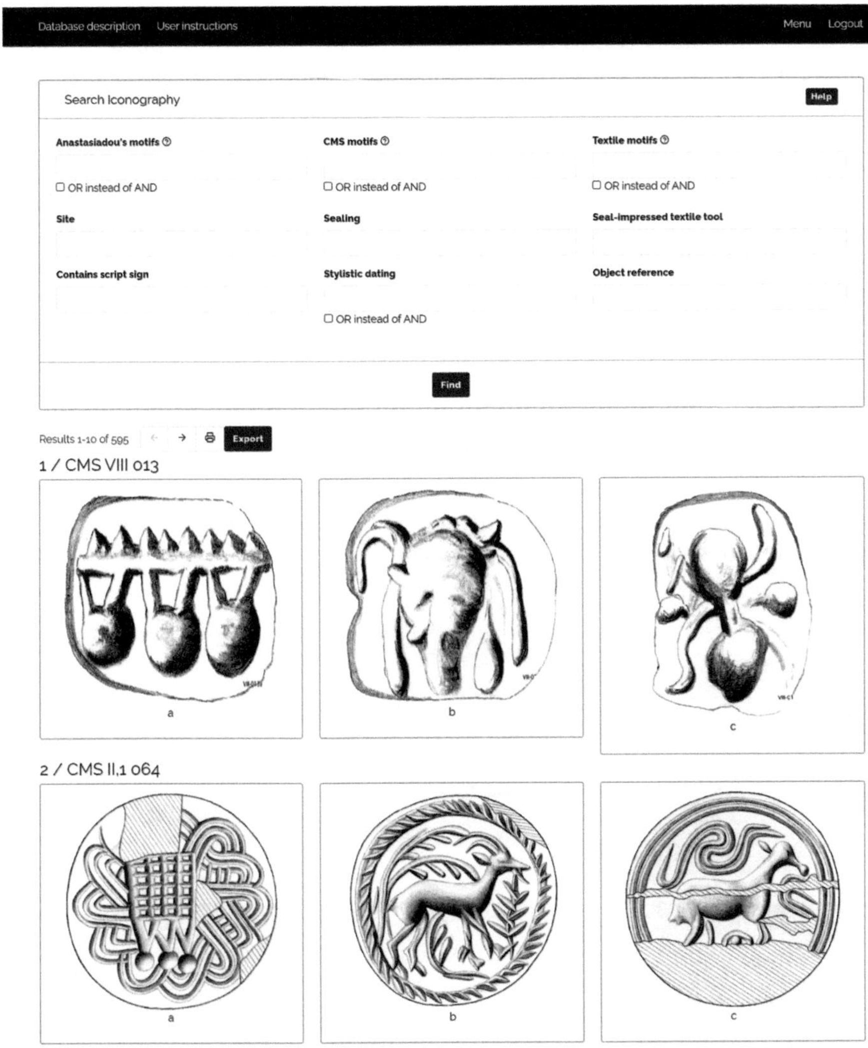

Fig. 17.4 Print screen of the search engine for the '*iconography of textile production*' module

17.4 Research Possibilities of the "Textiles and Seals" Database

Both published search engines have already proved to be useful tools for complex searches through data. The scope of data referring to imprints of textile products on the undersides of clay sealings is comprehensively recorded for the first time. The first statistical analyses of the data from Lerna and Phaistos show a certain uniformity in the spun products and textiles used in the sealing practices at both sites (cf.

Ulanowska 2020b). At Lerna, a total of 40 cords[14] was registered, including three dubious identifications (Fig. 17.5). Thirty seven are S-twisted or plied of single threads/cords twisted in z-direction or without observed twist. Diameter was measured for all cord imprints and it ranges from 2.41 to 7.03 mm, yet 29 cords fall into a cluster with diameter measures between 3–5 mm.[15] Twist angle, measured for 36 cord impressions, ranges from 19–53°. Both measurements suggest a certain standardisation of spun products used in the sealing practices at Lerna. This may possibly reflect technical uniformity of spun or drilled cords, but it can also imply the use of cords of similar quality for tying and wrapping. Analogous 'techno-functional' uniformity can be observed at Phaistos, however the range of products used for tying and wrapping at that side was wider. Cords still prevail with 85 examples, including seven dubious identifications, but alongside cords, finer threads (21 examples, including dubious identifications) and leather thongs (12 examples, including four dubious identifications) were also in use. As for the plied cords, S-twist still prevails (Fig. 17.5), however 11 examples are Z-plied. Seven impressions of cords and two impressions of threads possibly document braiding – a technique of production unknown from the Lerna evidence. Diameter was measured for 72 cord impressions and it ranges from 2.05 to 6.05 mm, however again, 62 cords fall into a

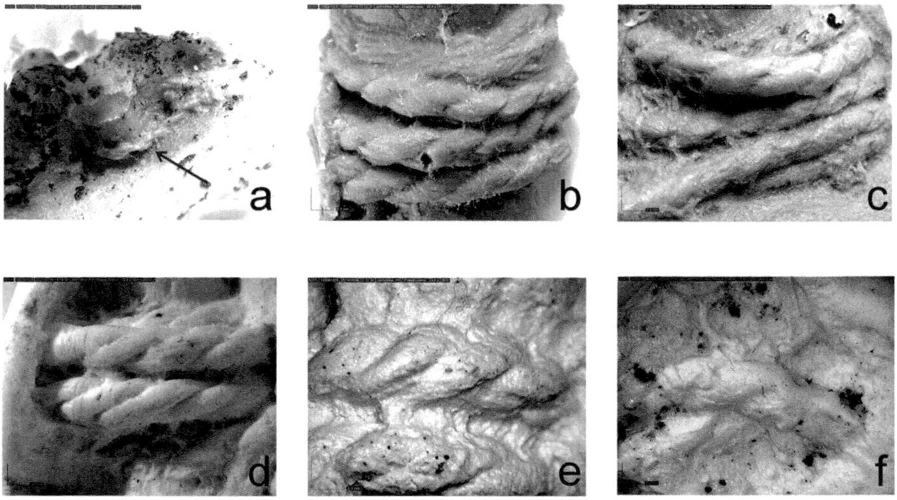

Fig. 17.5 Examples of impressions of threads and cords from Phaistos: (**a**) PH 855α?, diam. 0.75 mm; (**b**) PH 851δ′, diam. 3.01 mm; (**c**) PH 851γ, diam. 3.42 mm. Impressions of cords from Lerna: (**e**) LER 4.437, diam. 3.94 mm; (**f**) LER 4.375, diam. 4.31 mm; (**g**) L 4.374, diam. 5.24 mm. (Casts – CMS Heidelberg, photos by A. Ulanowska)

[14] Cords are spun products with a diameter between 2–8 mm. Threads are classified as products below 2 mm in diameter, while ropes have more than 8 mm in diameter (cf. Andersson Strand 2015: 48).

[15] For the accuracy of the measurements taken from textile impressions on clay, see Ulanowska 2020b.

cluster between 2–4 mm. Twist angle for cords was measured from 62 imprints ranging from 15–58°, but a cluster of narrower values between 25–45° has been recorded for 42 cord impressions. A comparison of the two clusters with the most frequent diameters implies that cords used at Lerna were generally thicker than the ones from Phaistos (Fig. 17.5).

More technical similarities are observed in textile impressions from Lerna (40 examples, including 24 dubious identifications) and Phaistos (119 examples, including 73 dubious identifications). The large number of dubious identifications is related to the nature of imprints preserved in small fragments, often impressed crumpled (Fig. 17.2). The only weave recognised on textile imprints is tabby. Thread dimeter was measured for 33 textile imprints from Lerna and 101 imprints from Phaistos, including three possible impressions of mats. The range of the diameters is wider at Phaistos ranging from 0.15–0.91 mm (0.19–0.56 mm at Lerna), however, on 73 possible textile impressions diameter of threads fall into a cluster between 0.2–0.45 mm. The thread count per cm could be estimated on the basis of short sections of 2–5 mm for 34 textile impressions from Phaistos and only four impressions from Lerna. Together, it ranges from 20–80 t/p/cm, but again, the majority of impressed fabrics (29 examples) falls into a narrower cluster between 30–50 t/p/cm. The first comparison of textile products used in the sealing practices at EBA Lerna and MBA Phaistos implies a more varied range of techniques and, generally, finer products in use at the latter site.

Moreover, the search possibilities of the module '*impressions of textiles on clay*' facilitate tracking potential correlations between the use of specific seals and a specific manner of tying or wrapping the sealed objects. This, in turn, could allow the verification of the original hypothesis of E. Fiandra about the sealing practices at Phaistos (1968, 1975). She has suggested a complex pattern in which a limited number of only 16 objects, such as pegs/knobs of doors, sides of chests, jars, etc., was subjected to repetitive sealing and opening over a short period of time, e.g. within a year, by, presumably, a limited number of seal owners. According to Fiandra (1968: 387), threads and cords used to secure these objects were very likely the same, i.e. the same cord was being used over and over again. Current evidence cannot exclude the possibility that several examples from Phaistos were indeed impressions of the same products. However, a brief comparison of cords characterised by the same diameter also demonstrates differences in their appearance, as well as in the manner of tying them around individual pegs/knobs (Fig. 17.6). Therefore, the preliminary analysis of data suggests a greater variety and number of textile products used in the sealing practices at Phaistos than previously assumed and, possibly, a greater variety of the sealed objects themselves.

The search engine for the module '*iconography of textile production*' allows searches focused on tracking textile production as a possibly distinct theme in the imagery of Aegean seals (cf. Nosch and Ulanowska forthcoming). It also facilitates the finding of possible repeating combinations of two textile production-related motifs on one seal face (e.g. '*comb*' and '*loom weights*' motifs, as on the seal published by Evans 1909: 132, fig. 71) and combinations of textile motifs with people (e.g. potential '*combers*', '*spinners*' and '*weavers*', cf. Ulanowska 2020b).

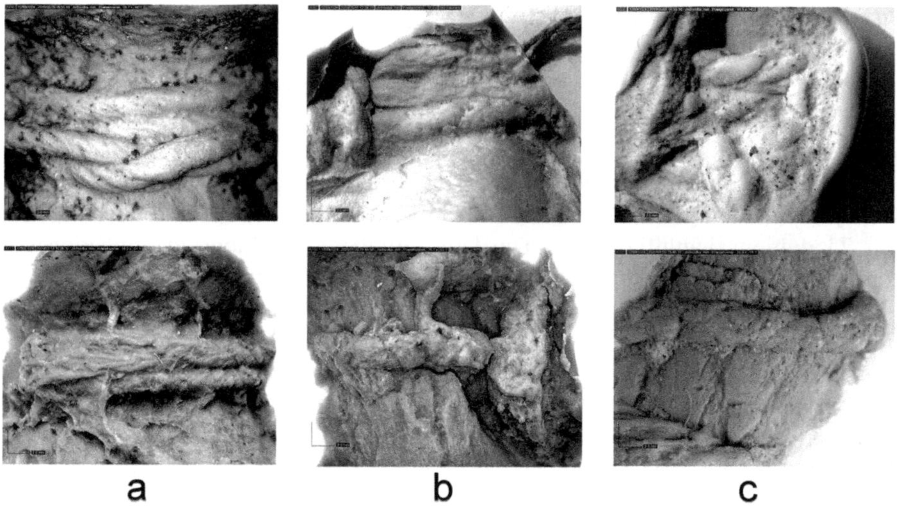

Fig. 17.6 Examples of cords impressions from Phaistos with the same diameters but different appearance and the manner of tying around the pegs/knobs or rims: (**a**) PH 735 (upper row) and PH 857γB, diam. 2.13 mm; (**b**) PH 674 (upper row) and PH 818γ?, diam. 2.65 mm; (**c**) PH 774 (upper row) and PH 789, diam. 2.91 mm. (Casts – CMS Heidelberg, photos by A. Ulanowska)

Additionally, it reveals repeating combinations of motifs that do not have an easily explicable relation to textile production, such as the so-called '*waterfowl*' motif.[16] The other possibility is a search for potentially repeating combinations of textile production-related motifs on different seal faces. Such combinations comprise e.g. depictions of '*loom weights*' and '*spiders*' on two different seal faces (nine examples, including three seals where a '*woolly animal*' motif is shown on the third face of a three-sided prism). It would be tempting to see such combinations as implying that textile production served as a 'meta-theme' for the imagery of an entire seal stone. However, it remains unclear whether multi-faced seals could be indeed interpreted as iconographic entities intended to convey a specific message or iconographic programme (cf. Anastasiadou 2011: 373–374).

17.5 Conclusions

Although still incomplete, the "Textiles and Seals" database and its search engines have already proved to be an efficient tool for the examination of the multifarious relations between textile production and seals and sealing practices in Bronze Age

[16]There are 12 recorded combinations of '*waterfowl*' with motifs such as '*loom weights*', '*spiders*', '*woolly animals*' and '*silk moth*' on one seal face, and 28 combinations where the '*waterfowl*' appears on a different seal-face. For the '*waterfowl*' motif, see Anastasiadou 2011: 184–185, Pl. 25–28.

Greece. By collecting various types of evidence (i.e. seal impressed textile tools, imprints of textile products on lumps of clay impressed by seals, and textile production-related imagery), the database facilitates comparative studies of published evidence (seal impressed tools and iconography) and new data coming from the first-hand analysis of the casts from the CMS Archive in Heidelberg (textile impressions). Several cross-references between the main modules of the database suggest the existence of possible semantic relations between the registered types of evidence. For example, textile tools that were occasionally impressed by seals potentially bearing textile-production related motifs, such as the spool-like tool from Malia, may suggest that a relationship between the image on a seal face and the use of a seal could, sometimes, exist. Similarly, the repetitive use of the same seals in direct object sealing, as observed at Lerna and Phaistos, could be related to the technique of wrapping or tying the sealed object and the parameters of impressed textile products. This can help to verify if the same objects were indeed sealed multiple times by individuals using similar or identical techniques for securing these objects by wrapping, tying, looping, etc. The comparison of textile impressions from Lerna and Phaistos implies different technical qualities of textile products at these two sites. This conclusion may seem obvious, given the distance in time and space between them. However, only now can it be supported by a significant quantity of data that reflects very similar technical uses of textiles in storage and in sealing practices.

Although several of the anticipated correlations and cross-references among the data are theoretically possible to track without a new database, in practice the amount of work required by such searches would potentially lead to important observations remaining unnoticed. The newly created search possibilities will open new paths for further questions and make the research undertaken in the ‘Textiles and Seals’ project more in-depth and specific. Moreover, open access to the database should allow further examination of data by scholars currently not involved in the project. Last but not least, creating an online database is a challenge that requires good communication between archaeologists and IT specialists: programmers and software developers. It also requires some elasticity in defining expectations on both sides, but the potential rewards for archaeological studies in and beyond any one particular project make such attempts and implementations worthwhile.

Acknowledgements I thank the Digital Competence Centre of the Univeristy of Warsaw team: Piotr Kasprzyk, Dominik Purchala, Ewa Serafin-Pursator, Tomasz Sporczyk and Wojciech Karpiński for their collaboration. I would also like to express my special thanks to Prof. Diamantis Panagiotopoulos and Dr Maria Anastasiadou for making the collection of casts in the CMS in Heidelberg available for my studies. The textile impressions on casts are published courtesy of the CMS in Heidelberg and Prof. Panagiotopoulos.

References

Alberti ME (2017) A measured world? Measures in Minoan daily life. In: Jasink AM, Weingarten J, Ferrara S (eds) *Non-scribal communication media in the Bronze Age Aegean and surrounding areas*. Periploi 9. Firenze University Press, Firenze, p 3–39.

Anastasiadou M (2011) *The Middle Minoan three-sided soft stone prism: A study of style and iconography*. CMS Beiheft 9. Philipp Von Zabern, Mainz.

Andersson Strand E (2015) The basics of textile tools and textile technology – from fibre to fabric. In: Andersson Strand E, Nosch M-L (eds) *Tools, textiles and contexts: Investigating textile production in the Aegean and Eastern Mediterranean Bronze Age*. Ancient Textiles Series 21. Oxbow Books, Oxford and Philadelphia, p 39–60.

Andersson Strand E, Nosch M-L (2015a) *Tools, textiles and contexts: Investigating textile production in the Aegean and Eastern Mediterranean Bronze Age*. Ancient Textiles Series 21. Oxbow Books, Oxford and Philadelphia.

Andersson Strand E, Nosch M-L (2015b) Introduction to the CTR database. In: Andersson Strand E, Nosch M-L (eds) *Tools, textiles and contexts: Investigating textile production in the Aegean and Eastern Mediterranean Bronze Age*. Ancient Textiles Series 21. Oxbow Books, Oxford and Philadelphia, p 145–152.

Breniquet C (2008) *Essai sur le tissage en Mésopotamie des premières communautés sédentaires au milieu du IIIe millénaire avant J.–C.* Travaux de la Maison René-Ginouvès 5. De Boccard, Paris

Burke B (2010) *From Minos to Midas: Ancient cloth production in the Aegean and in Anatolia*. Ancient Textiles Series 7. Oxbow Books, Oxford and Oakville.

CHIC, Olivier J-P, Godart L (1996) *Corpus hieroglyphicarum inscriptionum Cretae*. Études crétoises 32. De Boccard, Paris.

CMS, *Corpus der minoischen und mykenischen Siegel*. Volumes 1–13. Beihefte 1–10, Accessed 17 Apr 2020.

Cutler J (2016) Fashioning identity: weaving technology, dress and cultural change in the Middle and Late Bronze Age southern Aegean. In: Gorogianni E, Pavúk P, Girella L (eds) *Beyond thalassocracies. Understanding processes of Minoanisation and Mycenaeanisation in the Aegean*. Oxbow Books, Oxford and Philadelphia, p 172–185.

Del Freo M, Nosch M-L, Rougemont F (2010) The terminology of textiles in the Linear B tablets, including some considerations on Linear A logograms and abbreviations. In: Michel C, Nosch M-L (eds) *Textile terminologies in the Ancient Near East and Mediterranean from the third to the first millennia BC*. Ancient Textiles Series 8. Oxbow Books, Oxford and Oakville, p 338–373.

Evans A (1909) *Scripta Minoa: the written documents of Minoan Crete with special reference to the archives of Knossos*, vol. I. The Claredon Press, Oxford.

Fiandra E (1968) A che cosa servivano le cretule di Festòs. In: Panagiotaki NM (ed.) *Pepragména tou B′ diethnoús kritologikoú synedríou*. En Athínais, Athína, p 383–397.

Fiandra E (1975) Ancora a proposito delle cretule di Festòs: connessione tra i sistemi amministrativi centralizzati e l'uso delle cretule nell'età del bronzo. *Bollettino d'Arte* 60:1–25.

Heath MC (1958) Early Helladic clay sealings from the House of the Tiles at Lerna. *Hesperia* 27(2):81–121.

Heath Wiencke M (1969) Further seals and sealings from Lerna. *Hesperia* 38(4):500–521.

Jasink AM (2009) *Cretan hieroglyphic seals: A new classification of symbols and ornamental/filling motifs*. Biblioteca di Pasifae VII. Fabrizio Serra, Pisa and Roma.

Jones BR (2015) *Ariadne's threads: The construction and significance of clothes in the Aegean Bronze Age*. Aegaeum 38. Peeters, Leuven and Liège.

Karnava A (2019) The inscribed loom weight. In: Nikolakopoulou I. (ed.) *Akrotiri, Thera. Middle Bronze Age: Pottery and stratigraphy, vol. I. Stratigraphy, ceramic typology and technology, weaving equipment*. The Archaeological Society at Athens Library series 318. The Archaeological Society Foundation, Athens, p 501–504.

Killen JT (1984) The textile industries at Pylos and Knossos. In: Shelmerdine CW, Palaima TG (eds) *Pylos comes alive: Industry and administration in a Mycenaean palace*. Archaeological Institute of America, New York, p 49–63.

Killen JT (2007) Cloth production in Late Bronze Age Greece: the documentary evidence. In: Gillis C, Nosch M-LB (eds) *Ancient textiles: production, craft and society. Proceedings of the first international conference on ancient textiles, held at Lund, Sweden, and Copenhagen, Denmark, on March 19–23, 2003*. Ancient Textiles Series 1. Oxbow Books, Oxford, p 50–58.

Krzyszkowska O (2005) *Aegean seals: An introduction*. Bulletin of the Institute of Classical Studies supplement 85. Institute of Classical Studies, School of Advanced Study, University of London, London.

Krzyszkowska O (2012) Seals from the Petras cemetery: a preliminary overview. In: Tsipopoulou M (ed.) *Petras, Siteia – 25 years of excavations and studies. Acts of a two-day conference held at the Danish Institute at Athens, 9–10 October 2010*. Monographs of the Danish Institute at Athens 16. The Danish Institute at Athens, Athens, p 145–160.

Krzyszkowska O (2017) Further seals from the cemetery at Petras. In: Tsipopoulou M (ed.) *Petras, Siteia. The Pre- and Proto-Palatial cemetery in context. Acts of a two-day conference held at the Danish Institute at Athens, 14–15 February 2015*. Monographs of the Danish Institute at Athens 21. The Danish Institute at Athens, Athens, p 143–157.

Manning SW (2010) Chronology and terminology. In: Cline EH (ed.) *The Oxford handbook of the Bronze Age Aegean*. Oxford University Press, Oxford, p 11–28.

Maran J, Kostoula M (2014) The spider's web: Innovation and society in the Early Helladic 'Period of the Corridor Houses'. In: Galanakis Y, Wilkinson T, Bennet J (eds) *AΘYPMATA. Critical essays on the archaeology of the Eastern Mediterranean in honour of E. Susan Sherratt*. Archaeopress, Oxford, p 141–158.

Michailidou A (1990) The lead weights from Akrotiri: The archaeological record. In: Hardy DA, Doumas CG, Sakellarakis JA, Warren PM (eds) *Thera and the Aegean World III. Proceedings of the third international congress, Santorini, 3–9 September 1989*. Thera Foundation, London, p 407–419.

Militello PM (2018) An einer Stange hängene Gefässe. Nota sun un motivo sfragistico. In *Pepragména IA' diethnoús kritologikoú synedríou (Réthymno 21–27 Oktovríou 2011)*, vol. A1.3. Istorikí kai Laografikí Etaireía Rethýmnis, Réthymno, p 321–333.

Müller W (1999) Die Tonplomben und andere gestempelte Tonobjekte. In: Müller W, Pini I, Platon N (eds) *Iraklion Archäologisches Museum: die Siegelabdrücke von Aj. Triada und anderen zentral- und ostkretischen Fundorten, unter Einbeziehung von Funden aus anderen Museen*. CMS II,6. Gebr. Mann Verlag, Berlin, p 339–400.

Müller W (2010) The CMS-Database in Arachne. In: Müller W (ed.) *Die Bedeutung der minoischen und mykenischen Glyptik. VI. Internationales Siegel-Symposium, Marburg, 9.–12. Oktober 2008*. CMS Beiheft 8. Philipp von Zabern, Mainz am Rhein, p 427–446.

Müller W, Pini I (1997) Die ‚Schnüre' in den Plomben und die Gegenstandsabdrücke. In: Müller W, Olivier J-P, Pini I, Sakellariou A (eds) *Die Tonplomben aus dem Nestorpalast von Pylos*. Verlag Philipp Von Zabern, Mainz, p 67–69.

Nosch M-L (2012) The textile logograms in the Linear B tablets: les idéogrammes archéologiques des textiles. In: Carlier P, de Lamberterie C, Egetmeyer M, Guilleux N, Rougemont F, Zurbach J (eds) *Études Mycéniennes 2010. Actes du XIIIe colloque international sur les textes égéens, Sèvres, Paris, Nanterre, 20–23 septembre 2010*. Biblioteca di Pasiphae 10. Fabrizio Serra, Pisa and Rome, p 303–346.

Nosch M-L (2014) Mycenaean wool economies in the later part of the second millennium BC Aegean. In: Breniquet C, Michel C (eds) *Wool economy in the Ancient Near East and the Aegean: From the beginnings of sheep husbandry to institutional textile industry*. Ancient Textile Series 17. Oxbow Books, Oxfrord and Philadelphia, p 371–400.

Nosch M-L, Laffineur R (eds) (2012) *KOSMOS. Jewellery, adornment, and textiles in the Aegean Bronze Age. Proceedings of the 13th international Aegean conference, University of Copenhagen, Danish National Research Foundation's Centre for Textile Research, 21–26 April 2010*. Aegaeum 33. Peeters, Leuven and Liège.

Nosch M-L, Ulanowska A (2021) The materiality of the Cretan Hieroglyphic script: Textile production-related referents to hieroglyphic signs on seals and sealings from Middle Bronze Age Crete. In: Boyes PJ, Steele PM, Astoreca NE (eds) *The social and cultural contexts of historic writing practices*. Contexts of and relations between writing systems 2. Oxbow Books, Oxford and Philadelphia, p 73–100.

Nosch M-L, Ulanowska A, Żebrowska K, Bigoraj K, Gręzak A (2021) Sheep – 'a Factory without Waste'. Comparative, Interdisciplinary and Diachronic Views on Sheep in the Aegean Bronze Age. In: Laffineur R, Palaima TG (eds) *ZOIA. Animal-human interactions in the Aegean Middle and Late Bronze Age*. Aegaeum 45, Peeters, Leuven and Liège, p 35–49.

Palaima T (ed.) (1990) *Aegean seals, sealings and administration. Proceedings of the NEH-Dickson conference of the program in Aegean scripts and prehistory of the Department of Classics, University of Texas at Austin, January 11–13, 1989*. Aegaeum 5. Université de Liège, Liège.

Panagiotakopulu E, Buckland PC, Day PM, Doumas C, Sarpaki A, Skidmore P (1997) A lepidopterous cocoon from Thera and evidence for silk in the Aegean Bronze Age. *Antiquity* 71:420–429.

Rougemont F (2014) Sheep rearing, wool production and management in Mycenaean written documents. In: Breniquet C, Michel C (eds) *Wool economy in the Ancient Near East and the Aegean: From the beginnings of sheep husbandry to institutional textile industry*. Ancient Textiles Series 17. Oxbow Books, Oxford and Philadelphia, p 340–370.

Shaw MC, Chapin AP (2016) *Woven threads: Patterned textiles of the Aegean Bronze Age*. Ancient Textile Series 22. Oxbow Books, Oxford and Philadelphia.

Siennicka M (2015–2016) Étude d'empreintes de textile. In: Travaux de l'Ecole Française d'Athènes en 2014 et en 2015. Philippa-Touchais A, Papadimitriou N, Touchais G (eds) La Deiras. *Bulletin de Correspondance Héllenique* 139–140:837–838.

Siennicka M, Rahmstorf L, Ulanowska A (eds) (2018) *First textiles: The beginnings of textile manufacture in Europe and the Mediterranean. Proceedings of the EAA session held in Istanbul (2014) and the 'First Textiles' conference, Copenhagen (2015)*. Ancient Textiles Series 32. Oxbow Books, Oxford and Philadelphia.

Skals I, Möller-Wiering S, Nosch M-L (2015) Survey of archaeological textile remains from the Aegean and Eastern Mediterranean. In: Andersson Strand E, Nosch M-L (eds) *Tools, textiles and contexts: Investigating textile production in the Aegean and Eastern Mediterranean Bronze Age*. Ancient Textiles Series 21. Oxbow Books, Oxford and Philadelphia, p 61–74.

Spantidaki Y, Moulherat Ch (2012) Greece. In: Gleba M, Mannering U (eds) *Textiles and textile production in Europe: From Prehistory to AD 400*. Ancient Textiles Series 11. Oxbow Books, Oxford and Oakville, p 185–200.

Ulanowska A (2016) Representations of textile tools in Aegean glyptic: Cuboid seal from the tholos tomb A in Aghia Triada. In: Militello P, Żebrowska K (eds) *Sympozjum Egejskie. Proceedings of the 2nd students' conference in Aegean archaeology: methods – researches – perspectives, Institute of Archaeology, University of Warsaw, Poland, April 25th, 2014*. Syndesmoi 4. Università degli Studi di Catania, Catania, p 109–125.

Ulanowska A (2017) Textile technology and Minoan glyptic: Representations of loom weights on Middle Minoan prismatic seals. In: Żebrowska K, Ulanowska A, Lewartowski K (eds) *Sympozjum Egejskie. Papers in Aegean Archaeology*, vol. 1. Institute of Archaeology, University of Warsaw, Warsaw, p 57–66.

Ulanowska A (2020a) Textiles and seals: Recognising the relationship between textile production and seals, and sealing practices in Bronze Age Greece. *Revue de l'Archéologie du Vêtement et du Costume* 1:1–25.

Ulanowska A (2020b) Textile uses in administrative practices in Bronze Age Greece: New evidence of textile impressions from the undersides of clay sealings. In: Bustamante-Álvares M, Sánchez López EH, Jiménez Ávila J (eds) *Redefining ancient textile handcraft, structures, tools and production processes*. PURPUREAE VESTES VII. Universidad de Granada, Granada, p 413–424.

Ulanowska A (2020c) Towards engendering textile production in Middle Bronze Age Crete. In: Michel C, Harlow M, Quillien L (eds) *Textiles and gender in Antiquity: From the Orient to the Mediterranean*. Bloomsbury Academic, London, p 59–79.

Vakirtzi S, Georma F, Karnava A (2018) Beyond textiles: Alternative uses of twisted fibres and evidence from Akrotiri, Thera. *Światowit* LVI:75–88.

Vlasaki M (2010) *Iris cretica* and the Prepalatial workshop of Chamalevri. In: Krzyszkowska O (ed.) *Cretan offerings. Studies in honour of Peter Warren*. British School at Athens 18. The British School at Athens, London, p 359–366.

Weingarten J, Crouwel JH, Prent M, Vogelsang-Eastwood G (1999) Early Helladic sealings from Geraki in Lakonia, Greece. *Oxford Journal of Archaeology* 18(4):357–376.

Agata Ulanowska is Assistant Professor at the Faculty of Archaeology, University of Warsaw, Poland. Her research interests focus on textile production and technology, experimental archaeology, and seals and sealing practices in Bronze Age Greece. She was awarded two grants of the National Science Centre of Poland, for the projects: "Textile production in Bronze Age Greece – comparative studies of the Aegean weaving techniques" (2015–2017), and "Textiles and Seals. Relations between textile production and seals and sealing practices in Bronze Age Greece" (2018–2021). She is a chair of the COST Action CA 19131 "EuroWeb. Europe Through Textiles: Network for an integrated and interdisciplinary Humanities".

Chapter 18
3D Visualization of the 2400-Year-Old Garments of Salt Man 4 from Chehrābād, Iran

Aleksei Moskvin, Karina Grömer, Mariia Moskvina, Victor Kuzmichev, Thomas Stöllner, and Abolfazl Aali

Abstract This article presents a digital reconstruction of Salt man 4 (405–380 BCE), a salt miner's mummy found in the Douzlākh salt mine (Chehrābād, northern Iran). The mummy and its costume are relatively well preserved because the salt mine offers good preservation conditions for organic finds. Salt man 4 is unique in that his garments are a complete set of worker's gear. This offers an exceptional opportunity to study garments worn during heavy physical work in ancient times. An international research team has been conducting an in-depth study of the mummy since 2004. The results include measurements, photographs, MRI (magnetic resonance imaging) scans and descriptions, which taken together provide a good basis for a digital reconstruction. This article describes technical features of the reconstruction process and challenges that were solved to reproduce the garments. The reconstruction opens new ways of studying the costume in virtual reality using computer graphics and computer-aided design tools to obtain further insights into ancient mining clothes within the broader area of digital humanities.

A. Moskvin · M. Moskvina
Department of Clothes Design and Technology, Saint Petersburg State University of Industrial Technologies and Design, Saint-Petersburg, The Russian Federation

K. Grömer (✉)
Department of Prehistory, Naturhistorisches Museum, Vienna, Austria
e-mail: karina-groemer@nhm-wien.ac.at

V. Kuzmichev
Department of Clothes Design, Ivanovo State Polytechnic University, Ivanovo, The Russian Federation

T. Stöllner
Institute for Archaeological Science Ruhr-University Bochum, German Mining Museum Bochum, Bochum, Germany
e-mail: thomas.stoellner@bergbaumuseum.de

A. Aali
Archaeological Museum of Zanjan, Zanjan, Iran

A. Ulanowska et al. (eds.), *Ancient Textile Production from an Interdisciplinary Perspective*, Interdisciplinary Contributions to Archaeology,
https://doi.org/10.1007/978-3-030-92170-5_18

Keywords Chehrābād salt mine · Achaemenid period · Salt men of Iran · Archaeological textiles · Digital reconstruction

18.1 Introduction

The Douzlākh salt mine Chehrābād in northern Iran offers good preservation conditions for organic finds, such as wood, botanical remains, human faeces, textiles, ropes, leather and fur objects and human mummies (Aali and Stöllner 2015). More than 15% of all artefacts found in the waste layers of the salt mine are textiles, most of them covering roughly a timespan between c. fifth century BCE to the sixth century CE, which refers to the Achaemenid and Sassanid period of the Persian Empires.

Within the framework "Saltmen Iran 2" (Deutsche Forschungsgesellschaft) and "Patrimonies Project Gerda Henkel Stiftung", textile analysis was done with the expertise of salt mine textile research from the Natural History Museum Vienna (Grömer and Bagherpour 2018). Also, within this framework an interdisciplinary cooperation to make virtual 3D reconstructions of the salt mine garments was carried out.

Computer technologies provide researchers with a wide array of versatile tools for reproducing and studying archaeological artifacts in virtual reality (VR). Virtual archaeology improves our understanding of archaeological textiles (Cybulska 2012; Cybulska et al. 2013; Cybulska and Mianowska-Orlińska 2015) and offers valuable insights into features and construction of ancient clothing (Kuzmichev et al. 2018b; Moskvin et al. 2019; Wijnhoven and Moskvin 2020).

This study aims to reconstruct the garments and accessories worn by Salt man 4 in VR. Salt man 4 (14C-date: 405–380 cal. BCE, after Pollard et al. 2008: 140–141) was found during regular excavations in 2004 (Aali et al. 2012: 73–74). Six salt mummies have been found to date, of which Salt man 4 is the most well-preserved. As he died accidently in the mine, he still wears the garments and all items he had on his person during his work in the mine. To provide a basis for further study of the costume, we reproduced the process of weaving, dyeing, sewing and putting on the garments in VR. Computer graphics (CG) and computer-aided design (CAD) tools were applied to create a lifelike reconstruction which behaves under the influence of external forces, such as gravity and the avatar's movements, practically the same as real garments. Because the Douzlākh salt mine offers unique conditions for organic finds, special attention was given to textile materials.

18.2 Case Study: Salt man 4

18.2.1 Overview

The Achaemenid garments here are exemplified by the garments worn by Salt man 4 (Fig. 18.1a), as they represent the most complete ones. He was a young man at the age of c. 16 years. His completely intact clothing is accompanied by further objects, such as two small pottery jars worn on his belt, a fur cloak and a knife in a leather sheath. The salt man (Fn 987) had trousers, an upper garment ('tunic') and leather shoes (Aali et al. 2012: 73–74). He had no head gear to cover his short cut brown hair, but he wore two silver earrings. There is a woven belt and a bast cord under the upper garment; they might be in functional use together with the trousers, but as the belt is completely covered by the upper garment, this is not clear.

Similar fabrics have been used to make the upper garment and the trousers of salt man 4. In both cases, it is a dense, weft-faced tabby of natural light sheep wool, woven with 0.4–0.7 mm single yarn and thread count of 7–8 threads per cm in the warp and up to 22–30 in the weft. Table 18.1 shows the technical details of the fabrics, which are published for the first time in this chapter. Measurements of the technical details of the textile were carried out with a DinoLite Digital Microscope at the Archaeological Museum Zanjan, Iran, in October 2015. There, thread diameter, fine structures of the weave and details of patterns and seams were documented

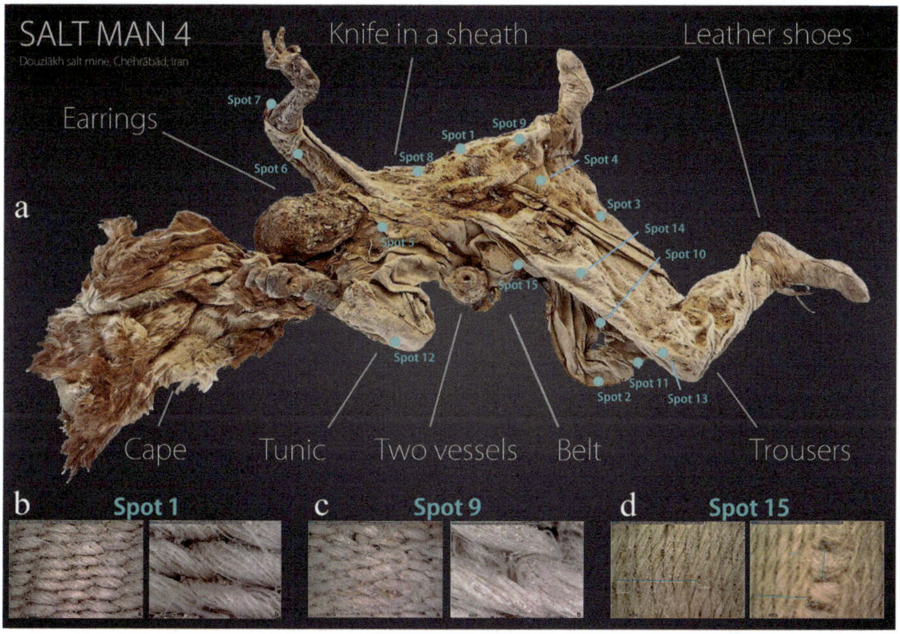

Fig. 18.1 Salt man 4: (**a**) the mummy and location of spots for measuring parameters of textiles. (Photo by A. Aali); (**b**) details of the fabrics. (DinoLite Microscope 50x and 200x, by K. Grömer)

Table 18.1 Parameters of Salt man 4 textiles[a]

	Tunic		Trousers		Belt	
	Warp	Weft	Warp	Weft	Warp	Weft
Yarn (y)/plied yarn (pl)	y	y	y	y	pl	y
Twist direction	s	z	s	z	Z	z
Twist angle (degrees)	50	20	50	20–30	20	30–40
Thread thickness (mm)	0.7–0.8	0.5–0.6	0.5–0.6	0.5	0.5	1.8
Thread count (threads per cm)	8	30	7–8	22	c. 28	5
Weave type	Weft-faced tabby		Weft-faced tabby		Dense warp-faced repp, round woven	
Colour (NCS-Code)	Most part (presumably the original colour): S3010-Y40R		Most part (presumably the original colour): S1005-Y20R		Visible part: S1005-Y	
	Sleeves and chest area: S1515-Y20R		Lower legs area: S1515-Y20R			
	Red cords: S 2060-Y70R		Red cords: S 2060-Y70R			

[a]Measurements by K. Grömer

by K. Grömer. Fifteen spots on the surface of the tunic, trousers and belt were chosen to measure the parameters (see Fig. 18.1a). Thanks to the good preservation, different technical features can be observed very easily: weave type and technical details such as thread count, yarn diameter, use of plied or single yarn, twist direction (s or z), weaving errors and seams can be studied.

18.2.2 *Upper Garment*

The tunic of Salt man 4 (Fn 987-11) is knee-length and has a straight neckline and long sleeves that reach the middle of the lower arms. The size is difficult to determine, because the tunic is still worn by the mummy, and we cannot gain clear measurements. An estimated length of the garment is 0.95 m, the length of the sleeves is 34 cm. The width of the garment at the bottom cannot be identified. On the right shoulder, a cord was attached to the slit to close the garment at the neck opening. The tunic consists of light beige wool and is decorated with red cords along its edges and seams.

The construction of the garment could easily be identified as it consists of a large, cross-shaped fabric that was woven to shape (Fig. 18.2a). The start was one sleeve, where the fabric was 34.5 cm wide. After c. 34 cm, the fabric was made wider, reaching c. 190 cm wide. (Approximations due to measuring directly on the mummy.) After the addition of warp threads, the main part of the tunic was woven. For the neck-opening, a slit was left open while weaving the main part. The slit was

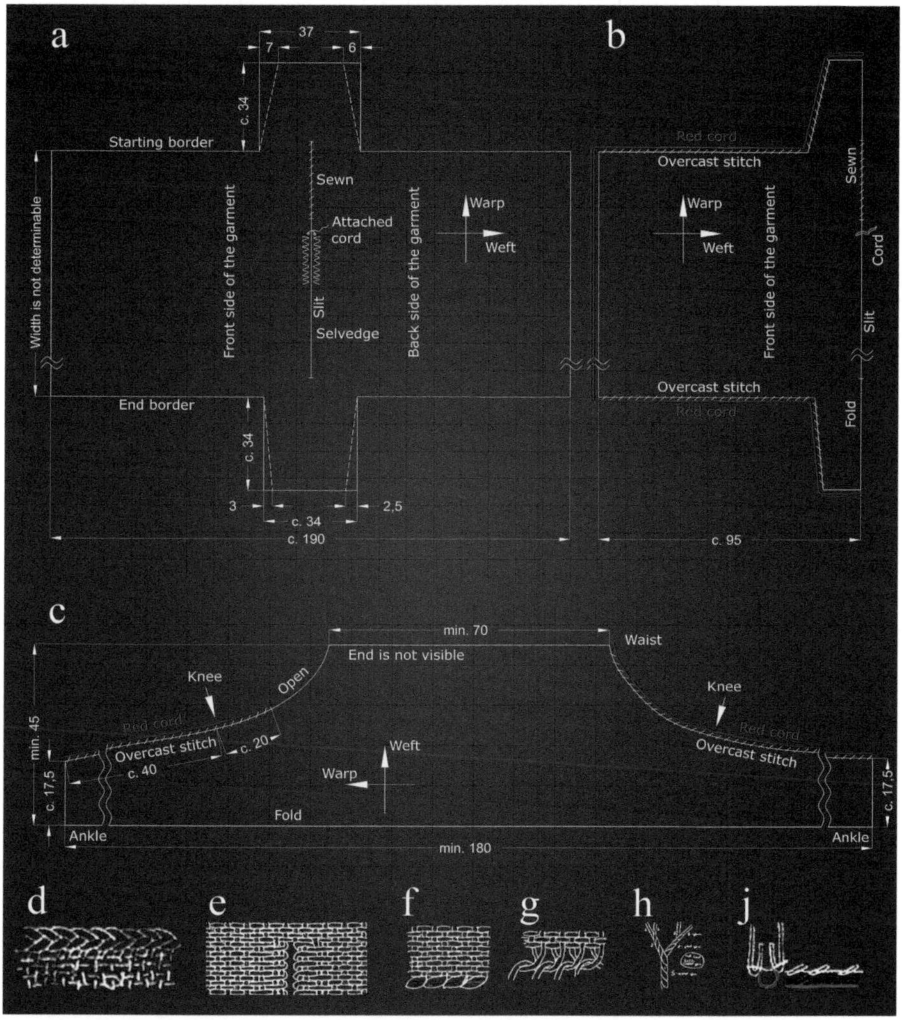

Fig. 18.2 Salt man 4, construction of the garments: (**a**) a woven-to-shape fabric of the tunic with constructive details which were designed during weaving; (**b**) folded and sewn tunic; (**c**) folded and sewn trousers; (**d**) starting border in the tunic; (**e**) woven-in slit in the centre of the tunic; (**f–g**) finishing border on the tunic; (**h**) attached red cord, schematic drawing of the cordage; (**j**) seam with overcast stitch and attached red cord. All measurements are given in cm. (Drawings by K. Grömer)

woven-in as the fabric was woven. On the fabric parts above and below, the weft thread was inserted into the shed along the whole width. To make the slit itself, the fabric was divided into two parts which were woven with separate weft threads and in the area of the slit, the weft threads formed a selvedge. To add strength to the edge, the four outer threads were taken together as paired. After the end of the slit, the weft threads were once again inserted through the full width of the fabric. To weave the second sleeve, the width of the whole fabric was reduced. In this case, the sleeve was 37 cm wide (i.e. a little wider than the first one). The slit was continued into the sleeve.

Finally, to sew the garment, the piece of fabric was folded and closed by two seams (Fig. 18.2b). The sleeves are not sewn strictly along the woven selvedge, but the width was reduced toward the wrist. Additionally, a decorative red cord was attached to the seams. The red cord coming from the sleeve seam was then also sewn onto the woven edge of the sleeve.

The tunic was repaired several times. On several parts of the garment damages can be found. Due to mechanical stress, the surface of the textile was rubbed off, the low spun and thinner weft threads faded and the stronger warp threads survived. Heavy use-wear is recognizable on the sleeves, especially on the part that face the palm of the wearer's hand. The use-wear on the lower part of the backside is not as intense. There is also use-wear on the neckline, caused by mechanical stress from taking the garment on and off. This damage has been met by repairing.

18.2.3 Trousers

The trousers of Salt man 4 (Fn 987-10) were made of beige weft-face woollen tabby of a very dense kind. Similar to the upper garment, the seams of the trousers are covered by a red cord. It is again difficult to gain clear measurements, because the garment is still worn by the salt mummy. The width at the waist is c. 70–75 cm, the length is about 80 cm (c. 40 cm from the knees downward).

The trousers are very wide in the pelvis region and between the legs. The ends of the ankle-long trousers are stuffed into the shoes, but there are no stockings directly attached to the trousers. In the slit of the shoe, the end of the trousers could be seen. The construction details of the trousers in the waist region (on the belt loop) could not be determined, because it is completely covered by the upper garment. The woven belt band, worn under the upper garment, might be motivated by a functional context, i.e. serving to hold the trousers.

The cut pattern and construction type of the trousers (Fig. 18.2c) are also interesting (Grömer and Aali 2016). The overall shape was long rectangular with an extended middle part. As the items are still worn by the mummy, it could not be determined if the edges of the fabric were cut or woven. Therefore, we cannot estimate if this garment was also woven to this distinctive shape or cut. The item was folded along the long side (Fig. 18.2b) and the fold forms the part of the inner feet. The seams have been carried out as top sewing in overcast stitch. During carrying out the seam, the red cord was attached to the sewn edge. The sewing thread went through the attached red cord, so the seam was covered by the cord. On the right side only c. 60 cm was sewn, the rest was left open to ensure the instep. On the other leg the whole side seam was closed completely. The trousers do not have any marks of repair, but some pilling can be seen, especially in the area between the legs.

18.2.4 Belt

Although the cloth types and the construction types of the garments associated with salt man 3–5 seem to be quite similar, the belts are all different. Salt man 4 had a 3.5 cm wide band (Fn 987-12) around the waist (Fig. 18.6). It might have been used for holding the trousers as well as the two pots, as the belt has a connection to the pot via the dark brown cord. The belt was made in very dense wool repp, but round-woven (as a tube). On its front side, a line of stretched weft threads appeared, which also has a decorative effect. The tube of the belt was filled at the end with some substance to get a bobbin-like three-dimensional end. Salt man 4 also wore a bast cord (Fn 987-7) under the upper garment, which might be in functional use together with the trousers, but since the belt is completely covered by the upper garment, this is not clear.

18.3 3D Reconstruction

18.3.1 Approach

In this study we applied CAD and CG tools to reproduce the process of clothes making in VR (Kuzmichev et al. 2017, 2018a; Moskvin et al. 2020). The key steps were as follows:

1. Digital reconstruction of textile materials. In this step, the appearance of the fabrics was reproduced. To achieve this, we generated texture maps for each garment. Texture maps are a set of special images which contain information about the characteristics of the surface, such as colour, reflectivity and glossiness. We used procedural textures, which were created using a mathematical description and modifiable parameters;
2. Reproduction of the garments laid out flat. The cut patterns of the tunic, trousers and belt are relatively simple. Each of these garments can be laid out flat. Thus, we recreated the three items of salt man's clothing laid out flat prior to the subsequent virtual try-on;
3. Modelling of an avatar based on dimensions of the mummy (an avatar is a virtual representation of the human body);
4. Virtual try-on. In the last step of the reconstruction, we put the trousers, tunic and belt on the avatar using the physically based cloth simulation.

18.3.2 Software

Four software programs, Substance Designer, Make Human, Clo3D and Blender, were used to reconstruct Salt man 4. The former, Substance Designer (Adobe 2020a), was applied to generate texture maps. Texture maps can be created in a wide range of raster graphics editors (Adobe Photoshop, GIMP and Krita) and vector graphics editors (Adobe Illustrator, CorelDraw and Inkscape). The main reason for using Substance Designer is that the software is a specialized CG material creation tool. The program allows users to create procedural textures with a real-time preview of the material.

MakeHuman (The MakeHuman team 2020) was employed to create an avatar of Salt man 4. Many software programs based on 3D morphing technology, such as iClone, Daz Studio, Poser and Adobe Fuse, can be utilized as a substitute to MakeHuman. However, MakeHuman uses up to 250 body and head morphs (a morph is a special command that seamlessly changes one shape into another) alongside 20 anthropometric measurements to generate anatomically correct virtual human bodies.

Clo3D (CLO Virtual Fashion 2020) was used to reproduce the garments laid out flat and put them on the avatar. The program can be substituted with Marvelous Designer, Optitex, vStitcher and Vidya.

And lastly, Blender (Blender Foundation 2020) was used to model the accessories and generate photorealistic images of the reconstructed costume. As an alternative, other 3D graphics programs, such as 3DSMax, Maya and Houdini, can be employed.

18.3.3 Reconstructing Textile Materials

Substance Designer uses ‘Atomic Nodes’ to create simple 2D images and transform them into high-resolution texture maps. According to the user’s guide, Atomic Nodes are “the lowest-level most basic building blocks of Substance Designer” (Adobe 2020b). We developed a graph (Fig. 18.3) that automatically generated texture maps of textile materials required for the reconstruction. The graph consists of 86 nodes.

In the first stages of texture generation (steps 1 and 2 in Fig. 18.3), images of two basic components used in weaving, namely warp and weft yarns, were created. In these steps we used such nodes as ‘Shape’, ‘Bevel’, ‘Safe Transform’, ‘Messy Fibers’, ‘Blend’, ‘Levels’ and ‘Perlin Noise’. In the third step, the two systems of yarns were overlapped by using the ‘Blend’ node. Then, to make the fabric look realistic, we added special effects (steps 4–7 in Fig. 18.3). In step 8, we dyed the fabric by using the ‘Uniform Colour’, ‘Blend’ and ‘Levels’ nodes. Diffuse, normal, glossiness, specular and height maps were generated in step 9.

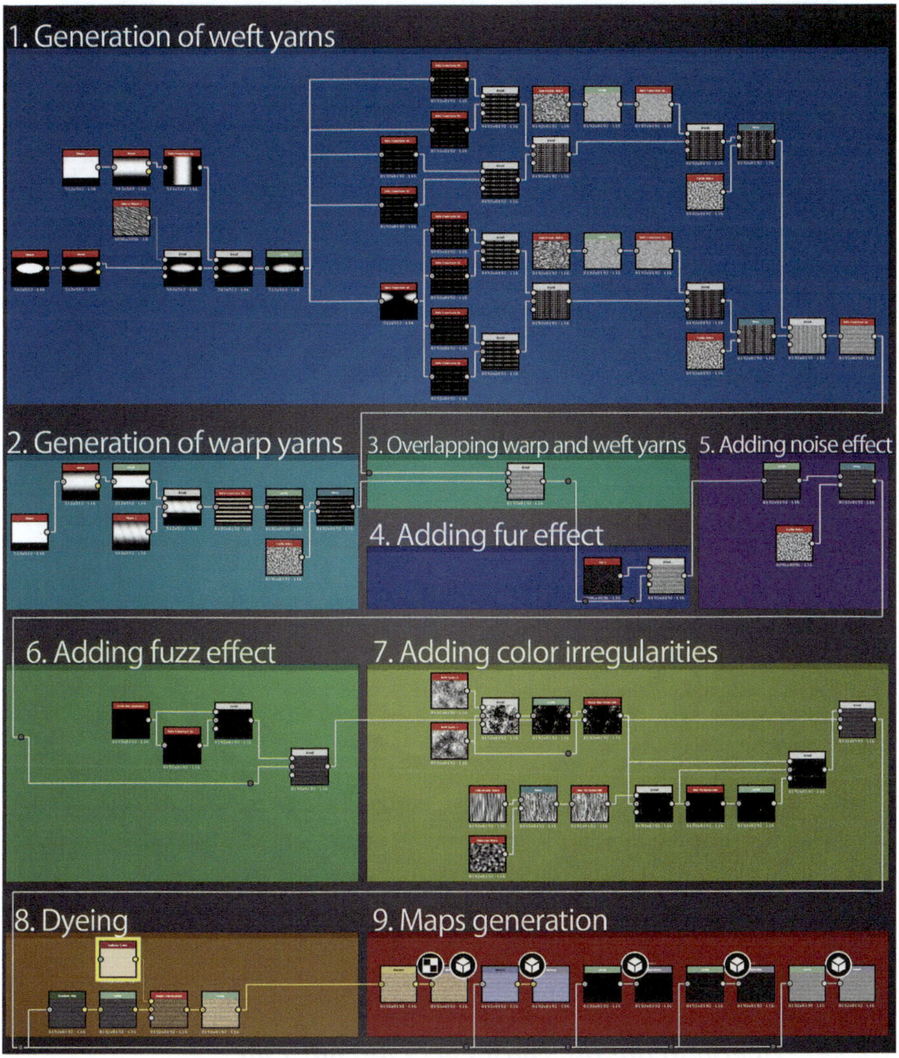

Fig. 18.3 The graph used to generate texture maps of the reconstructed textile materials in Substance Designer

In the software, the parameters of texture maps depend on the parameters of nodes. To ensure that the digital reconstruction matches the actual fabrics, we used the parameters of the Salt man's textiles to control the nodes. In order to accurately reproduce the structure of the Salt man's fabrics, we used actual values of twist direction, twist angle, thread thickness, thread count and weave type (see Table 18.1) in steps 1 and 2. The use of NCS-Codes allowed the colours of the textile materials to be reproduced at step 8.

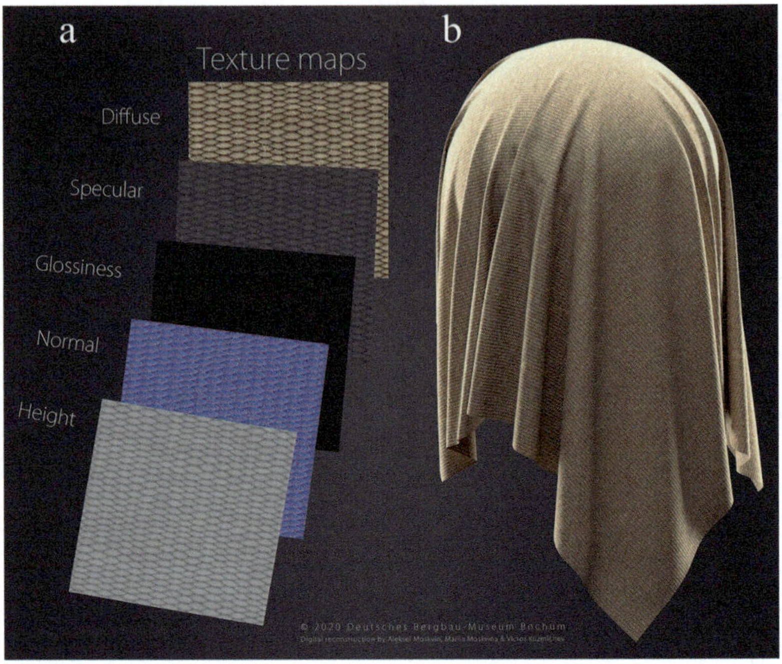

Fig. 18.4 Reproduction of the woven fabric from the Douzlākh salt mine (tunic of Salt man 4): (**a**) fragments of texture maps; (**b**) computer visualization of the fabric. (Image: A. Moskovin and V. Kuzmichev)

The output size of the texture maps was 8192 × 8192. The maps were made for small fragments of the fabrics (10 × 10 cm, 80 warp yarns). These small fragments were subsequently used to tile the whole surface of the garments. Figure 18.4 shows texture maps, the ones generated in Substance Designer and subsequently used to reconstruct the tunic of Salt man 4, and a computer visualization of the fabric.

18.3.4 Reproducing the Garments Laid out Flat

The garments laid out flat were modelled in Clo3D. First, we reproduced the outlines of the cut patterns by using the 'Polygon' command. To control the lengths and shapes of the lines we used measurements shown in Fig. 18.2. Then, we used the 'Internal Polygon/Line' command to draw internal seam lines on the tunic and trousers. To make the head opening (slit) we used the 'Cut' command. The ends of the belt were filled up by using the 'Pressure' command. To ensure the tunic will be placed over the trousers and the belt will be placed over the tunic, we applied the 'Set Sublayer' command. Figure 18.5 shows the digital tunic and trousers laid out flat.

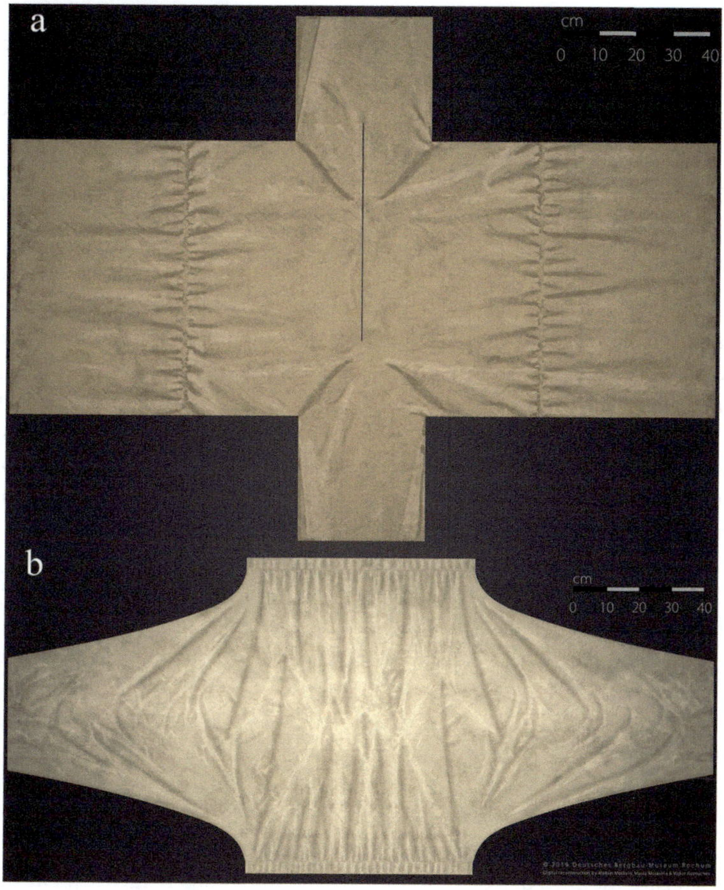

Fig. 18.5 Digital reconstruction of the salt man's garments laid out flat: (**a**) tunic; (**b**) trousers. (Image: A. Moskovin and V. Kuzmichev)

18.3.5 Avatar Modelling

Two types of anthropometric measurements were involved in the avatar modelling, namely body circumferences (girths) and linear measurements (heights and lengths). The circumferences of the mummy do not represent the actual pristine body shape of Salt man 4. The muscle and adipose tissue decreased in size due to dehydration resulting from being buried in salt. However, despite the damage caused by the mine collapse, most of the bones have virtually the same lengths as 2400 years ago. Thus, we measured heights and lengths on the mummy (Table 18.2, upper row).

A complete set of 16 body measurements is needed to generate an avatar. To fill in the gaps, we used a 2010 anthropometric dataset of 12,635 Iranian men (Pourtaghi et al. 2014). First, we evaluated how the linear measurements of the mummy differ from the ones presented in the dataset. We estimated that the differences between the salt man's measures and contemporary mean values do not exceed 3.4%

Table 18.2 Body measurements that were used to generate an avatar of Salt man 4

Source, date, units	Values[a]															
	Height	Chest girth	Chest height	Natural waist girth	Waist height	Hip girth	Hip height	Neck girth	Arm length	Inside leg length	Knee height	Biceps girth	Wrist girth	Calf girth	Ankle girth	Back width
The mummy, 405–380 BCE (cm)	**173.1**	–	**125.7**	–	**105.2**	–	**88.3**	–	**57.4**	**78.1**	**50.2**	–	–	–	–	–
Contemporary anthropometric dataset, 2010 (cm)	174.1	**93.6**	126.5	**83.8**	101.6	**97.2**	–	**36.3**	57.3	75.5	50.5	**32.5**	**17.5**	**36.8**	**26.1**	**42.1**
Difference (%)	0.6	–	0.6	–	3.4	–	–	–	0.2	3.3	0.6	–	–	–	–	–

[a]Bold font indicates values that were used to generate an avatar of Salt man 4

(Table 18.2, lower row). Secondly, we obtained the missing values of body circumferences from the dataset (Table 18.2, middle row).

It is interesting to note that, according to the National Non-Communicable Disease Risk Factor Surveillance (Haghdoost et al. 2008), the mean stature of 20-year-old Iranian men in 2005 was 173.39 cm. The same source states that the height of Iranian men among those who were born between 1940 and 1984 has increased by around 1.53 cm per decade. This allows one to assume that the salt man (173.1 cm height) would have been considered comparatively tall in his time.

Distinctive features of the salt man's appearance, such as his head proportions, were reproduced by using head and face morphs in MakeHuman. The hairstyle was replicated in Blender. We used the 'Particle System' command to generate strands of hair. Then, we applied the 'Comb', 'Length' and 'Cut' brushes to edit the hair.

18.3.6 Virtual Try-on

Mechanical behavior of fabrics in computer simulations is controlled by the physical and mechanical properties of textile materials. The library of Clo3D contains 70 presets of these properties. We chose the presets of the contemporary fabrics that have the weave type, content and thread thickness similar to the ones of the salt man's textiles (see Table 18.1). Table 18.3 shows the physical and mechanical properties of digital textile materials used for the reconstruction.

The virtual try-on was carried out in the following steps:

1. We set the parameters of the simulation (gravitational acceleration of 9.8 m/s2 and time step of 0.03 s), the garments (particle distance of 3 mm and additional thickness – collision of 0.1 mm) and the surface of the avatar (static friction coefficient of 0.8, kinetic friction coefficient of 0.05 and skin offset of 0.1 mm);
2. The chosen properties of textile materials (Table 18.3) were assigned to the tunic, trousers and belt;
3. We folded the tunic and trousers as Fig. 18.2(b, c) shows by using the 'Fold Arrangement' command;
4. The garments were placed around the avatar by using the 'Select/Move' command;
5. The side seams of the tunic and trousers were marked by using the 'Segment Sewing' command;
6. We put the trousers, tunic and belt on the avatar, one after another, using the 'Fast (GPU) Simulation' command;

The 'Fitting (Accurate Fabric)' command were applied to additionally improve the quality of the 3D models. The 3D models were exported to Blender. We applied polygonal modelling tools of the software to place red cords along the edges of the garments as Fig. 18.2 shows. The same tools were applied to model the accessories, namely the pottery jars, the knife, the leather sheath and two earrings. Figure 18.6 shows the digital reconstruction of the Iranian salt miner and his costume.

Table 18.3 Physical and mechanical properties of digital textile materials used for the reconstruction

	Properties[a]										
Garment	Content	Density (g/m^2)	Stretch-weft stiffness (g/s^2)	Stretch-warp stiffness (g/s^2)	Bending-weft stiffness ($g/mm^2/s^2/grad$)	Bending-warp stiffness ($g/mm^2/s^2/grad$)	Bending-bias stiffness ($g/mm^2/s^2/grad$)	Buckling ratio-weft (0–1)	Buckling ratio-warp (0–1)	Buckling stiffness-weft (0–1)	Buckling stiffness-warp (0–1)
Tunic	100% Wool	345	368706	489034	1300	1300	1300	0	0	0.20	1.0
Trousers	100% Wool	345	368706	489034	1300	1300	1300	0	0	0.20	1.0
Belt	100% Wool	427	512383	269877	2683	5616	3181	0	0	0.20	0.20

[a]The properties of fabrics are shown with the names and units used in Clo3D and Marvelous Designer software

Fig. 18.6 Complete digital reconstruction of the tunic, trousers, belt and all the accessories of Salt man 4 except the fur cloak. The rope that connects the pots to the belt was left uncolored because its presence is an educated guess. (Image: A. Moskovin and V. Kuzmichev)

18.4 Results, Discussion and Outlook

Contemporary methods of digitization of archaeological artifacts are based mostly on 3D scanning technology. That technology has certain limitations. It does not allow inner layers of clothes to be digitized. Nor can parts of human body covered by clothes be accessed by 3D scanners. Furthermore, items of clothing reveal their shape when garments are put on the human body and influenced by gravity. 3D scanning data cannot reproduce the mechanical behavior of textiles under the influence of external forces. In case of Salt man 4, these limitations are of crucial importance. The mummy is damaged by the influence of time and the inner layers of the costume along with most of the body surface are hidden from the observer.

Other approaches to digital reconstruction of archaeological textiles do exist (Cybulska 2010, 2012, 2020; Cybulska et al. 2013). These methods replicate textile materials at yarn level in a 3D environment. However, the process of 3D modelling is time-consuming and not yet completely automated. Moreover, the 3D models of textiles are applied to garments as 2D textures. The maximum size of 2D textures in many of contemporary software programs for viewing 3D models (Sketchfab,

Unreal Engine 4, Unity, CryEngine, Substance Painter, etc.) is limited to 8192 × 8192 pixels or, in some cases, to 16384 × 16384 pixels. If the garment laid out flat is 190 cm in length, pixel density of the texture map ranges between 109 pixels per inch (8192 × 8192 pixels) and 219 pixels per inch (16384 × 16384 pixels). Such densities are relatively small. Thus, the 3D–2D conversion blurs the small details of yarns and fabrics, which are the main benefit of 3D reconstructions. This in turn makes it difficult for an observer to access the benefits of 3D reconstructed textiles. Due to hardware limitations, it is difficult to apply large fragments of 3D textiles to garments directly. This would require a huge amount of RAM (Random Access Memory) to process the 3D garment. Physically based simulation of such fragments, which could reproduce the draping behavior of archaeological textiles, is limited by CPU performance. In this regard, the innovative methods by Cybulska are still ahead of their times. In addition, these methods are focused on textiles and do not provide researchers with tools for generating avatars of the human body.

In this study we used another approach and reproduced the entire process of clothes making and try-on in VR. We used a node-based texture generation software to transform parameters of ancient textiles into its texture maps. Then, we recreated the Salt man's tunic, trousers and belt laid out flat in accordance with 2D measurements of the actual garments. An avatar of the mummy was created by using two sources of information, namely the mummy itself and a contemporary anthropometric dataset of Iranian men. Lastly, the physically based cloth simulation was applied to put all the garments on the avatar.

The reconstructed garments are dynamic 3D clothing models. In contrast to 3D scans and models created by using polygonal modelling, these virtual garments behave like actual clothing. In other words, the 3D shapes of the garments can be generated automatically in terms of the physical forces acting on them. The reconstruction is fully compatible with a wide range of contemporary software programs (3D modelling packages, computer simulation software and game engines), which can be used to obtain further insights into the behaviour and features of ancient mining clothes.

One of the interesting facts relating to the garments found in the salt mine is that they do represent a specific working environment. Usually, more or less complete garments from a period between 600 BCE and 600 CE in the Eastern Mediterranean, Near East and Central Asia derive from funeral contexts (e.g., Barber 1991; Rudenko 1970; Falck and Lichtwark 1996; Wieczorek and Lind 2007). Often, they are from persons of high status and as such represent a specific selection according to status and funeral practices. In Chehrābād, the garments are worker's gear, worn during daily activities.

Recent theories on Functional Clothing Design (e.g., Watkins and Dunne 2015) could provide further insights in our study of prehistoric and ancient garments. The salt mine Chehrābād offers an especially valuable basis for such research, as its finds of garments and fully dressed people – salt miners – derive from a specific working context. They died as the mine collapsed while they were at work; this is a human tragedy, but it offers a unique opportunity to study garments worn during heavy physical work in ancient times.

Contemporary software allows researchers to visualize and analyze contact points between the body and clothes (Liu et al. 2010; Jevšnik et al. 2017), the strain of textile materials (Bogović et al. 2019) and heat transfer processes (Puszkarz et al. 2020). Since the digital reconstruction of Salt man 4 is a dynamic 3D clothing model, it can be comprehensively studied in VR. Further research will help us understand what kinds of cloth and garment shapes have been chosen by ancient salt miners in order for them to be dressed for work, to be able to move in a proper way (as their work required) and to be protected from the rough surroundings of a salt mine.

Acknowledgements The reconstruction would not have been possible without valuable contributions of all the individuals and organizations involved in Salt man projects. The authors would like to thank Gabriela Russ-Popa (Austrian Academy of Sciences), Lena M. Öhrström (University of Zurich) and Natascha Bagherpour Kashani (Archäologisches Museum Frankfurt).

References

Aali A, Stöllner T (eds) (2015) *The archaeology of the salt miners. Interdisciplinary research 2010–2014, Metalla 21, 1–2/2014*. Deutsches Bergbau Museum, Bochum.

Aali A, Stöllner T, Abar A, Rühli F (2012) The salt men of Iran. The salt mine of Douzlakh, Chehrābād. *Archäologisches Korrespondenzblatt* 42(1):61–81.

Adobe (2020a) *Substance designer official website*. https://www.substance3d.com/products/substance-designer/, accessed 20 Aug 2020.

Adobe (2020b) *Documentation for the Substance products. Substance Painter. Nodes reference*. https://docs.substance3d.com/sddoc/nodes-reference-129368078.html, accessed 28 Jul 2020.

Barber EJW (1991) *Prehistoric textiles: The development of cloth in the Neolithic and Bronze Age with special reference to the Aegean*. Princeton University Press, Princeton.

Blender Foundation (2020) *Blender. Open source 3D creation official website*. https://www.blender.org/, accessed 17 Sept 2020.

Bogović S, Stjepanovič Z, Cupar A, Jevšnik S, Rogina-Car B, Rudolf A (2019) The use of new technologies for the development of protective clothing: Comparative analysis of body dimensions of static and dynamic postures and its application. *AUTEX Research Journal* 19(4):301–311. https://doi.org/10.1515/aut-2018-0059.

CLO Virtual Fashion (2020) *Clo3d official website*. https://www.clo3d.com/, accessed 28 Jul 2020.

Cybulska M (2010) Reconstruction of archaeological textiles. *FIBRES & TEXTILES in Eastern Europe* 18, No. 3(80):100–105.

Cybulska M (2012) To see the unseen. Computer graphics in visualization and reconstruction of archaeological and historical textiles. In: Mukai N (ed.) *Computer graphics*. InTech, Rijeka, p. 213–228.

Cybulska M (2020) Woman's costume in the territories of Poland during the Roman Period. Reconstruction based on finds from Nowy Łowicz in Pomerania. *FIBRES & TEXTILES in Eastern Europe* 28, No. 5(143):124–129.

Cybulska M, Mianowska-Orlińska E (2015) Analysis, reconstruction and interpretation of two early Medieval embroideries from Kruszwica. In: Grömer K, Pritchard F (eds) *Aspects of the design, production and use of textiles and clothing from the Bronze Age to the Early Modern Era. The North European symposium of archaeological textiles, 21 May 2014 in Hallstatt, Austria*. NESAT XII. Archaeolingua 33, Budapest, p 311–321.

Cybulska M, Kuberski S, Maik J, Orlińska-Mianowska E (2013) Figural embroidery from Tum Collegiate Church – analysis, reconstruction and identification. In: Banck-Burgess J, Nübold C (eds) *The North European symposium for archaeological textiles XI, 10–13 May 2011 in Esslingen am Neckar.* NESAT XI. Verlag Marie Leidorf GmbH, Rahden, p 185–191.

Falck M, Lichtwark F (eds) (1996) *Ägypten. Schätze aus dem Wüstensand. Kunst und Kultur der Christen am Nil. Katalog zur Ausstellung*. Ludwig Reichert Verlag, Wiesbaden.

Grömer K, Aali A (2016) Durchdacht gemacht. Die Kleidung eines Salzmannes aus Chehrabad, Iran. *Universum Magazin* 3:108–109.

Grömer K, Bagherpour N (2018) The saltmummies of Zanjan. Textile research in Iran in a conservation and exhibition project. *Archaeological Textiles Review* 60:110–113.

Haghdoost AA, Mirzazadeh A, Alikhani S (2008) Secular trend of height variations in Iranian population born between 1940 and 1984. *Iranian Journal of Public Health* 37(1):1–7.

Jevšnik S, Stjepanovič Z, Rudolf A (2017) 3D virtual prototyping of garments: Approaches, developments and challenges. *Journal of Fiber Bioengineering and Informatics* 10(1):51–63. https://doi.org/10.3993/jfbim00253.

Kuzmichev V, Moskvin A, Moskvina M (2018a) Virtual reconstruction of historical men's suit. *AUTEX Research Journal* 18(3):281–294. https://doi.org/10.1515/aut-2018-0001.

Kuzmichev V, Moskvin A, Moskvina M, Pryor J (2018b) Research on 3D reconstruction of late Victorian riding skirts. *International Journal of Clothing Science and Technology* 30(6):790–807. https://doi.org/10.1108/IJCST-12-2017-0192.

Kuzmichev V, Moskvin A, Surzhenko E, Moskvina M (2017) Computer reconstruction of 19th century trousers. *International Journal of Clothing Science and Technology* 29(4):594–606. https://doi.org/10.1108/IJCST-12-2016-0139.

Liu Y-J, Zhang D-L, Yuen M (2010) A survey on CAD methods in 3D garment design. *Computers in Industry* 61(6):576–593. https://doi.org/10.1016/j.compind.2010.03.007.

Moskvin A, Kuzmichev V, Moskvina M (2019) Digital replicas of historical skirts. *The Journal of the Textile Institute* 110(12):1810–1826. https://doi.org/10.1080/00405000.2019.1621042.

Moskvin A, Moskvina M, Kuzmichev V (2020) Parametric modeling of historical mannequins. *International Journal of Clothing Science and Technology* 32(3):366–389. https://doi.org/10.1108/IJCST-06-2019-0093.

Pollard AM, Brothwell DR, Aali A, Buckley S, Fazeli H, Hadian Dehkordi M, Holden T, Jones AKG, Shokouhi JJ, Vatandoust R, Wilson AS (2008) Below the salt: A preliminary study of the dating and biology of five salt-preserved bodies from Zanjān province, Iran. *Iran* 46:135–150.

Pourtaghi G, Valipour F, Sadeghialavi H, Lahmi M (2014) Anthropometric characteristics of Iranian military personnel and their changes over recent years. *The International Journal of Occupational and Environmental medicine* 5(3):115–124.

Puszkarz AK, Machnowski W, Błasińska A (2020) Modeling of thermal performance of multilayer protective clothing exposed to radiant heat. *Heat and Mass Transfer* 56:1767–1775. https://doi.org/10.1007/s00231-020-02820-1.

Rudenko S (1970) *Frozen tombs of Siberia*. Dent, London.

Shahbazi S (2012) Clothing. ii. In the Median and Achaemenid periods. In: *Encyclopædia Iranica,* V(7):723–737. http://www.iranicaonline.org/articles/clothing-ii, accessed 10 Sep 2020.

The MakeHuman team (2020) *Make Human Official Website*. http://www.makehumancommunity.org/, accessed 20 Aug 2020.

Watkins S M, LE Dunne (2015) *Functional clothing design*. Bloomsbury, London.

Wieczorek A, Lind C (2007) *Ursprünge der Seidenstraße. Sensationelle Neufunde aus Xinjiang, China. Begleitbuch zur Ausstellung*. Thesis, Stuttgart and Mannheim.

Wijnhoven MA, Moskvin A (2020) Digital replication and reconstruction of mail armour. *Journal of Cultural Heritage*. Advance online publication. https://doi.org/10.1016/j.culher.2020.04.010.

Aleksei Moskvin is Associate Professor at the Department and Technology, Saint Petersburg State University of Industrial Technologies and Design, Russian Federation. He studied apparel design, pattern making and garment manufacturing technology at the same university in 2006–2011. His PhD thesis (2015) is entitled "Menswear design based on retrospective analysis of historical costumes' construction". He specializes in 3D digital reconstruction of historical and archaeological garments dated from the Iron Age to the Late Modern Period. Currently, he is working on a habilitation thesis entitled "Methodological basis of computer reconstruction of historical costumes". He is also a Senior Research Fellow at Ivanovo State Polytechnic University, Russian Federation.

Karina Grömer is the Head of the Department of Prehistory, Natural History Museum Vienna. She studied prehistoric archaeology, ethnology and anthropology at the University of Vienna in Austria. Her habilitation thesis was entitled "Archaeological Textile Research – Technical, economic and social aspects of textile production and clothing from Neolithic to the Early Modern Era". She is specialized in interdisciplinary and integrated analysis of textiles, research on textile tools and reconstruction of prehistoric costume. Her research covers a timespan from c. 2500 BCE to 1000 CE and a geographical area from Central Europe to Iran, with research projects like "DressID – Clothing and Identity in Roman Empire", "CinBA – Creativity in the Bronze Age" and "Chehrābād Saltmummy and Saltmine Exploration Project". She also teaches at the Universities of Vienna (Austria) and Southampton (UK).

Mariia Moskvina studied apparel design, pattern making and engineering drawing in 2006–2011 at Saint Petersburg State University of Industrial Technologies and Design, Russian Federation. Since 2007, she held the positions of Laboratory technician, Assistant Professor and Associate Professor (currently) at the Department of Clothes Design and Technology at the same university. Her PhD thesis (2016) is entitled "Providing anthropometric compliance in computer-aided design of clothing with specified silhouette shapes". She specializes in physically based computer simulation of cloth and virtual try-on technologies. She is also a Research Fellow at Ivanovo State Polytechnic University, Russian Federation.

Victor Kuzmichev is the Head of the Clothing Design Department, Ivanovo State Polytechnic University, Russian Federation, and Professor of Wuhan Textile University, China. His research interests include clothing engineering, fashion, simulation and visualisation of clothing and virtual prototyping of garments. He is also the head of the Laboratory for IT in Fashion. His current projects are devoted to virtual reconstruction of historical costume.

Thomas Stöllner holds the chair for Pre- and Protohistory at the Ruhr-University of Bochum, Germany and directs the Department of Mining Archaeology at the Deutsches Bergbau-Museum Bochum (DBM). Since 2014, he is also Head of the research department at the DBM and member of the directorate of the DBM. His main area of research is the social and economic development of mining communities throughout pre- and protohistory with a focus on mining, the archaeometry of mining, the archaeology of technology and societal interrelation by the help of re-search on settlements and graveyards. His research spans from ld World archaeology including Central and East Europe to the Caucasus, the Middle East and Central Asia as well as South America.

Abolfazl Aali is the Head of the Department of Museums and Cultural Properties in the General Directorate of Cultural Heritage, Handicrafts and Tourism of Zanjan Province in Iran. He studied archaeology at the University of Tehran. His research interest is the archaeology of Iran, with a special focus on the Achaemenid, Parthian and Sassanid periods, which covers a time-frame between 500 BC and 600 AD. In the last decade, he has jointly headed the International Chehrābād Salt Mine research project with Professor Thomas Stöllner from the Ruhr-University of Bochum and the Deutsches Bergbau-Museum Bochum, Germany.

GPSR Compliance

The European Union's (EU) General Product Safety Regulation (GPSR) is a set of rules that requires consumer products to be safe and our obligations to ensure this.

If you have any concerns about our products, you can contact us on ProductSafety@springernature.com

In case Publisher is established outside the EU, the EU authorized representative is:

Springer Nature Customer Service Center GmbH
Europaplatz 3
69115 Heidelberg, Germany

Batch number: 10732090

Printed by Printforce, the Netherlands